User Performance 2018

Transportation Research Record: Journal of the Transportation Research Board Volume 2672, Issue 37

ISBN: 978-1-5108-8245-4

Printed from e-media with permission by:

Curran Associates, Inc.
57 Morehouse Lane
Red Hook, NY 12571

proceedings.com

Some format issues inherent in the e-media version may also appear in this print version.

Published by Sage Publications USA

ISBN (Print) 978-1-5108-8245-4
ISBN 2018 Printed Set (All Issues) 978-1-5108-7735-1

Printed by Curran Associates, Inc. (2019)

For permission requests, please contact sagepub.com/journals-permissions

Additional copies of this publication are available from:

Curran Associates, Inc.
57 Morehouse Lane
Red Hook, NY 12571 USA
Phone: 845-758-0400
Fax: 845-758-2633
Email: curran@proceedings.com
Web: www.proceedings.com

Volume 2672 Number 37 December 2018

TRR
JOURNAL OF THE TRANSPORTATION RESEARCH BOARD

Contents

User Performance 2018

Articles

---

* Membership as of February 2019.

TRB

TRANSPORTATION RESEARCH BOARD

CORE SPONSORS

Transportation Departments of the 50 States and the District of Columbia

Federal Government

U.S. Department of Transportation
- Federal Aviation Administration (FAA)
- Federal Highway Administration
- Federal Motor Carrier Safety Administration (FMCSA)
- Federal Railroad Administration (FRA)
- Federal Transit Administration
- National Highway Traffic Safety Administration (NHTSA)
- Office of the Assistant Secretary of Transportation for Research and Technology (OST-R)

U.S. Air Force Civil Engineer Center (AFCEC)
U.S. Coast Guard (USCG)
U.S. Department of Energy
U.S. Department of the Interior (DOI)
U.S. Environmental Protection Agency (EPA)

Nongovernmental Organizations

American Public Transportation Association (APTA)
Association of American Railroads (AAR)
California Air Resources Board

The National Academies of
SCIENCES • ENGINEERING • MEDICINE

The **National Academy of Sciences** was established in 1863 by an Act of Congress, signed by President Lincoln, as a private, nongovernmental institution to advise the nation on issues related to science and technology. Members are elected by their peers for outstanding contributions to research. Dr. Marcia McNutt is president.

The **National Academy of Engineering** was established in 1964 under the charter of the National Academy of Sciences to bring the practices of engineering to advising the nation. Members are elected by their peers for extraordinary contributions to engineering. Dr. C. D. Mote, Jr., is president.

The **National Academy of Medicine** (formerly the Institute of Medicine) was established in 1970 under the charter of the National Academy of Sciences to advise the nation on medical and health issues. Members are elected by their peers for distinguished contributions to medicine and health. Dr. Victor J. Dzau is president.

The three Academies work together as the National Academies of Sciences, Engineering, and Medicine to provide independent, objective analysis and advice to the nation and conduct other activities to solve complex problems and inform public policy decisions. The Academies also encourage education and research, recognize outstanding contributions to knowledge, and increase public understanding in matters of science, engineering, and medicine.

Learn more about the National Academies of Sciences, Engineering, and Medicine at **www.national-academies.org.**

The **Transportation Research Board** is one of seven major programs of the National Academies of Sciences, Engineering, and Medicine. The mission of the Transportation Research Board is to increase the benefits that transportation contributes to society by providing leadership in transportation innovation and progress through research and information exchange, conducted within a setting that is objective, interdisciplinary, and multimodal. The Board's varied activities annually engage about 7,000 engineers, scientists, and other transportation researchers and practitioners from the public and private sectors and academia, all of whom contribute their expertise in the public interest. The program is supported by state transportation departments, federal agencies including the component administrations of the U.S. Department of Transportation, and other organizations and individuals interested in the development of transportation.

Learn more about the Transportation Research Board at **www.TRB.org**.

Article

TRR
JOURNAL OF THE TRANSPORTATION RESEARCH BOARD

Transportation Research Record
2018, Vol. 2672(37) 1–10

DOI: 10.1177/0361198118791394
journals.sagepub.com/home/trr
SAGE

Prevalence of Engagement in Single versus Multiple Types of Secondary Tasks: Results from the Naturalistic Engagement in Secondary Task (NEST) Dataset

Martina Risteska[1], Birsen Donmez[1], Huei-Yen W. Chen[2], and Miti Modi[1]

Abstract

We investigated engagement in single vs. multiple types of secondary tasks in distraction-affected, safety-critical events (SCEs), i.e., crashes/near-crashes, and baselines reported in the Naturalistic Engagement in Secondary Tasks (NEST) dataset. NEST was created from Second Strategic Highway Research Program (SHRP2) data for studying distractions in detail. Early descriptive analysis on NEST found that most distraction-affected SCE and baseline epochs (10 s long) include more than one type of secondary task, suggesting that a considerable number of drivers may be engaging in multiple secondary activities within a relatively short time frame, potentially being exposed to increased demands brought on by multi-tasking and task-switching. We conducted inferential statistics on NEST focusing on engagement in single vs. multiple types of tasks across SCEs and baselines. A logit model was built to compare the odds of engaging in single vs. multiple types of tasks with the following predictors: event type (SCE, baseline), environmental demand, GPS speed, and driver age. The last three predictors were included to capture the driving demands experienced, which may have impacted drivers' task engagement behavior. Odds of engagement in multiple types of secondary tasks was higher in SCEs than baselines. Furthermore, with marginal statistical significance, drivers 65 years and over were less likely to engage in multiple types of secondary tasks than younger drivers. Overall, engagement in multiple secondary task types is more prevalent in SCEs. Most crash risk studies to date have reported the effects associated with one type of secondary task. However, it appears that these effects may be confounded by the presence of other secondary tasks.

Driver distraction is an important safety concern that affects all road users. The National Highway Traffic Safety Administration (NHTSA) found that approximately 3,477 crash-related fatalities (10% of all fatal crashes) and 391,000 injuries (15% of injury crashes) in 2015 were affected by distraction (*1*). In addition, simulator and on-road studies have found a considerable amount of evidence for driving performance degradation under distracted driving (*2*). Recent technological advances in personal communication devices and in-vehicle infotainment systems have generated a particular focus on the potential distracting effects of technology. For example, many controlled studies have examined the effects of prominent types of cell-phone usage: e.g., dialing (*3–5*) and talking on the phone (*6–8*). Previous research has also investigated the effects of in-vehicle technologies, such as radios (*9*) and navigation devices (*10*).

Unlike police reports and simulator/on-road studies, naturalistic driving studies provide real-time driver behavior data collected without the contrived settings utilized in experiments. Through sensors installed in the participating vehicles, naturalistic driving studies provide a unique opportunity to evaluate the prevalence of, and the crash risk associated with, potentially risky behaviors (e.g., engagement in a particular type of secondary task). The Second Strategic Highway Research Program (SHRP2) Naturalistic Driving Study (NDS) is the largest

[1]Department of Mechanical and Industrial Engineering, University of Toronto, Toronto, ON, Canada
[2]Department of Industrial and Systems Engineering, University at Buffalo, Amherst, NY

Corresponding Author:
Address correspondence to Birsen Donmez: donmez@mie.utoronto.ca

naturalistic study conducted to date that captured more than 35 million miles of continuous driving and 2 petabytes of video, kinematic, and audio data (*11*). A recent analysis on the SHRP2 data has found that drivers are distracted from their primary task of driving roughly 50% of the time, and that 68% of crashes involve some type of observable distraction (*12*). Furthermore, similar to the findings of earlier naturalistic studies (*13*), visual-manual secondary activities were found to be the most detrimental to safety compared with other types of secondary activities.

Although a considerable number of studies examined driver distraction both in controlled and naturalistic settings, most studies, if not all, focused on the effects associated with one type of secondary task compared with no task or with another task. Early descriptive analysis from a subset of the SHRP2 data, i.e., the Naturalistic Engagement in Secondary Tasks (NEST) dataset, suggests that drivers may be engaging in more than one type of secondary task in relatively short periods of time (i.e., within 10 s) (*14*), potentially being exposed to increased demands brought on by multi-tasking and task-switching. The NEST dataset (*15*) was created from SHRP2 data, for the specific purpose of investigating driver distraction. NEST includes a sub-sample of the SHRP2 safety-critical events (SCEs), i.e., crashes/near-crashes, where secondary task engagement was identified to be a potential contributing factor. It also includes baselines matched on the driver. Two 10 s epochs were coded for each SCE prior to the precipitating event, and for each baseline event. Driver actions related to distraction were coded in detail, including secondary task descriptions. The descriptive analysis mentioned earlier (*14*) revealed that the majority of distraction-affected SCE and baseline epochs reported in NEST include more than one type of secondary task, suggesting that engagement in multiple types of secondary tasks does occur. This finding raises questions about the overall prevalence of this phenomenon. Most crash risk studies to date have reported the effects associated with one type of secondary task, when it appears that these effects may be confounded by the presence of other secondary tasks.

In this paper, through inferential statistics, we compare the prevalence of engagement in single vs. multiple types of secondary tasks in distraction-affected SCEs and baselines reported in the NEST dataset. A logit model was built to compare the odds of engaging in single vs. multiple types of secondary tasks with event type (SCE, baseline) as a predictor. NEST provides detailed information about the driving environment, which provides the opportunity to investigate and control for the effects of environmental demand on secondary task engagement type. Findings from naturalistic studies suggest that drivers exhibit self-regulating behaviors when engaging in secondary activities. For instance, Funkhouser and Sayer (*16*) found that drivers initiated cell-phone conversations more frequently during low-speed conditions and engaged in visual or manual phone tasks more frequently when stopped. Likewise, Tivesten and Dozza (*17*) reported that drivers were more likely to initiate visual-manual phone tasks while stationary compared with while driving at high speeds. Therefore, environmental demand, GPS speed, and driver age were included in the model as a way of capturing the driving demands experienced by the driver, which may have impacted their secondary task engagement behavior, in particular, whether they engaged in more than one type of secondary task in a given epoch. Driver age was used as a proxy for driver skill and ability.

NEST Dataset

The SHRP2 NDS monitored approximately 3,400 drivers between 2010 and 2013. About 50 million miles of naturalistic driving data were collected from six sites across the United States: Seattle, Washington; Tampa, Florida; Buffalo, New York; Durham, North Carolina; State College, Pennsylvania; and Bloomington, Indiana. To date, SHRP2 data identified 1,549 crashes and 2,705 near-crashes (*18*). Crashes have been categorized into four severity levels: most severe crashes are level I (n = 112), moderate severity police-reportable crashes are level II (n = 159), minor severity crashes are level III (n = 633), and low-risk tire strikes are level IV (n = 637). Near-crashes were defined as safety-critical situations with successful but rapid evasive maneuvers exhibited by the drivers.

The NEST dataset was reduced from the SHRP2 data, administered and coded by Virginia Technical Transportation Institute, contracted by and under supervision of Toyota Collaborative Safety Research Center (*15*). NEST data reduction took place in 2014 and was completed in October 2014. NEST includes only a subset of SHRP2 data as SHRP2 data was not entirely ingested by the time the contract ended.

Overall, NEST contains 236 SCEs in total from 204 drivers. Only SCEs where secondary task engagement was identified to be a potential contributing factor were included in NEST. Four baseline events were also included per SCE by matching on the driver, leading to a total of 944 baseline events (4 baselines per SCE × 236 SCEs). While all SCEs included in NEST were associated with secondary task engagement, the matched baseline events did not necessarily involve secondary task engagement. SCEs and baseline epochs in NEST were coded beyond the standard SHRP2 data reduction process. Specifically, NEST coding for SCEs was performed for 30 s: 20 s prior to and 10 s after the precipitating event, i.e., the state of environment or action that initiated the

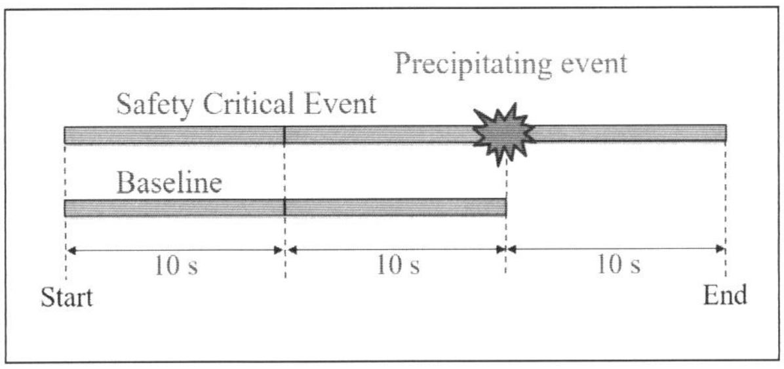

Figure 1. Data coding epochs for SCEs and baseline events in NEST.

crash or near-crash event sequence (Figure 1). Baseline epochs, which did not have a precipitating event by definition, were coded for 20 s. In contrast, SHRP2 data was coded for 6 s (5 s prior to and 1 s after the precipitating event) for both SCEs and baselines. NEST coded secondary task engagement information (e.g., whether the driver talked on the cell-phone) in blocks of 10 s epochs as presented in Figure 1. Moreover, every 10 s epoch contains information of up to a maximum of 6 different secondary tasks. Thus, compared to SHRP2, NEST provides the opportunity to investigate a longer time period leading up to an SCE, as well as the ability to study driver behavior within two 10 s epochs leading up to an event. This additional information is valuable given that the length of engagement in a secondary task can be considerably longer than a few seconds, especially if drivers are engaging in multiple secondary tasks.

Method

In this section, we first present our statistical modeling approach along with the outcome and predictor variables used in the model. NEST provides an extensive list of variables. Given the relatively small sample size, it was infeasible to include all these variables separately in our model. Hence, aggregations were performed to create meaningful groupings of NEST variables, with reasonable sample sizes within each group. After presenting an overview of our model, we describe in detail how our model variables were created.

We should also note here that our analysis excluded the third 10 s epoch coded in NEST for SCEs, which represents the period from the precipitating event to the crash/near-crash. It is reasonable to assume that this period is inherently different than the 20 s preceding the precipitating event and may represent significant shifts in secondary task engagement, which is a point for future research.

Statistical Model

A logistic regression model was built to compare the odds of engaging in single vs. multiple types of secondary tasks for different event types (i.e., SCEs and baselines). Given that NEST only included SCEs that involved secondary task engagement, baselines included in our model were those that also involved secondary task engagement (n = 704). The outcome variable was whether the driver engaged in multiple types of secondary tasks in either or both of the first two 10 s epochs of an event, or not. We also conducted our analysis on the two 10 s epochs independently but obtained similar results. Therefore, we only report our results that considered the two 10 s epochs together.

The predictor variables that were included in the model are as follows: event type (baseline, lower severity SCE, higher severity SCE), environmental demand (lower, higher), GPS speed at the start of the event, driver age (16–19, 20–24, 25–34, 35–64, 65 and over), and all two-way interactions of these variables. Repeated measures were accounted for through generalized estimating equations. Backward selection was used to drop non-significant interaction terms from the final model. The model was built in SAS University Edition using the GENMOD procedure. Figure 2 presents the model covariates tested along with how the environmental demand variable was created.

Secondary Task Types

NEST provides an extensive list of secondary tasks (Table 1), with each 10 s epoch providing information of up to a maximum of 6 different secondary tasks. Some of these tasks appear to be complementary to each other and may not suggest a shift in task type when they occur together in the same epoch (e.g., viewing PDA vs. operating PDA). Instead, the drivers may be expected to perform some of these tasks together or in sequence as they complete a higher-level task (e.g., to operate a PDA, the driver may first have to locate/reach the PDA). Thus, we aggregated the NEST secondary tasks into nine more general secondary task types as presented in Table 1. These nine secondary task types relate to generally distinct objects and/or activities.

To create the outcome variable used in our model, we identified whether the driver engaged in multiple types of secondary tasks in a given epoch or not. The event was then labeled as "engagement in multiple types of secondary tasks" if the driver engaged in multiple types of secondary tasks in at least one of the 10 s epochs. All other events were labeled as "engagement in single secondary task type". Events that included unknown and N/A were excluded from our analysis. Further, events that involved writing/reading and pet interaction were too infrequent for these tasks to be in their own categories. In addition, these tasks did not fit well within the other nine task types, so they were also excluded from our analysis.

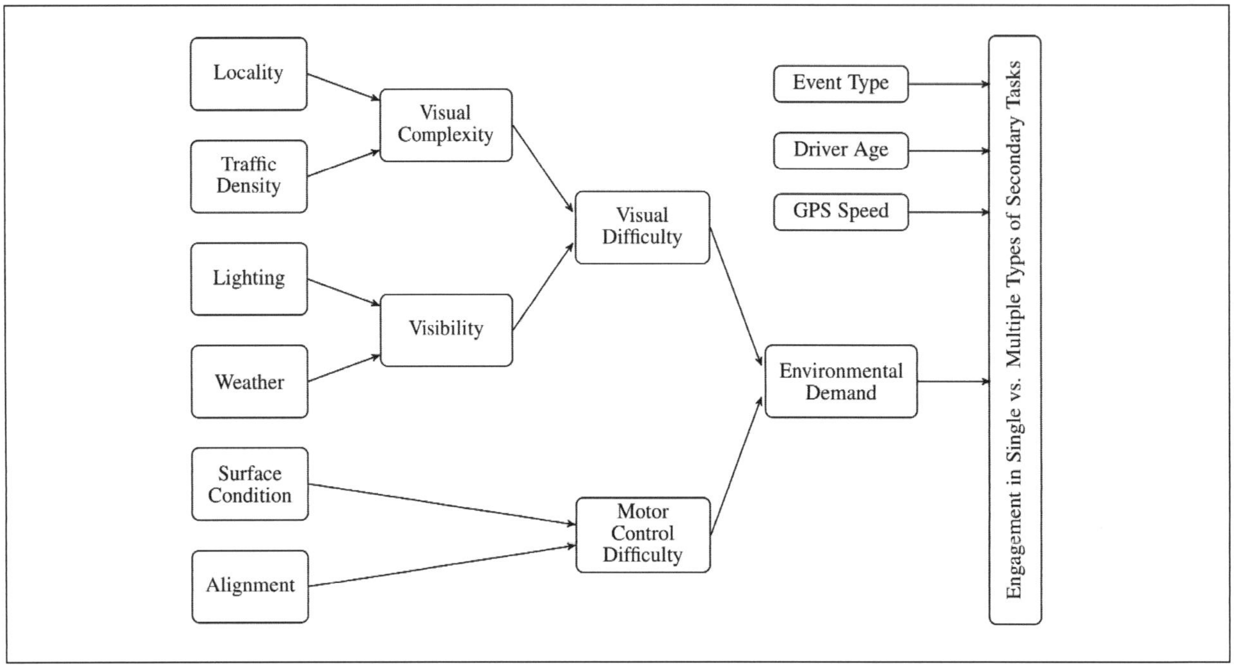

Figure 2. Model covariates and the creation of the environmental demand variable.

Finally, although NEST considered the presence of passengers or children with no interaction as secondary tasks, we did not. Thus, SCEs (n = 11) and baselines (n = 228) that were associated with only these NEST secondary tasks were excluded from our analysis. Further, three baseline events and one SCE were removed due to missing information about the environment given that environmental demand was a variable included in our model. In total, 218 SCEs and 704 baseline events were included in our analysis from a total of 204 drivers.

Event Type

SCEs and baselines represent the events of interest in our analysis. However, we wanted to investigate whether the severity of an SCE made a difference as well. Thus, the predictor "event type" was assigned to three levels: baseline (n = 704), lower severity SCE (n = 107), and higher severity SCE (n = 111). The categories of lower and higher severity SCE were generated from the SHRP2 SCE severity levels by taking sample size into account (Table 2).

Environmental Demand

Six NEST variables describing the environment around the time of an event were used to capture environmental demand. Each of these variables had various levels and was coded in NEST once to three times per event. Therefore, these variables and their levels needed to be grouped to have a reasonable sample size within each environmental demand level used in our model.

In creating the environmental demand construct, two demand types that the environment may impose on the drivers were considered: motor control difficulty and visual difficulty. Figure 2 presents an overview of how these constructs were created from the following NEST variables: locality, traffic density, lighting, weather, surface condition, and alignment. Motor control difficulty describes the lateral and longitudinal vehicle control demands imposed on the driver by the environment and is expected to be mainly affected by alignment (e.g., road curvature is expected to increase lateral control demand) and surface condition (e.g., road surface being wet is expected to increase longitudinal control demand). Visual difficulty consists of two sub-constructs: visual complexity and visibility. Visual complexity describes the density of static and dynamic elements of the driving scene that the driver needs to observe and track (i.e., traffic signs, vehicles, pedestrians, cyclists, intersections), and is expected to be affected by locality (e.g., "open country" vs. "urban") and traffic density. Visibility describes the lighting and weather conditions that may influence driver's ability to perceive those driving scene elements.

Motor control difficulty and visual difficulty constructs were assigned two levels each: lower and higher. Environmental demand variable created from these two constructs was also assigned two levels: lower and higher. Environmental demand was labeled to be lower

Table 1. Secondary Task Types Used in Our Analysis and Corresponding NEST Secondary Tasks

Secondary task type	NEST secondary tasks
Visual/manual interactions with vehicle integrated device/control	Adjusting/monitoring (a) climate control, (b) radio, (c) other devices integral to vehicle Inserting/retrieving CD
Personal hygiene	Applying make-up Biting nails/cuticles Brushing/flossing teeth Combing/brushing/fixing hair Removing/adjusting jewelry Removing/inserting/adjusting contact lenses or glasses Other personal hygiene
Interaction with carried-in device	Dialing hand-held (a) cell phone, (b) cell phone using quick keys Texting on cell phone Locating/reaching/answering cell phone Cell phone, other Operating PDA/other hand-held device Locating/reaching PDA/other handheld device Viewing PDA/other handheld device PDA/other handheld device, other
Talking on hand-held cell phone	Talking/listening on hand-held cell phone
Drinking/eating	Drinking Eating
Dancing/singing	Dancing Talking/singing
Outside distraction	Looking at (a) an object external to the vehicle, (b) pedestrian, (c) previous crash or incident Other external distraction Distracted by construction
Reaching/manipulating object	Moving object in vehicle Object (a) in vehicle, other, (b) dropped by driver Reaching for (a) food-related or drink-related item, (b) object that is a manufacturer-installed device, (c) object, other, (d) personal body-related item, (e) lighting/smoking/extinguishing cigar/cigarette
Passenger interaction	Passenger in adjacent or rear seat – interaction Child in adjacent or rear seat – interaction

Table 2. NEST SCE Severity Levels and Severity Categories Used in Our Analysis

NEST SCE severity levels	Frequency	SCE severity categories
Crash Level I: Most severe	34	Higher
Crash Level II: Police-reportable crash	38	(n = 111)
Crash Level III: Minor crash	39	
Crash Level IV: Low-risk tire strike	38	Lower
Near-crash	69	(n = 107)

if both motor control difficulty and visual difficulty were deemed to be lower (n = 461, e.g., industrial locality with free traffic flow in normal weather conditions on a straight road), it was labeled to be higher otherwise (n = 461, e.g., busy urban areas with medium traffic density during rainy weather). Cases where the road surface was dry, and the road was straight were deemed to correspond to lower motor control difficulty; all other cases were labeled as higher motor control difficulty. As for visual difficulty, the categorization was more complex given that the NEST variables which made up this construct had significantly more levels. An example for lower visual difficulty is interstate/bypass/divided highway with no traffic lights, level of service C, daylight, and normal weather. An example for higher visual difficulty is interstate/bypass/divided highway with no traffic lights, level of service A, non-daylight, and poor weather. The grouping and labeling performed were largely based on the NEST and SHRP2 variable descriptions;

Table 3. NEST Age Groups and Age Categories Used in Our Analysis

NEST age groups	Frequency	Age categories
16–19	49	16–19 (*n* = 49)
20–24	69	20–24 (*n* = 69)
25–29	20	25–34 (*n* = 31)
30–34	11	
35–39	2	35–64 (*n* = 26)
40–44	1	
45–49	3	
50–54	4	
55–59	7	
60–64	9	
65–69	5	65 and over (*n* = 26)
70–74	3	
75–79	7	
80–84	8	
85–89	3	
Missing age data	3	Missing

however, two of the authors also reviewed a subsample of the SHRP2 Insight Data Access Website video sequences to assess the level of visual complexity associated with different localities and traffic levels of service.

It should be noted that the same level of environmental demand is not expected to induce the same level of workload on each driver. The perceived workload would depend on the driver skills, their chosen speed, and the interactions among the two along with environmental demand level. Thus, in addition to environmental demand, GPS speed at the beginning of an event (the very first frame coded for the event) and driver's age (a proxy of skill and ability) were also included in our model as predictor variables.

Driver Age

As discussed earlier, driver age was used in our model to capture skill and ability. Table 3 presents the original NEST age groups, as well as the five categories we created to obtain a reasonable minimum sample size within each category.

Descriptive Analysis of Secondary Task Types

In addition to our statistical model, we conducted preliminary descriptive analysis to investigate how frequently different task types occurred within an epoch as well as how frequently they occurred together with another task type within an epoch.

Results

The final statistical model included event type, $\chi^2(2) = 30.75$, $p < .0001$, environmental demand, $\chi^2(1) = 1.28$, $p = .26$, and driver age, $\chi^2(4) = 8.13$, $p = .09$. Overall, GPS speed information was found to be missing from 42 crashes, 10 near-crashes, and 85 baseline events. Thus, the non-significant main effect of GPS speed was excluded from the final model in order not to lose any additional statistical power. None of the interaction terms were significant and were dropped from the model ($p > .05$).

The results showed that the likelihood of engagement in multiple types of secondary tasks (as opposed to in a single secondary task type) was more likely to occur during SCEs compared to baselines (Figure 3); lower severity SCEs vs. baselines: odds ratio (OR) = 2.10, 95% confidence interval (95% CI) = 1.36, 3.24, $p = .0008$; higher severity SCEs vs. baselines: OR = 2.75, 95% CI = 1.84, 4.11, $p < .0001$. No significant difference was observed between lower severity and higher severity SCEs, $p > .05$.

Even though from Figure 3 it appears that there might be an interaction effect between event type and driver age, the interaction was not statistically significant as mentioned earlier. The main effect of driver age was marginally significant as reported above. With a marginal statistical significance, drivers 65 years old and over were less likely to engage in multiple types of secondary tasks (as opposed to a single secondary task) compared to other age groups; 16–19 year olds: OR = 0.45, 95% CI = 0.26, 0.78, $p = .005$; 20–24 year olds: OR = 0.56, 95% CI = 0.33, 0.95, $p = .03$; 25–34 year olds: OR = 0.60, 95% CI = 0.33, 1.09, $p = .09$; 35–64 year olds: OR = 0.56, 95% CI = 0.29, 1.07, $p = .08$.

Preliminary Descriptive Analysis of Secondary Task Types

Table 4 presents the number of times different secondary task types were observed alone within a 10 s epoch, as well as the number of times they were observed together with other task types within a 10 s epoch. For the latter number, an epoch could have been included more than once in the total count, as a 10 s epoch could have more than two secondary task types.

The most common task type observed alone in the baseline epochs were "passenger interaction", followed by "dancing/singing", and "interaction with carried-in device". For SCEs, the list consisted of "interaction with carried-in device", "passenger interaction", and "reaching/manipulating object". The most common task types that were observed together in baseline epochs were the same as the most common pairs observed together in SCE epochs: dancing/singing + interaction with

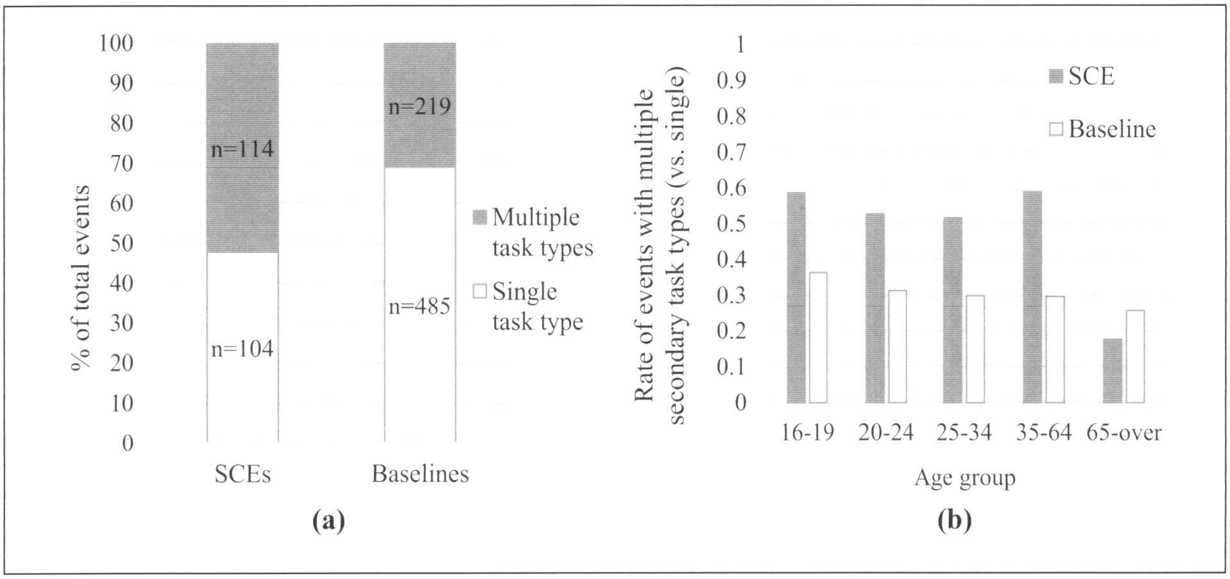

Figure 3. Showing (*a*) frequency and percentage of SCEs and baselines with single vs. multiple types of secondary task engagement; and (*b*) rate of events (type: SCEs and baselines) with multiple types of secondary task engagement (as opposed to single) across different age groups.

carried-in device, dancing/singing + outside distraction, and outside distraction + passenger interaction.

Discussion of Results

This paper investigated engagement in single vs. multiple types of secondary tasks in distraction-affected SCEs and baselines reported in the NEST dataset. Engagement in multiple types of secondary tasks was observed in 52% of the distraction-affected SCEs and 31% of the baselines involving secondary task engagement. Moreover, our statistical model revealed that engagement in multiple types of secondary tasks was significantly more likely (as opposed to engagement in a single secondary task) during SCEs compared to baselines. The severity of the SCEs was also considered but no statistical difference between the lower and higher severity levels was found.

Overall, it appears that engagement in multiple secondary task types is prevalent in both distraction-affected SCEs and baselines, but is more likely in SCEs than baselines. Most crash risk studies to date reported the effects associated with one type of secondary task when it appears that in reality these effects may be confounded by the presence of other secondary tasks. For example, the prevalence rates of different distraction types and the associated crash/near-crash risks reported in Dingus et al. (*12*) do not consider whether drivers engaged in more than one distraction type. Therefore, the numbers reported in Dingus et al. (*12*) and other naturalistic studies of crash risk only capture part of the story. The NEST dataset is a small portion of the SHRP2 data, and based on its sampling methodology, is not appropriate for crash risk calculations. Future research can use SHRP2 data or other naturalistic driving data to conduct similar analysis to the ones we report in our paper and calculate crash risks associated with engagement in single vs. multiple types of secondary tasks.

Although we have presented some preliminary descriptive analysis in this paper, future research should also further investigate which types of secondary tasks are more likely to occur together, and whether there are patterns that may explain such groupings (e.g., the type of attentional resources claimed by different secondary tasks). The limited sample size of the NEST database prevented us from making conclusions about the different multiple task combinations that the drivers engaged in. It is possible for drivers at times to engage in multiple secondary tasks without a significant degradation in their driving performance. Drivers may be able to self-regulate their attention allocation efficiently across the driving task and multiple types of secondary tasks, especially if these secondary tasks claim attentional resources that are not central to the driving task at hand (*19*, *20*). SHRP2 data can be used to assess prevalence of different multiple secondary task combinations, but also crash risks associated with the more prevalent ones. The prevalence and crash risk findings from SHRP2 data can also inform controlled studies (e.g., simulator or instrumented vehicle) that investigate how driving performance is affected by multi-tasking situations that are commonly

Table 4. Number of Times Different Secondary Task Types Were Observed Within a 10 s Epoch Alone and Together with Another Task Type

			# of times task type observed together with another task type							
Secondary task type	Event type	# of times task type observed alone	Visual/ manual vehicle center stack	Personal hygiene	Interaction with carried-in device	Talking on hand-held cell phone	Drinking/ eating	Dancing/ singing	Outside distraction	Reaching/ manipulating object
Visual/manual vehicle center stack	Baseline	58	–							
	SCE	12								
Personal hygiene	Baseline	54	8	–						
	SCE	12	7							
Interaction with carried-in device	Baseline	189	13	4	–					
	SCE	79	4	3						
Talking on hand-held cell phone	Baseline	76	3	2	9	–				
	SCE	32	1	1	8					
Drinking/eating	Baseline	29	1	2	3	0	–			
	SCE	5	3	0	0	0				
Dancing/singing	Baseline	215	16	14	25	0	3	–		
	SCE	20	7	7	18	0	1			
Outside distraction	Baseline	96	15	2	17	4	5	29	–	
	SCE	33	6	7	14	5	3	25		
Reaching/ manipulating object	Baseline	59	7	5	12	0	15	15	9	–
	SCE	34	6	5	11	3	9	11	7	
Passenger interaction	Baseline	243	14	10	16	1	7	7	28	23
	SCE	53	10	4	6	3	3	2	20	14

Note: Prevalent cases highlighted.

experienced by drivers or that are associated with increased crash risks.

Our results also showed that, although marginally significant, drivers 65 years old and over were less likely to engage in multiple types of secondary tasks compared to younger age groups. Risk-reducing compensatory behaviors for older drivers have been reported in previous studies (*21*, *22*). Further research is needed to identify whether the reduction observed in our analysis is intentional to compensate for decreasing cognitive and motor abilities, the result of a cognitive saturation with multiple types of secondary tasks that older drivers experience more than other age groups, or an effect of generational differences in driving styles and technology use. Further, raw data seemed to indicate that there might be an interaction effect between event type and driver age. An increased sample size may reveal stronger effects for age-related factors. Sample size may have also played a role in the lack of significance for the environmental demand variable. It might also be that environmental demand may not have an influence on engagement in single vs. multiple types of secondary tasks or that our categorization of environmental demand needs to be improved. We could not identify any previous literature systematically classifying environmental demand in driving. Thus, our categorization was a first attempt that was exploratory in nature. Further research is needed for quantifying environmental demand.

The results presented in this paper should be interpreted considering the fact that NEST data only includes drivers who have had at least one distraction-affected crash/near-crash. Thus, they are not necessarily representative of the entire driving population, but a particularly at-risk one for distraction-affected crashes.

Acknowledgments

Funding for this work was provided by the Toyota Collaborative Safety Research Center (CSRC). We would like to thank Joshua E. Domeyer, Heishiro Toyoda, James Foley, and Chuck Gulash from Toyota CSRC for their feedback. We would like to also acknowledge VTTI (Julie Cook, Suzie Lee, Justin Owens, and Miguel Perez) for their help with our inquiries about the data. Special thanks to MIT AgeLab (Bryan Reimer and Joonbum Lee) for sharing their reduction code with us.

Author Contributions

The authors confirm contribution to the paper as follows: study conception and design: Martina Risteska, Birsen Donmez, Huei-Yen W. Chen; analysis and interpretation of results: Martina Risteska, Birsen Donmez, Miti Modi; draft manuscript preparation: Martina Risteska, Birsen Donmez, Huei-Yen W. Chen. All authors reviewed the results and approved the final version of the manuscript.

References

1. National Center for Statistics and Analysis. *Distracted Driving 2015*. Traffic Safety Facts Research Note. Report No. DOT HS 812 381. National Highway Traffic Safety Administration, Washington, D.C., 2017.
2. Regan, M. A., J. D. Lee, and K. L. Young. *Driver Distraction: Theory, Effects, and Mitigation*. CRC Press, Boca Raton, 2009.
3. Alm, H., and L. Nilsson. Changes in Driver Behaviour as a Function of Handsfree Mobile Phones - A Simulator Study. *Accident Analysis & Prevention*, Vol. 26, No. 4, 1994, pp. 441–451.
4. Brookhuis, K. A., G. de Vries, and D. de Waard. The Effects of Mobile Telephoning on Driving Performance. *Accident Analysis & Prevention*, Vol. 23, No. 4, 1991, pp. 309–316.
5. Reed, M., and P. Green. Comparison of Driving Performance On-road and in a Low-cost Simulator Using a Concurrent Telephone Dialling Task. *Ergonomics*, Vol. 42, No. 8, 1999, pp. 1015–1037.
6. McKnight, A. J., and A. S. McKnight. The Effect of Cellular Phone Use Upon Driver Attention. *Accident Analysis & Prevention*, Vol. 25, No. 3, 1993, pp. 259–265.
7. Strayer, D. L., and F. A. Drews. Profiles in Driver Distraction: Effects of Cell Phone Conversations on Younger and Older Drivers. *Human Factors*, Vol. 46, No. 4, 2004, pp. 640–649.
8. Strayer, D. L., and W. A. Johnston. Driven to Distraction: Dual-Task Studies of Simulated Driving and Conversing on a Cellular Telephone. *Psychological Science*, Vol. 12, No. 6, 2001, pp. 462–466.
9. Sodhi, M., B. Reimer, and I. Llamazares. Glance Analysis of Driver Eye Movements to Evaluate Distraction. *Behavior Research Methods*, Vol. 34, No. 4, 2002, pp. 529–538.
10. Lee, J., J. Forlizzi, and S. E. Hudson. Iterative Design of MOVE: A Situationally Appropriate Vehicle Navigation System. *International Journal of Human-Computer Studies*, Vol. 66, No. 3, 2008, pp. 198–215.
11. Dingus, T. A., J. M. Hankey, J. F. Antin, S. E. Lee, L. Eichelberger, K. E. Stulce, D. McGraw, M. Perez, and L. Stowe. *SHRP 2 Report S2-S06-RW-1: Naturalistic Driving Study: Technical Coordination and Quality Control*. Transportation Research Board of the National Academies, Washington, D.C., 2015.
12. Dingus, T. A., F. Guo, S. Lee, J. F. Antin, M. Perez, M. Buchanan-King, and J. Hankey. Driver Crash Risk Factors and Prevalence Evaluation Using Naturalistic Driving Data. *Proceedings of the National Academy of Sciences*, Vol. 113, No. 10, 2016, pp. 2636–2641.
13. Klauer, S. G., T. A. Dingus, V. L. Neale, J. D. Sudweeks, and D. J. Ramsey. *The Impact of Driver Inattention on Near-Crash/Crash Risk: An Analysis using the 100-Car Naturalistic Driving Study Data*. Report No. DOT HS 810 594. National Highway Traffic Safety Administration, Washington, D.C., 2006.
14. Domeyer, J. E., S. Seaman, L. Angell, J. Lee, B. Reimer, C. Zhang, and B. Donmez. SHRP2 NEST database: Exploring Conditions of Secondary Task Engagement in Naturalistic Trip Data. *Adjunct Proc., 8th International Conference*

on Automotive User Interfaces and Interactive Vehicular Applications, 2016, pp. 185–190.
15. Owens, J. M., L. Angell, J. M. Hankey, J. Foley, and K. Ebe. Creation of the Naturalistic Engagement in Secondary Tasks (NEST) distracted driving dataset. *Journal of Safety Research*, Vol. 54, 2015, pp. 33–40.
16. Funkhouser, D., and J. Sayer. Naturalistic Census of Cell Phone Use. *Transportation Research Record: Journal of the Transportation Research Board*, 2012. 2321: 1–6.
17. Tivesten, E., and M. Dozza. Driving Context Influences Drivers' Decision to Engage in Visual-Manual Phone Tasks: Evidence from a Naturalistic Driving Study. *Journal of Safety Research*, Vol. 53, 2015, pp. 87–96.
18. Hankey, J. M., M. A. Perez, and J. A. Mcclafferty. *SHRP 2 Report S2-S08A-RW-1: Description of the SHRP 2 Naturalistic Database and the Crash, Near-Crash, and Baseline Data Sets*. Transportation Research Board of the National Academies, Washington, D.C., 2016.
19. Wickens, C. D. Multiple Resources and Performance Prediction. *Theoretical Issues in Ergonomics Science*, Vol. 3, No. 2, 2002, pp. 159–177.
20. Wickens, C. D. Multiple Resources and Mental Workload. *Human Factors*, Vol. 50, No. 3, 2008, pp. 449–455.
21. Donorfio, L. K. M., L. A. D'Ambrosio, J. F. Coughlin, and M. Mohyde. To Drive or Not to Drive, That isn't the Question—The Meaning of Self-Regulation Among Older Drivers. *Journal of Safety Research*, Vol. 40, No. 3, 2009, pp. 221–226.
22. Reimer, B., B. Donmez, M. Lavallière, B. Mehler, J. F. Coughlin, and N. Teasdale. Impact of Age and Cognitive Demand on Lane Choice and Changing Under Actual Highway Conditions. *Accident Analysis & Prevention*, Vol. 52, 2013, pp. 125–132.

The Users Performance Section (AND00) peer-reviewed this paper (18-06316).

Article

TRR
JOURNAL OF THE TRANSPORTATION RESEARCH BOARD

Transportation Research Record
2018, Vol. 2672(37) 11–18

Article reuse guidelines:
sagepub.com/journals-permissions
DOI: 10.1177/0361198118790328
journals.sagepub.com/home/trr
SAGE

Bicycling Simulator Calibration: Proposed Framework

Dylan Horne[1], Masoud Ghodrat Abadi[1], and David S. Hurwitz[1]

Abstract

Bicycling simulation allows for the low-risk experimental study of human factors within transportation environments. A cyclist pedals on a stationary bike trainer, which is instrumented to detect the speed of the wheel and the steering angle of the bicycle. This paper proposes a speed calibration procedure to increase the validity of the simulator results, by using an independent bicycle computer for comparing the simulator speed. The speed ratio, defined as the simulator speed divided by the bike computer speed, approaches one when the simulator is properly calibrated. The effect of tire pressure was analyzed by examining the speed ratio for various tire pressures. The optimal tire pressure was selected as the one that provided a speed ratio closest to one when all other factors were held constant. In the final calibration, a gain factor was used to modify the simulator speed calculation that was embedded in the simulator's bicycle dynamics model. Following calibration, the final simulation speed was within 99.5% of the bicycle computer speed, indicating that the physical speed of the wheel was accurately modeled in the simulation environment. The calibration procedure uses general equations and techniques that can be applied to other bicycling simulators to calibrate speed measurements and improve the consistency of experimental data worldwide.

Bicycling simulation allows for the careful examination of bicyclist behaviors and interactions with various elements in the built environment in a controlled experimental setting. Novel or existing infrastructure can be analyzed to determine the effectiveness of traffic control devices. Interactions between conflicting modes of travel can be evaluated with surrogate safety measures to understand crash risk. The controlled and repeatable nature of human-in-the-loop simulator experimentation provides a means to develop explanatory mechanisms for transportation user behavior, which is difficult to extract from naturalistic experiments (*1*). The virtual reality environment significantly reduces the risk for participants, who can be exposed to risky scenarios while avoiding potential harm (*2*).

The ability to extrapolate conclusions from simulation studies to real-world practice requires that the simulation and real-world performances be matched. Thus, calibration, measurement accuracy, and validation must be given careful attention. Simulated environments may not yet be able to emulate every nuance of real-world experiences, but they are sufficient to create environments in which user responses are similar to those they give in the real world (*1*). Thus, these simulations include relative validity—meaning that users respond in the same direction as in the real world—but do not include absolute validity—meaning that the simulation response is not yet identical to the real-world response in both magnitude and direction (*2*). In fact, reducing some of the variability that is experienced in the real world contributes to the power of simulation in controlled experiments, as almost all environmental factors are administered. However, because of the limited number of bicycle simulators worldwide, results from such simulators have been considered less rigorous than similar results from the comparatively more mature field of driving simulation (*3*).

The Oregon State University Bicycle Simulator is the focus of this calibration effort, although the procedure could be applied to other bicycling simulators (Figure 1). The simulator uses SimCreator (Realtime Technologies Inc.) as the simulation software package, which manages the vehicle dynamics and visual field. For the bicycle simulation, the vehicle dynamics are modified to create a vehicle that has the characteristics of a bicycle. The user's vehicle has a narrower, shorter wheelbase and reduced

[1]School of Civil and Construction Engineering, Oregon State University, Corvallis, OR

Corresponding Author:
Address correspondence to David S. Hurwitz: david.hurwitz@oregonstate.edu

Figure 1. Calibrated Oregon State bicycle simulator.

speed to emulate the performance of a bicycle. The simulator has two input devices: a cradle for the front wheel to capture the steering angle, and an instrumented stationary bike trainer to capture the speed of the bicycle (Figure 2). The visual field is projected on a screen in front of the cyclist, and a surround sound system provides audio. The platform is adjustable for various bicycle sizes, including a children's bike.

Figure 2. Bicycle computer used to collect the physical speed of the wheel.

Methodology

Calibrating the wheel speed input between real-world and virtual bicycling will increase the validity of the bicycling simulator. Wheel speed was calibrated through an independent bike computer (Figure 2), which calculated the physical speed of the wheel based on the wheel size and a spinning magnet attached to a spoke (Figure 3). The physical speed of the wheel was transferred through the bike trainer onto a rotational sensor, which the bicycle simulator used to calculate the simulation speed.

Speed data from the bike computer were exported and compared with the speed data recorded by the simulation computer (Figure 4). Two factors, gain and tire pressure, were investigated to understand the operational interface between the bicycle and the simulator. The gain factor, embedded in the vehicle dynamics package in the simulation software, was used to calculate the wheel speed based on the angular speed from the rotation sensor input. The gain factor can be defined by the operator and used to calibrate the simulated speed measurement.

The interface between the tire and the bike trainer effected the transfer of motion from the wheel and the input device to the computer. The tire pressure and tightness of the bike trainer on the tire are related to the amount of friction or rolling resistance. If the rolling resistance is too low, the tire will slip past the trainer, especially if the cyclist is exerting high torque. Previous research on bicycle tire pressure has found an

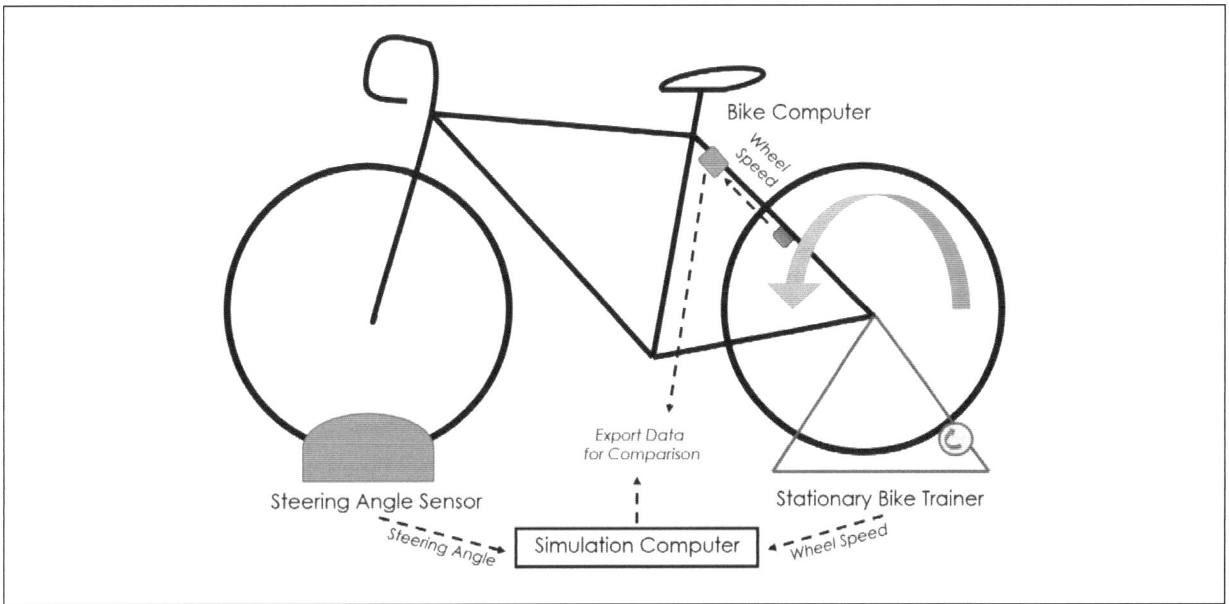

Figure 3. Wheel speed calibration diagram.

inconsistent relationship between tire pressure and rolling resistance, with tire diameter being a better indicator of rolling resistance (*4*).

Various tire pressures between 40 and 60 pounds per square inch (psi) were tested with one cyclist using consistent gearing to determine the effect of tire pressure on speed measurements. The manufacturer's recommended tire pressure for the bicycle was 50–60 psi. Higher pressure was expected to reduce slip, decreasing the variance of the simulation measurement. Equation 1 shows how the tire pressure factor and gain factor influence the simulated speed measurement.

$$\text{Speed}_{\text{Sim}} = \text{Speed}_{\text{Wheel}} * \text{Tire Pressure Factor} * \text{Gain Factor} \quad (1)$$

where

$\text{Speed}_{\text{Sim}}$ is the observed wheel speed calculated from the simulation;

$\text{Speed}_{\text{Wheel}}$ is the actual speed of the physical wheel;

Tire Pressure Factor accounts for losses caused by the tire/bike trainer interface; and

Gain Factor is a variable in the simulation for speed calibration.

Equation 2 shows the relationship between the bike computer and the wheel speed, where $\text{Speed}_{\text{Bike Computer}}$ is the speed recorded by the bike computer.

$$\text{Speed}_{\text{Bike Computer}} \approx \text{Speed}_{\text{Wheel}} \quad (2)$$

The goal was to minimize the difference between speed observations. Speed measurements of the bike computer and the simulator were converted to common units (mph), and the delta speed was calculated as the difference between the simulation speed and the bike computer measured speed, as

$$\Delta\text{Speed} = \text{Speed}_{\text{Sim}} - \text{Speed}_{\text{Bike Computer}} \quad (3)$$

To account for variations in the magnitudes of speed measurements, a ratio of the two speed measurements was used (Equation 4). The bike computer speed, which directly represents the wheel speed, was used as the denominator. Ratio values greater than one indicate overestimates of bike speed, whereas values less than one indicate underestimates. Calibration occurs as the speed ratio approaches one.

$$\text{Speed Ratio} = \frac{\text{Speed}_{\text{Sim}}}{\text{Speed}_{\text{Bike Computer}}} \quad (4)$$

The simulation computer collects data during the whole simulation, but the bike computer only collects data while the rear bicycle wheel is moving. At the startup and termination of the simulation, several seconds of data are collected that do not have corresponding bike computer data. The time series for the simulation was trimmed to make it equal to the time series of the bike computer (Equation 5). A similar process was used to trim the end of simulation data by removing all zero-speed data recorded during simulation shut down.

$$\text{Sim Time}_{\text{Modified}} = \text{Sim Time}_{\text{Original}} - \text{Time Step of First Non Zero Speed} \quad (5)$$

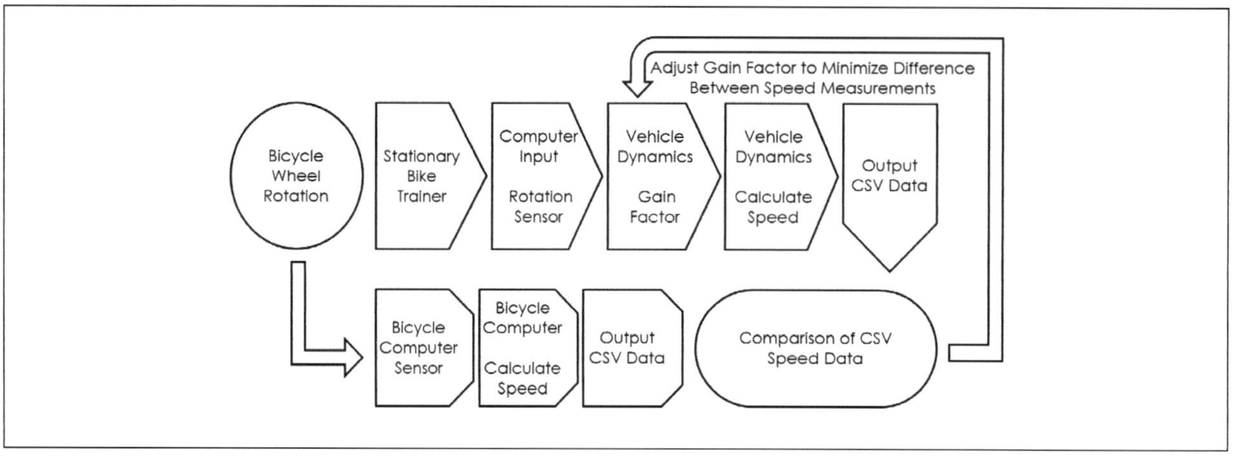

Figure 4. Flowchart of the comparison of two speed measurements.

The time intervals must be set to the same frequency. Instead of aggregating the higher-frequency simulation data, the bike computer data were disaggregated to match the higher frequency of the simulation computer. Each of the time intervals was rounded to the tenths place, and then the corresponding bike computer speed was matched to each time step. Under this framework, each bike computer speed corresponded to around 850 simulation speed measurements. This disparity in data resolution resulted in error, especially following large changes in speed. For example, when a participant stopped for a traffic signal in the simulation, the bike computer data were slow to respond to the change in speed, and then again slow to respond to the acceleration on the green indication. A potential improvement would be to use a geometric estimate of the higher resolution speeds based on the current and next bicycle computer speed measurements, as shown by

$$\begin{aligned} \text{Speed}_{\text{Est}} &= m(x - x_0) + y_0 \\ &= \frac{\text{Speed}_{n+1} - \text{Speed}_n}{\text{Time}_{n+1} - \text{Time}_n} * (\text{Time}_{\text{Est}} - \text{Time}_n) + \text{Speed}_n \end{aligned} \tag{6}$$

where

$\text{Speed}_{\text{Est}}$ is the estimated speed in the high-resolution interval;

Speed_n is the previous speed measurement (from the bike computer);

Speed_{n+1} is the next speed measurement;

Time_{Est} is the high-resolution interval;

Time_n is the previously measured interval (from the bike computer); and

Time_{n+1} is the next measured interval.

Inversely, the simulation data could be aggregated by using a moving average or a Kalman filter, to remove much of the noise and decrease the computational effort. These methods create a local average based on nearby data points, reducing data variability caused by short-term fluctuations, while maintaining the general trend of the data. A moving average formula is shown by

$$\begin{aligned} \overline{\text{Speed}_{\text{MA}}} &= \overline{\text{Speed}_{\text{Previous MA}}} \\ &+ \frac{\text{Speed}_{\text{Current Interval}}}{n} - \frac{\text{Speed}_{\text{Previous Interval}}}{n} \end{aligned} \tag{7}$$

where

$\overline{\text{Speed}_{\text{MA}}}$ is the simple moving average speed estimate;

$\text{Speed}_{\text{Current Interval}}$ is the simulation speed measurement at current time; and

n is the number of intervals included in the average.

Results

A graphical comparison of the two speed measurements was used to check the initial alignment of the data, as shown in Figure 5. Large variance between speeds followed large changes of speed because of the large difference in sampling rate, as evident around the 16,000th step when the participant stopped the bicycle. The much higher data resolution for the simulation contributed to some of the noise in the simulator speed. However, in general, the speeds followed the same trends. In Figure 5, the noise in the simulation speeds indicates that the system does occasionally over- or underestimate the speed for short durations. These events are typically very short, as the simulator records data at 85 Hz.

Tire Pressure

One participant rode the bicycle simulator while data on tire pressure were collected. In general, the ride was at least 5 min long using the same gearing for each run.

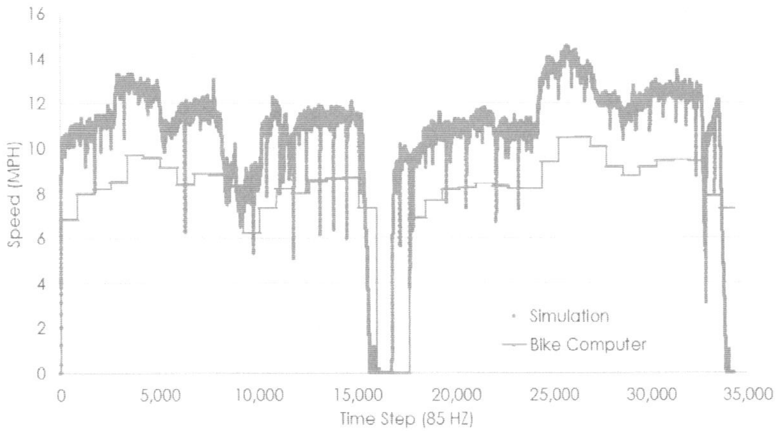

Figure 5. Comparison of speed data from typical participant.

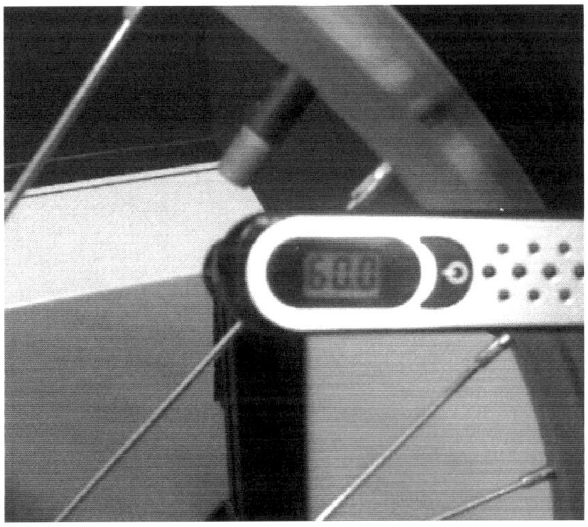

Figure 6. Digital tire pressure gauge during tire pressure analysis.

The tire pressure of the rear tire was set at 40 psi and increased by 5 psi increments up to 60 psi. A digital tire pressure gauge was used to ensure accurate pressure measurements, as shown in Figure 6.

Figure 7 shows the distribution of speed ratios for each tire pressure. A speed ratio of one indicates that the bike computer speed and the simulator speed were perfectly calibrated. The high variance for the 40 and 55 psi data was caused by the low bike computer speeds during the startup and termination portions of the run. During these 10-s intervals, the bike computer speed was much lower than the simulation speed, resulting in large speed ratios. Although the speed ratio means were similar, an analysis of variance (ANOVA) test indicated that they were statistically different (F value 1165.75, $P < 0.000$). A Tukey honest significant difference (HSD) test was used for multiple comparisons of the tire pressure, with all pairs except 45 and 50 psi ($P = 0.711$) being statistically different. This result indicates that tire pressure had a statistically significant influence on the speed ratio.

Descriptive statistics for the various tire pressure speed ratios are shown in Table 1. The speed ratios indicate the relative difference in measured speeds. For example, with a tire pressure of 40 psi and a bike computer speed of 10 mph, the simulator speed mean would be 13.49 mph. The lowest speed ratio corresponded to 40 psi, indicating that the simulator and bike computer were most calibrated at this tire pressure.

Gain Factor

Various gain factor settings were tested to analyze the effect of the gain factor on simulator speed. A single participant rode for 300 s at each setting, using the same gearing and tire pressure for all runs. The participant pedaled the bike to a steady-state speed before the simulation began to minimize variance during the startup period. The steady-state speed was maintained through the end of the simulation to reduce variance further.

Tire pressure was set to 60 psi during the gain factor analysis. This tire pressure is not the optimal tire pressure determined in the tire pressure experiment. Both data sets were collected before the data were analyzed, and it was incorrectly assumed that the higher tire pressure would reduce the slip the most. The factors were tested independently, however, and it is not expected that a significant interaction between the factors exists, as one is a physical relationship and the other is a software setting. The final

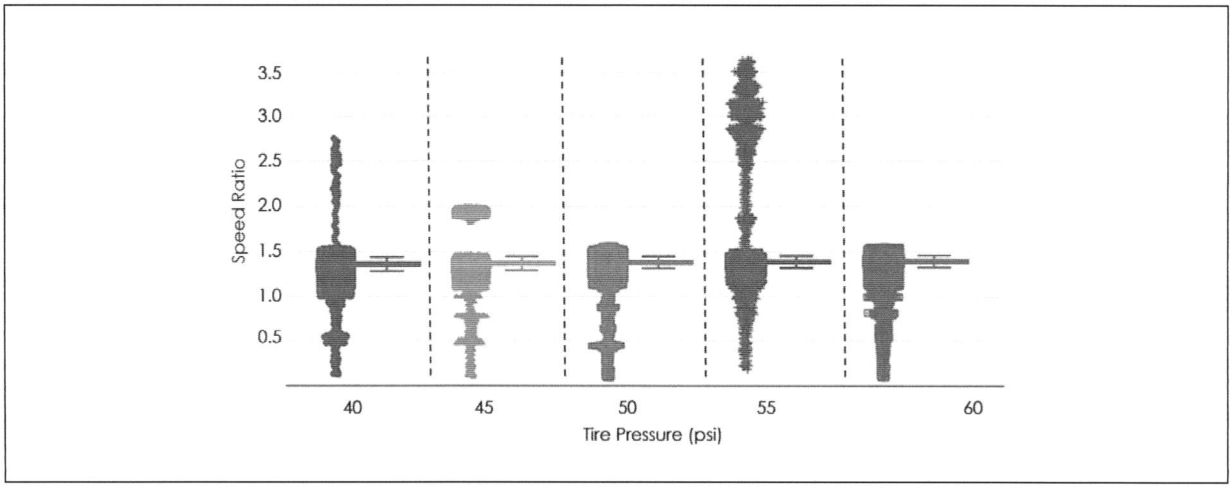

Figure 7. Distribution of speed ratios for each tire pressure.

Table 1. Descriptive Statistics for Various Tire Pressure Speed Ratios

					95% CI			
Tire pressure	*N*	Speed ratio mean	SD	SE	Lower	Upper	Min	Max
40	35,375	1.349	0.0869	0.000462	1.349	1.350	0.1233	2.757
45	47,882	1.368	0.0990	0.000452	1.367	1.369	0.0975	1.998
50	39,332	1.369	0.0772	0.000389	1.368	1.370	0.0841	1.567
55	32,220	1.420	0.2854	0.001590	1.417	1.424	0.1846	3.626
60	34,964	1.380	0.0831	0.000444	1.379	1.381	0.0914	1.549
Total	189,773	1.376	0.1439	0.000330	1.375	1.376	0.0841	3.626

Note: SD = standard deviation; SE = standard error; CI = confidence interval; Min = minimum; Max = Maximum.

Table 2. Gain Factor Settings and Speed Ratio Descriptive Statistics for Each Simulation Run

					95% CI			
Gain factor	*N*	Speed ratio mean	SD	SE	Lower	Upper	Min	Max
0	23,487	−3.1E-7	3.44E-5	2.25E-7	−7.5E-7	1.29E-7	−0.003	0.002
0.1	22,556	0.995	0.140	0.00093	0.993	0.996	−6.88E-7	1.275
0.1	22,837	0.975	0.126	0.00084	0.973	0.976	−2.43E-8	1.252
Default*	22,147	1.585	0.194	0.00130	1.582	1.587	−8.12E-9	1.894
0.2	21,651	2.070	0.244	0.00166	2.066	2.073	−2.26E-8	2.391
1	23,487	8.190	3.191	0.02082	8.149	8.230	−9.454	11.792
Total	136,165	2.328	3.056	0.00828	2.311	2.344	−9.454	11.792

Note: SD = standard deviation; SE = standard error; CI = confidence interval; Min = minimum; Max = Maximum.
*Default gain factor is 0.15707963267949.

calibration using 40 psi should be even more accurate than described here.

Table 2 shows the various gain factor levels and descriptive statistics for each of the runs. An ANOVA analysis indicated that the differences between gain factors were statistically significant (F value = 115845, $P <$ 0.000). A Tukey HSD test indicated that each of the factors was significantly different except for the two 0.1 runs. The slightly different speed ratios with the same gain factor of 0.1 were evidence of random system variability. Setting the gain factor to zero reduced the simulation speed of the bicycle to zero, and setting the gain

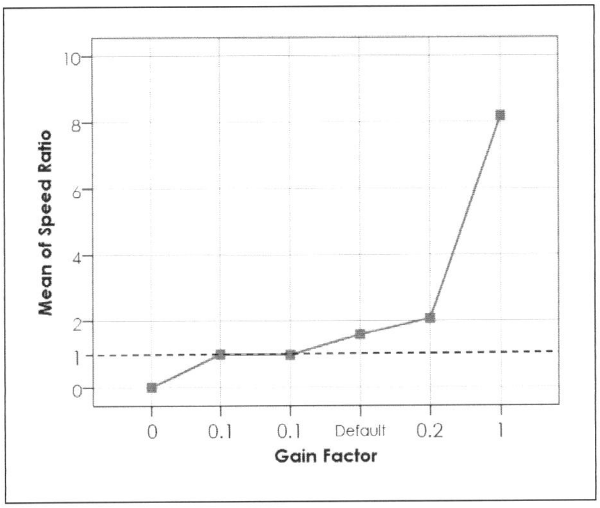

Figure 8. Speed ratios for each gain factor (speed ratio of one corresponds to proper calibration).

factor to one dramatically increased the simulation speed of the bike (8.19 times faster than the real speed of the wheel). With a gain factor of one, the simulated bike would reach speeds of 65 mph, and then become unstable and crash with any steering input. Setting the gain factor to 0.1 produced the most calibrated results, with the simulated speed within 97.5% to 99.5% of the bike computer speed.

Figure 8 graphically shows how the gain factor relates to the speed ratio from the data in Table 2. The *X*-axis shows the values that were tested during the sensitivity experiment, which should be interpreted as categorical variables. Theoretically, any speed ratio could be achieved by adjusting the gain factor; therefore, a solid line was used between the data points. The target value was a speed ratio of one, as this was the indicator of good calibration. During the experiment, the gain factor was adjusted following the principles of Newton's Method, adjusting the value in an iterative fashion to approach the goal of a speed ratio of one. Additional steps could have been performed (0.095 or 0.15), but 99.5% was determined to be acceptable for demonstration purposes.

Conclusions

Bicycle simulator studies provide an experimental framework to evaluate novel and existing infrastructure and human factors while controlling for environmental factors and reducing risk to participants. Calibrating the inputs of the bicycle simulator improves the authenticity of the user experience. The calculated speed of the rear wheel was compared with an independent speed observation from a bike computer to minimize the difference between measurements. Calibrating the observed simulator speed and the actual speed of the bicycle wheel should make the simulation more representative of real cycling, thereby improving the user's experience and the applicability of the results.

The speed ratio, or the simulated speed divided by the bike computer speed, was used to evaluate the influences of tire pressure and gain factor. Various tire pressures were tested based on the manufacturer's recommended tire pressure range, with 40 psi having the most accurate and statistically significant speed ratio measurement. A gain factor of 0.1 brought the simulation to within 99.5% of the bike computer speed, indicating good calibration. The calibration could be further improved by additional refinement after testing tire pressures outside of the manufacturer's recommendations and additional gain factor settings.

The general procedure describe here can be applied to other bicycling simulators around the world. The use of a commercially available bike computer allows for the comparison of simulator speeds against an independent speed measurement. The calibration of speed measurements could increase the repeatability of experimental data across different simulators. The speed ratio framework enables discussion of the difference between the real speed of the bike and the simulated speed, which is especially important when validating simulator results to real-world experiences.

Future Research

The experiment as described only explored the steady-state speed, to minimize speed variance. Evaluating acceleration or deceleration would be difficult using the current bike computer because of data resolution, as the system only records data every 10 seconds. Acceleration events are likely to be much shorter than this interval. Deceleration events are a potential performance measure during experimentation, as they reflect braking as a response to simulated conflicts. These events could be used as a measure of reaction time, specifically involving stopping situations. The calibration effort only focused on speed, but speed is a fundamental property of any human-in-the-loop simulator.

This research creates a standard procedure for bicycle simulator speed calibration. Applying this methodology to other bicycle simulations will help to improve fidelity of bicycling simulation in general, as speed measurements will have a common calibration procedure. The future of this research thread includes applying this procedure to other bicycle simulators, developing a procedure to analyze the calibration of steering input through observation of visual latency, and validation studies to match simulator performance to field experiments.

Future research for validation of the simulator will help to answer questions about the human perception of the simulation. The focus of this research was to calibrate the calculated speed of the simulator to a physical measurement of the speed of the wheel, rather than to calibrate the human perception of speed. The implication is that a calibrated simulator will better emulate the real-world experience. However, because of the relative validity of simulation research, the participant's perception of the simulation speed is arguably at least as important, if not more so. Because of the limits of what sensor information can be presented in a simulator environment, the simulator may seem much faster or slower to participants than the real-world experience. Therefore, a study of user perceptions of bicycling simulation should be performed with the research question, "Does the bicycle simulator match user expectations from riding a real bicycle?" This feedback mechanism should be used to validate the simulator to match user expectations.

Author Contributions

The authors confirm contribution to the paper as follows: study conception and design: Horne, Hurwitz; data collection: Horne; analysis and interpretation of results: Horne, Ghodrat Abadi; draft manuscript preparation: Horne, Ghodrat Abadi, Hurwitz. All authors reviewed the results and approved the final version of the manuscript.

References

1. O'Hern, S., J. Oxley, and M. Stevenson. Validation of a Bicycle Simulator for Road Safety Research. *Accident Analysis & Prevention*, Vol. 100, 2017, pp. 53–58. http://dx.doi.org/10.1016/j.aap.2017.01.002.
2. Kaptein, N., J. Theeuwes, and R. Van Der Horst. Driving Simulator Validity: Some Considerations. *Transportation Research Record: Journal of the Transportation Research Board*, 1996. 1550: 30–36.
3. Nehaoua, L., H. Arioui, and S. Mammar. Review on Single Track Vehicle and Motorcycle Simulators. *Proc., 19th Mediterranean Conference on Control and Automation*, Corfu, Greece, 2011, pp. 940–945.
4. Akanda, F., D. Ahmad, O. Jamarie, S. Sulaiman, and A. Fashina. Empirical Determination of the Motion Resistance of Pneumatic Bicycle Wheels for On and Off-Road Performance. *African Journal of Agricultural Research*, Vol. 5, 2010, pp. 3322–3332.

The Users Performance Section (AND00) peer-reviewed this paper (18-02150).

Article

TRR
JOURNAL OF THE TRANSPORTATION RESEARCH BOARD

Transportation Research Record
2018, Vol. 2672(37) 19–30

DOI: 10.1177/0361198118792995
journals.sagepub.com/home/trr
SAGE

Field Study of Driver Exiting Behavior at Complex Interchanges

Bryan J. Katz[1], Scott O. Kuznicki[1,2], Nicholas Kehoe[1], and Jim Shurbutt[3]

Abstract

A field study explored driver behavior at complex interchanges throughout the United States in order to better understand geometric, signing, and marking characteristics that influence driving behavior and lane selection. Two types of data were collected at 13 interchanges spanning four states: videos from fixed-location cameras, and videos from unmanned aerial vehicles. Findings from the study highlight common behavior as drivers negotiate complex interchanges. For each interchange studied, video was reduced to track vehicle paths throughout the study site. Although data were captured on both through and exiting traffic, the focus of this review was on the exiting vehicles. No major safety issues were observed through the study. One common finding across all sites was that exiting traffic was found to most commonly use the EXIT ONLY lane rather than the option lane. In addition, few common behaviors identified through the video show last-minute lane changes; drivers typically entered their target lane well upstream of the interchange.

The modern urban freeway interchange can present a demanding environment for drivers. Drivers approaching an interchange must undertake the navigation task while workloads from the guidance and control tasks are particularly high. In addition to the task of selecting a lane appropriate to the desired route, drivers also experience workload demands associated with conflicting traffic from lane changes associated with entering and exiting maneuvers. These workload demands are intensified by other factors such as roadway geometric design, short distances between critical points, traffic volumes and density, and speed. The present field study was part of a multi-phase project intended to address safety and operational issues related to complexity within interchange areas. For the purposes of this project, complexity was defined as follows: complexity occurs when the choice of more than one movement is available from a lane or group of lanes in which the decision points occur successively in close proximity. The following project activities were completed:

- Identification of attributes influencing interchange complexity.
- Evaluation of current geometric design, signing, and marking practices.
- Simulator study investigating driver behavior at different interchange layouts.

The present paper describes a fourth activity, a field study in which video data were collected at 13 interchanges spanning four states through the use of both fixed-position cameras and unmanned aerial vehicles (UAVs). The purpose of this field study was to investigate how roadway geometry and signing may affect driving behavior, in terms of making unnecessary lane changes or unsafe maneuvers, in the real world with traffic.

Literature Review

The Federal Highway Administration (FHWA) has sponsored recent research related to complex interchanges (*1–3*) with the purpose of identifying factors that contribute to complexity and characteristics that increase driver confusion, understanding driver expectations, gathering practitioner experience and input, and quantifying interchange complexity.

Richard and Lichty conducted a review of previous work on driver navigation problems and expectations at

[1]Toxcel, LLC, Blacksburg, VA
[2]Transpo Group, Kirkland, WA
[3]Federal Highway Administration, McLean, VA

Corresponding Author:
Address correspondence to Bryan J. Katz: Bryan.Katz@toxcel.com

interchanges and intersections (*2*). Various measures were identified that can be used to record driver behavior and expectations, including driver errors (e.g., number of missed exits, unnecessary lane changes, erratic maneuvers, etc.), operational measures (e.g., lane change distance, response time, vehicle speed, etc.), and subjective measures (e.g., sign expectations, certainty of choice selection, etc.). The researchers used such performance measures to infer the following about driver expectations (*2*):

- Drivers expect to need to be in the lane closest to the exiting direction to be able to exit (*4*).
- Drivers expect lane drops at exits (*5*).
- Signs and lane markings appear to be effective ways to set driver expectations at interchanges (*6*).
- Driver expectations impact driver information acquisition (*7*).

Complex interchanges do not typically have conventional layout patterns; instead, each interchange is unique and, thus, the current guidance and practices are not always sufficient for complex conditions (*8*). Because each interchange is unique, the challenges regarding complex interchanges are often the result of the unique interaction between multiple components, rather than a result of one particular scenario. These components may include roadway geometry variables, signing and markings, traffic volume, driver expectancy violations, and driver workload. Therefore, it is difficult to explicitly define a complex interchange. There are, however, a variety of resources that address the geometric and signing conditions that contribute to interchange complexity. For example, one resource discusses complex geometric situations such as ramp and interchange spacing, collector–distributor (C-D) roadways, ramp braids, and weave sections, and provides guidance in evaluating safety and operational considerations of potential solutions (*9*). Feedback from stakeholders representing 17 state departments of transportation and 32 different regions revealed that multiple routes converging or diverging within a short distance are a common indication of a complex interchange (*3*). Fitzpatrick et al. developed a tool to evaluate the complexity of an interchange, which included higher weights for factors considered to have a greater impact on complexity (*1*). Lane continuity violations and weaving sections (less than 0.5 mi in length) were considered to be the biggest contributors to driver workload and perceived complexity. Other factors with higher weights included number of entrance ramps per mile, ramp densities, number of general purpose lanes, loop ramps or curved approaches to ramps, left-side ramps, and ramps with multiple destinations. Additional research and crash data support that factors such as interchange spacing and ramp characteristics play a significant role in interchange complexity (*10–12*).

Previous literature was helpful in identifying important complex interchange characteristics, challenges that need to be addressed within these interchanges, and strategies to improve uniformity and decrease driver confusion.

Site Evaluation and Selection

The site selection process began with a list of 70 interchanges from throughout the United States and Canada. A variety of factors were used in selecting the final sites, including:

- Similarity to nearby interchanges.
- Geographic dispersion.
- Location in states that are probable candidates for working agreements.
- Number and location of attributes related to complexity.
- Presence of differing treatments for similar geometric designs within a single interchange or administrative region.
- Age of signing within the interchange.
- Type, extent, and age of modifications and improvements.
- Compliance with contemporary design criteria.
- Traffic congestion experience.

Four states were selected to serve as partners in this project. The project team identified seven characteristics of a suitable participating state:

- Freeway system in a large urban area or megaregion.
- Established, published sign design criteria.
- Mixture of central office and regional oversight of planning, design, and operations.
- At least one interchange with an unusual configuration.
- Interchanges exhibiting attributes from the preliminary attributes list.
- Agency commitment to working with FHWA.
- Availability of agency staff and resources to support a project.

The initial list of 70 interchanges was filtered to include only those interchanges within the four partner states, and the criteria listed above were used to refine the list to six sites identified for field evaluations:

1. Site 11: I-4 (US 192 to SR 535) in Kissimmee, FL.
2. Site 26: I-85 (I-285, northeast junction) in Atlanta, GA.

3. Site 27: I-20 (I-285, west junction) in Atlanta, GA.
4. Site 31: I-35W (TH 62) in Minneapolis, MN.
5. Site 41: I-5 (I-405 and Washington SR 518) in Tukwila, WA.
6. Site 43: I-5 (US 101 to SR 510) in Lacey, WA and Olympia, WA.

Data Collection

To ensure that data-collection methods would produce imagery without gaps or areas out of view, the team used satellite imagery to determine the ideal locations for fixed-position cameras and UAV hovering.

Data-Collection Methods

Fixed-location cameras were deployed at Site 11 (supplementing the aerial video), Site 27, and Site 31. For sites 11 and 27, the project team used high-resolution cameras with a weather-resistant housing and supplemental battery. For Site 31, a vendor provided equipment used for intersection traffic counts.

UAV video data were collected at sites 11 and 43 using high-resolution, 4K cameras mounted to professional quality UAVs. Battery technology limited each UAV to approximately 15-min data-collection intervals separated by approximately 5 min of transit and servicing time.

Data Reduction

Data reduction activities varied per site and focused on distilling driving behavior (e.g., route choice, lane selection, origin/destination, etc.) from the video. In general, the data reduction involved viewing each video, selecting vehicles based on a pre-determined sampling plan, and noting the lane position of each selected vehicle at screen lines along the route. Those screen lines were chosen based on the position of guide and regulatory signing, the location of pavement markings and pavement marking pattern changes, and other factors that could influence driver behavior relative to navigating to an exit. Any vehicle that could not be positively tracked across the entire study area was removed from the dataset. Where possible, qualitative observations concerning vehicle trajectory, speed, and sudden movements were also recorded to determine whether some locations exhibited a higher frequency of these movements.

Data Analysis

The primary data analysis approach discussed in this paper is a cluster analysis. Cluster analysis is a data mining technique that groups data into clusters, in which the data items within a single cluster are similar, but items between clusters are dissimilar. The project team used open-source data mining software (*13*) that included an algorithm for k-means clustering. All cluster analyses performed on the field data were completed using the simple k-means algorithm ranging from 2 to 10 clusters. The final number of clusters was selected by identifying the "elbow" of the curve when the numbers of clusters were plotted against the sum of error within each cluster; this is the point where additional clusters do little to better explain the dataset. This analysis provided groups of common driving behaviors (i.e., what lanes drivers selected at different locations) witnessed in the video data.

For this analysis, lane selection per location within a single study site was used as an input. Output of the cluster analysis showed what lane best represented driver behavior at each location within the study site. It is important to note that this method assigns vehicles to the cluster that most closely represents their driving behavior. In other words, while the driver behavior will be similar between drivers of a single cluster, not every driver within a given cluster has the exact same driving behavior.

Site Descriptions and Results

Site 11: Orlando

Description. This site included roughly 15,000-linear-ft (457.5 m) along I-4, between US 192 in the southwest and SR 536 in the northeast. This area is particularly complicated by the presence of braided ramps between World Drive and US 192, between US 192 and Osceola Parkway, and between Osceola Parkway and SR 536. This means that there is no I-4 access to US 192 from Osceola Drive in the westbound direction.

Attributes present at the site:

- Auxiliary lanes
- Exiting lanes
- Exit with downstream split
- Guide signs for option lanes.

Observations. Video collected from the eastbound and westbound approaches will be discussed separately below. Approximately 70 min of video per UAV were reviewed for the eastbound video and approximately 50 min of video per UAV were reviewed for the westbound video. Vehicles were sampled at a rate of one sample per minute, with each sample representing one vehicle per lane, resulting in a total sample size of 328 vehicles (182 eastbound; 146 westbound).

Results

Eastbound. At Exit 65, there were two typical exiting behaviors; either traffic in the right lane remaining in

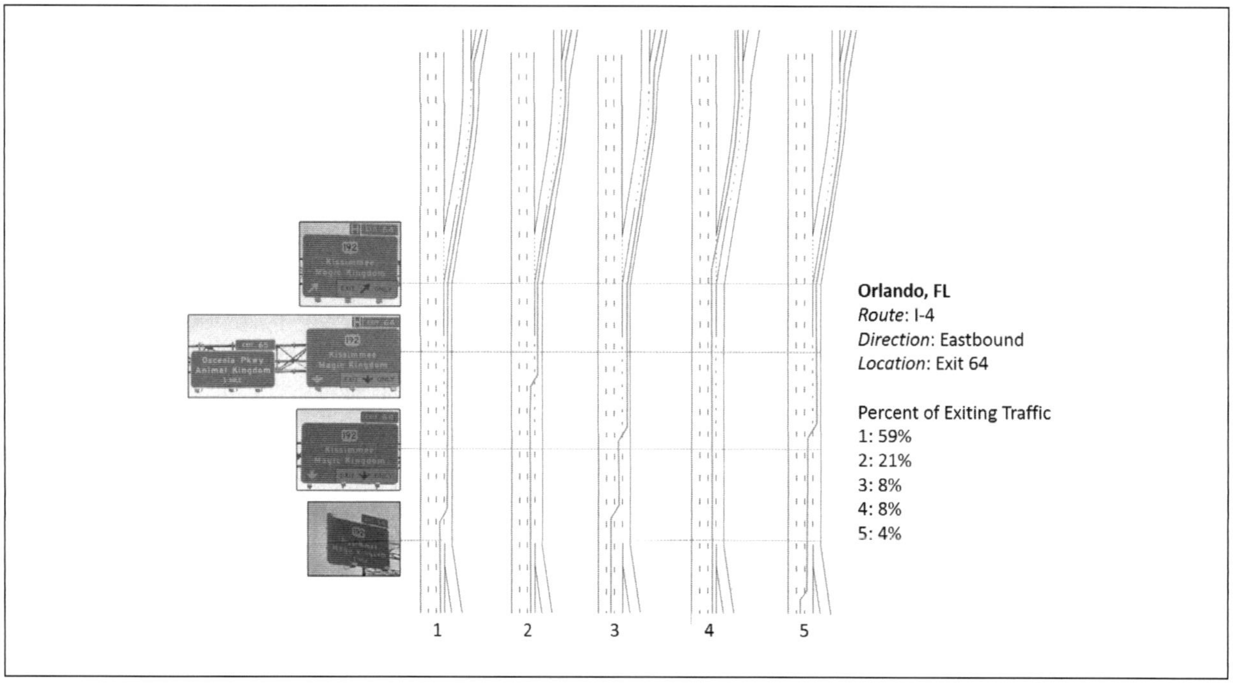

Figure 1. Common exiting driver behaviors at Site 11, eastbound, Exit 64 ($n = 24$).

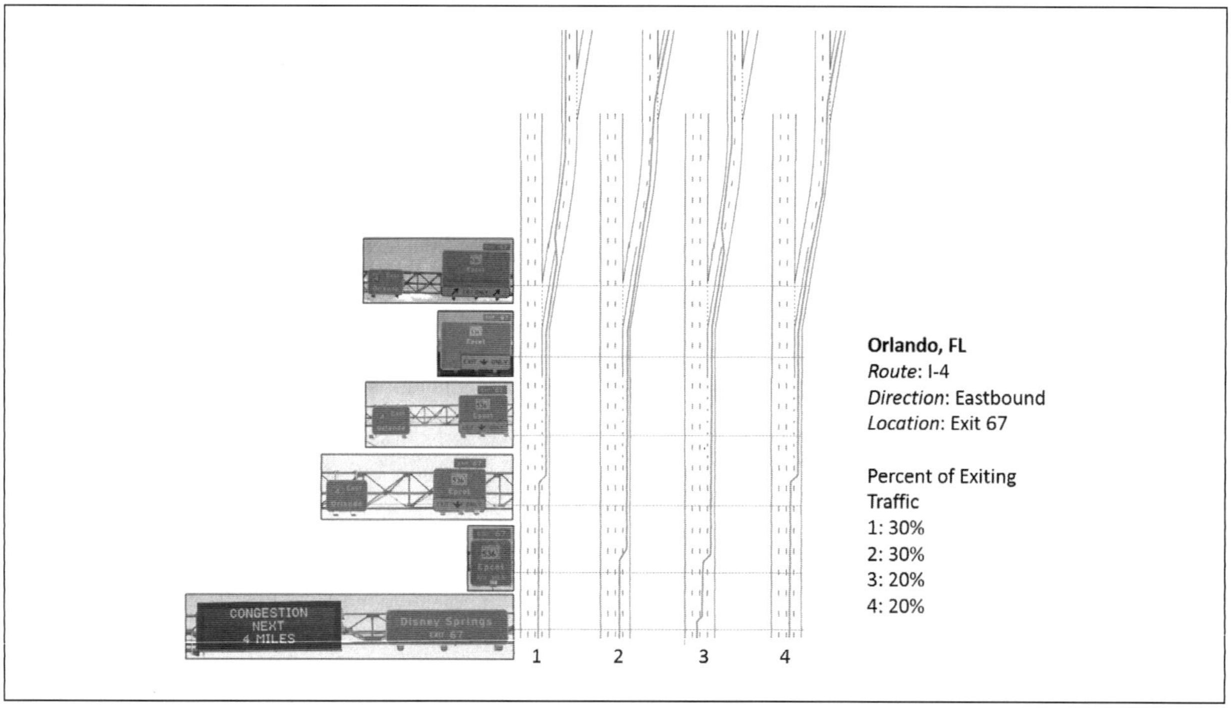

Figure 2. Common exiting driver behaviors at Site 11, eastbound, Exit 67 ($n = 10$).

that lane until they exited, or traffic in the middle lane moved into the right lane after viewing the first set of overhead guide signs, and then remained in the right lane until they exited. Figures 1 and 2 illustrate common driver behavior identified at exits 64 and 67, respectively on eastbound I-4. Depiction 3 in Figure 2 illustrates

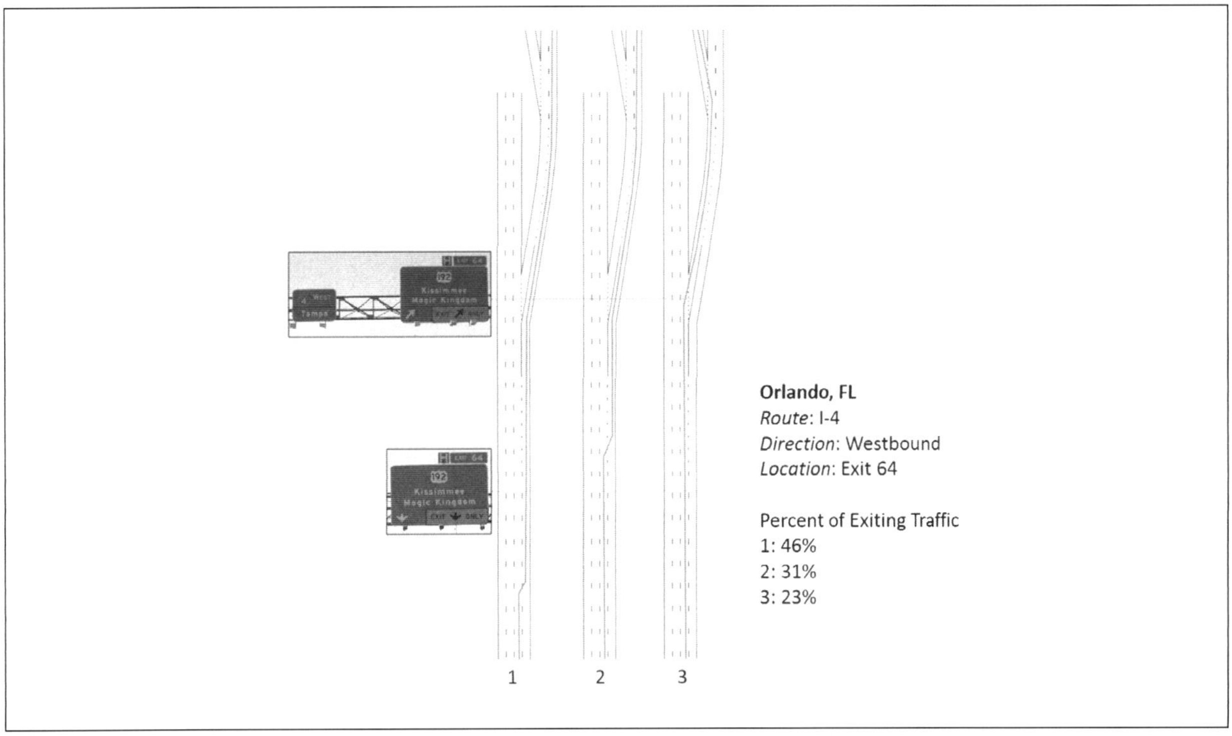

Figure 3. Common exiting driver behaviors at Site 11, westbound, Exit 64 ($n = 26$).

apparent unnecessary lane changes where drivers move out of the option lane to exit the freeway in the exit only lane, yet these drivers make a lane change to the left on the C-D roadway to exit left at the downstream split.

Westbound. Figures 3–5 illustrate common driver behavior at exits 64, 65, and 67, respectively on westbound I-4.

Site 27: Atlanta

Description. Site 27 includes an approximately 1.5-mi (2.42-km) stretch of I-20 eastbound leading up to exits 51A and 51B. In this location, I-20 eastbound consists of four lanes, with the rightmost lane an EXIT ONLY lane, and the adjacent lane an option lane at both the first (Exit 51A) and second (Exit 51B) exits. The attributes present at the site include:

- System interchange with sub-optimal geometry
- Exiting lanes
- Guide signs for option lanes.

Observations. Ninety minutes of video data were reviewed. The video was sampled at a rate of two samples per minute, with each sample representing one vehicle per lane, making a total of four vehicles per sample. The total sample size from the data reduction was 719 vehicles, with 51% of the vehicles exiting I-20.

Results. Figures 6 and 7 illustrate common driver behavior identified at Site 27. In addition, several instances of erratic behavior near Exit 51B were identified. Two passenger vehicles were observed standing in the exit gore before returning to the freeway, and two commercial vehicles were observed crossing the exit gore (one crossed over the exit gore to return to the mainline and the other came to a complete stop on the mainline adjacent to the exit gore for several seconds before crossing the gore to exit).

Site 31: Minneapolis

Description. Site 31 included three approaches to the interchange of I-35W with Minnesota State Trunk Highway 62. These approaches included southbound I-35W, westbound TH 62, and northbound I-35W, in Richfield and Minneapolis. The three interchanges (sites 31-1, 31-2, and 31-3) are described separately below. Site 31-1 was most complicated, with three total splits and a total of four downstream.

Site 31-1. Site 31-1 encompassed a distance of roughly 1.5 mi (.805 km) along southbound I-35W. The first downstream split involves five lanes diverging into six,

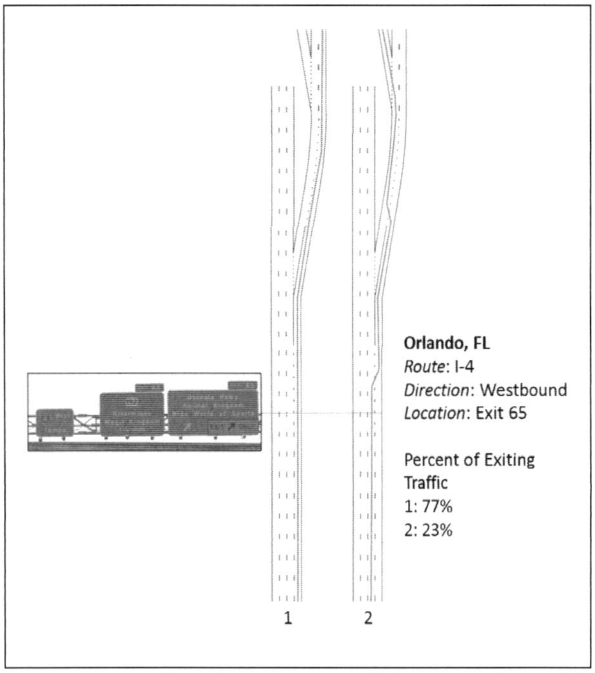

Figure 4. Common exiting driver behaviors at Site 11, westbound, Exit 65 ($n = 13$).

with the center lane serving as an option lane for the exit. The signing in advance of all these locations consisted of discrete arrow signing using a vertical divider above the option lane arrow. The arrows at the departure location are angled to better emphasize the point of departure. The three-lane exit ramp diverges downstream into two 2-lane ramps. This second divergence is referred to as Split 2. Each of those two ramps further splits (splits 3L and 3R), with exclusive-lane exits to Portland Avenue (left-hand ramp) and Lyndale Avenue (right-hand ramp) and are built to accommodate two-lane entrances to TH 62 as the Crosstown Freeway is expanded in the future. The attributes present at this site include:

- Auxiliary lanes
- Exit with downstream split
- Guide signs for option lanes

Site 31-2. Cameras captured both vehicles' behavior at the section of TH 62 where an additional lane is added on the right, creating three lanes, as well as drivers' destination selection (i.e., remain on TH 62 or exit to I-35W northbound). The attributes present at this site include:

- Auxiliary lanes
- Guide signs for option lanes

Site 31-3. Video captured behavior of vehicles as they exited I-35W and traveled on the exit ramp to TH 62. The single-lane exit ramp expands into two lanes after approximately 400 ft (122 m) from exiting I-35W, and the ramp

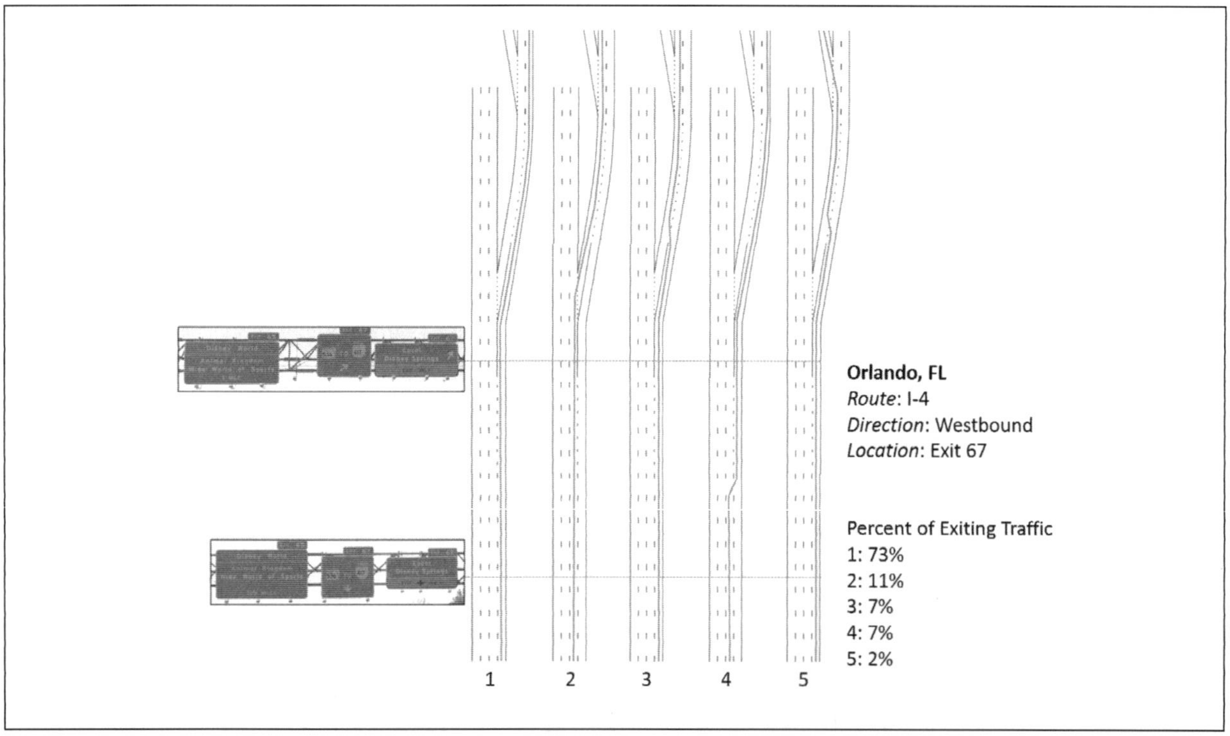

Figure 5. Common exiting driver behaviors at Site 11, westbound, Exit 67 ($n = 45$).

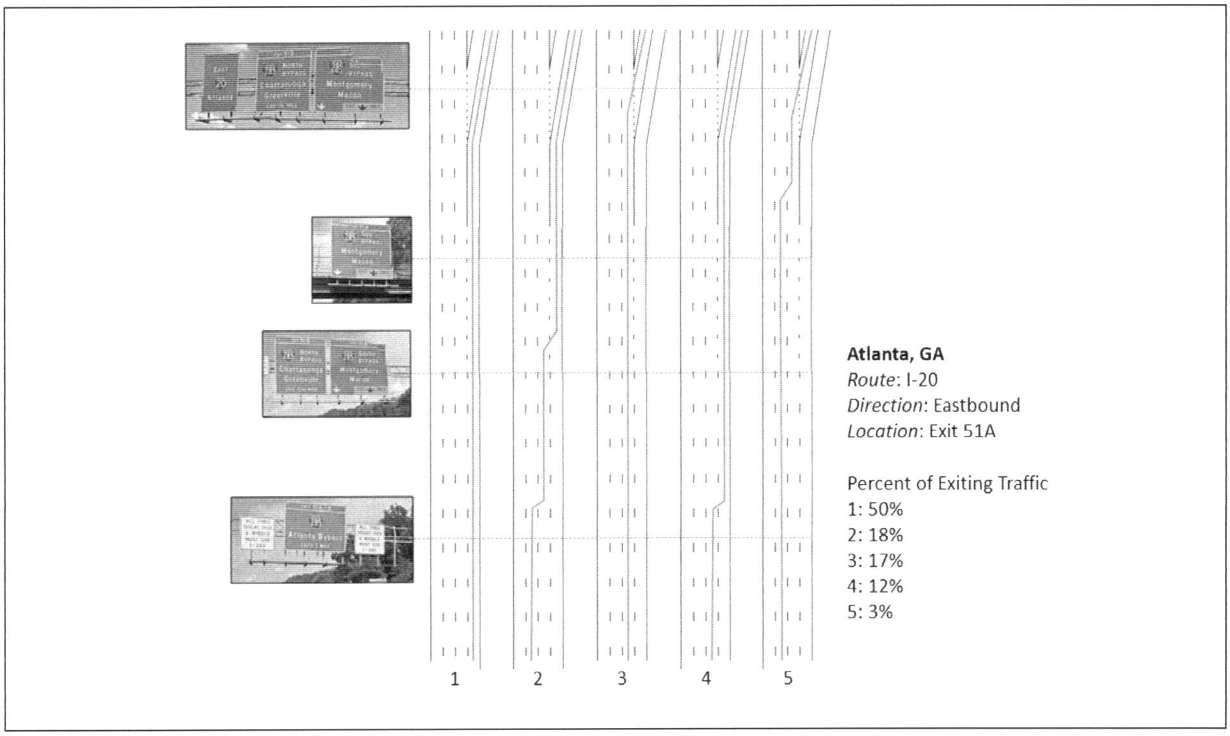

Figure 6. Common exiting driver behaviors at Site 27, Exit 51A ($n = 202$).

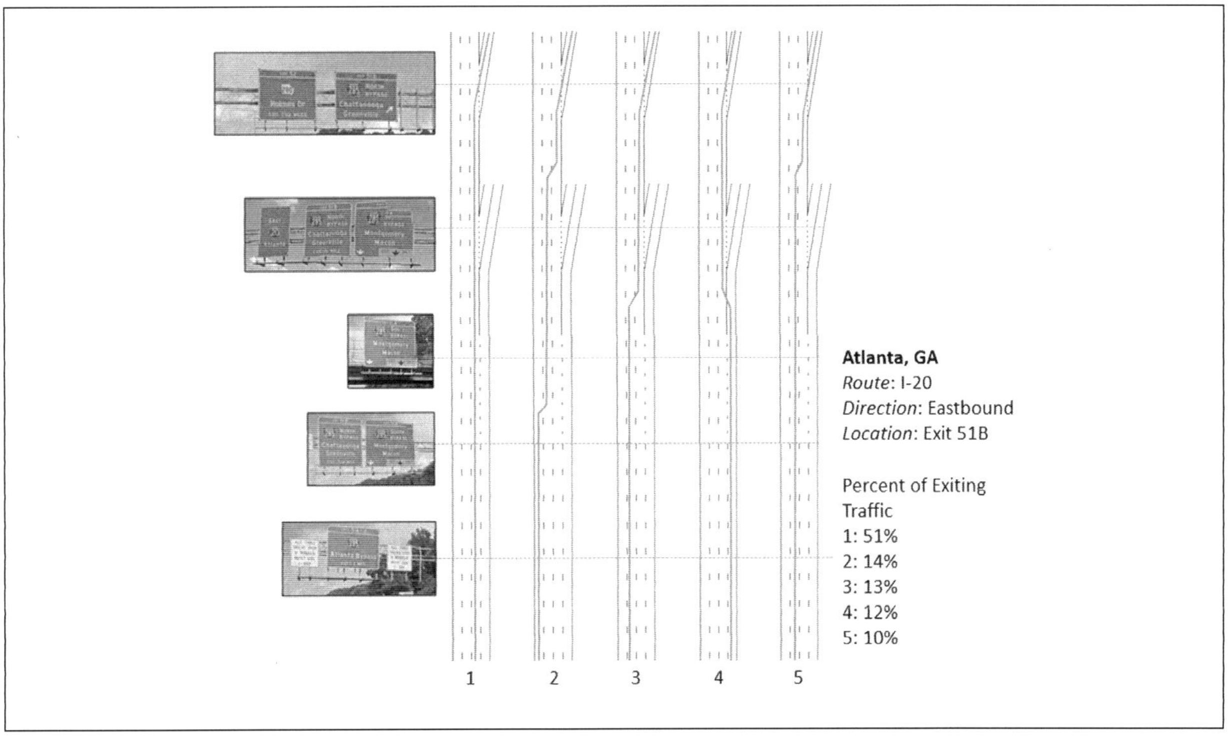

Figure 7. Common exiting driver behaviors at Site 27, Exit 51B ($n = 164$).

splits with the exit on the left bringing the driver to TH 62 westbound and the exit on the right bringing the driver to TH 62 eastbound. The attributes present at this site include:

- Exiting lanes
- Exit with downstream split.

Observations

Site 31-1. The video was sampled once per minute, collecting information on one vehicle per lane across the five lanes of I-35W, totaling 100 vehicles (33 exiting and 62 remaining on I-35W) from the approximately 20 min of video.

Site 31-2. The video was sampled twice a minute, and one vehicle from each lane was selected in each sample, resulting in the capture of driver behavior from 474 vehicles from the approximately 2 h of video. Approximately 1 h of video was recorded during the a.m. peak period (234 vehicles; 79 exiting), and approximately 1 h was recorded during the p.m. off-peak period (240 vehicles; 98 exiting).

This site focused on understanding behavior at a location upstream of an exit where the freeway expands from two to three lanes. The research team recorded vehicle position immediately prior to the addition of the third lane and after the third lane was fully established. Finally, the vehicles' destinations (e.g., remain on or exit the freeway) were recorded. The hypothesis was that drivers intending to exit to I-35W north would avoid using the right-hand lane until the overhead signing was visible. Some drivers may use the lane immediately in heavy traffic or because their intended destination is the next right-hand exit on I-35W north.

Site 31-3. One hour of video was reviewed during the a.m. peak (625 vehicles; 554 toward TH62 West; 71 toward TH62 East), 1 h was reviewed during the p.m. off-peak (781 vehicles; 637 toward TH62 West; 144 toward TH62 East), and 1 h was reviewed during the p.m. peak (738 vehicles; 566 toward TH62 West; 171 toward TH62 East). The p.m. off-peak and p.m. peak times were selected in an effort to maximize the percent of traffic exiting to TH62 East.

The focus at this location was to observe how vehicles behaved when approaching the split. The hypothesis was that vehicles exiting to TH 62 eastbound would move into the formed right-hand lane further down than the beginning of the lane addition taper, owing to the lack of upstream guide signing with route marking. Further, it was hypothesized that vehicles would drift toward the right, following the right white edge line, on account of the lack of dotted extension lines providing positive guidance and indicating to drivers the lane addition taper and the path for those wishing to remain in the left-hand lane.

Results. Figure 8 illustrates common driver behavior identified at Site 31-1.

Findings from Site 31-2 show that fewer drivers exiting the freeway entered the newly formed exit lane near the beginning of the lane (i.e., once it first became available) than those that entered it later. This finding could indicate that exiting drivers are changing lanes further downstream (i.e., closer to the exit) due to uncertainty in identifying the proper lane to navigate the exit.

Findings from Site 31-3 found that traffic exiting to the left (westbound) displayed more uniformity than the traffic exiting to the right (eastbound). The most common behavior for the eastbound traffic was to enter the study area in the center of the lane, maintain the position in the center of the lane, and then move to the right side of the lane near the exit. Fewer eastbound vehicles entered the study area on the right side of the lane and maintained that position before exiting to the right.

Some erratic or uncertain behaviors were identified at Site 31-3. During the p.m. off-peak period, five vehicles were observed exiting the freeway by crossing the exit gore and one vehicle was identified crossing the exit gore at the downstream split. During the p.m. peak period, one vehicle was observed crossing the exit gore, and another vehicle was observed missing the exit, pulling over to the shoulder, and backing up to take the exit.

Site 43: Olympia

Description. Similar to Site 11, Site 43 featured a length of urban freeway with longer spacing between exits, across a distance of 7 mi. Along I-5, three interchange locations were selected (sites 43-1, 43-2, and 43-3), which are described separately in the following sections. Each of these exits features an option lane from the mainline roadway. The attributes at each site included:

- Auxiliary lanes
- Exit with downstream split
- Guide signs for option lanes.

Site 43-1. Site 43-1 included slightly less than 1 mi of I-5 southbound focusing on Exit 105B and Exit 105A. In this location, I-5 southbound consists of four lanes, with the rightmost lane an EXIT ONLY lane, and the adjacent lane an option lane.

Site 43-2. Site 43-2 includes slightly less than 1 mi of I-5 southbound focusing on Exit 104. In this location, I-5 southbound consists of four lanes, with the rightmost lane an EXIT ONLY lane, and the adjacent lane an option lane.

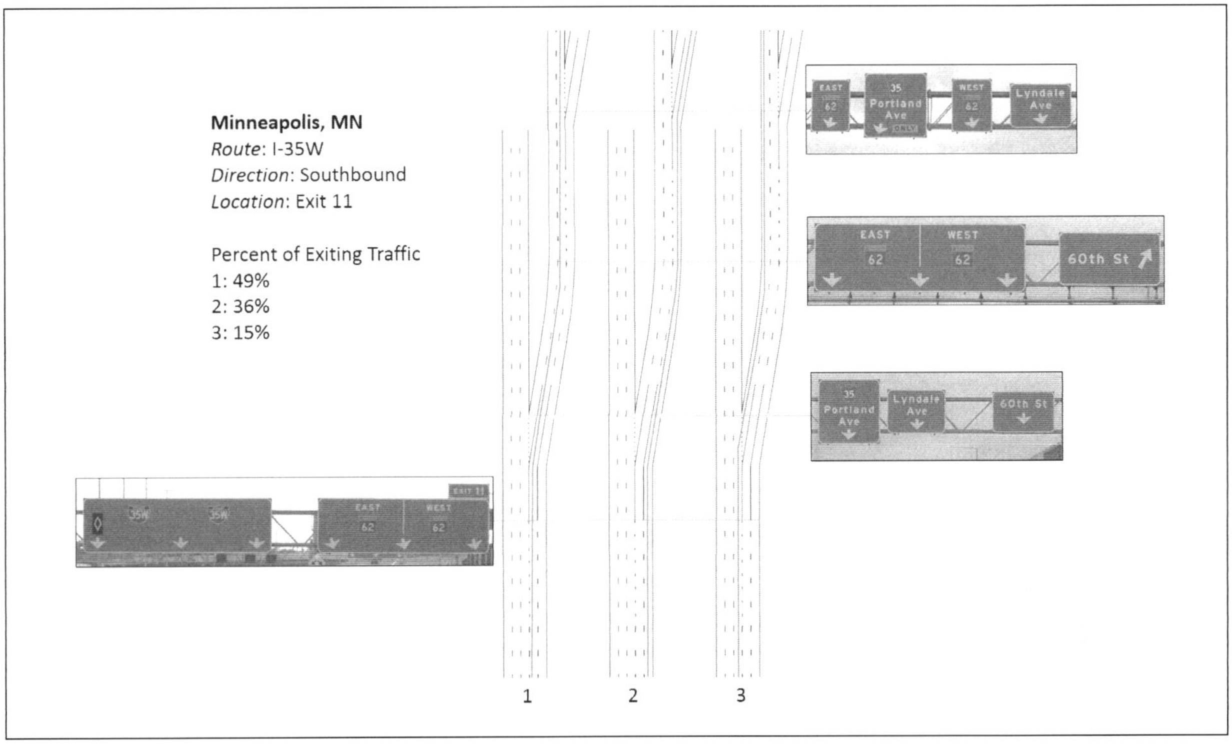

Figure 8. Common exiting driver behaviors at Site 31-1, Exit 11 ($n = 33$).

Site 43-3. Site 43-3 includes slightly less than 1 mi of I-5 northbound focusing on Exit 111. At this location, I-5 northbound consists of three lanes. Approximately halfway through the study area, a fourth lane is added with the rightmost lane an EXIT ONLY lane, and the adjacent lane an option lane.

Observations. At Site 43-1, there was 28 min of video reduced, which included observations of 434 vehicles (97 exiting; 337 remaining on I-5). At Site 43-2, there was 20 min of video reduced, which included observations of 334 vehicles (141 exiting; 193 remaining on I-5). At Site 43-3, there was 30 min of video reduced, which included observations of 346 vehicles (84 exiting; 262 remaining on I-5).

Results. Figures 9–11 illustrate common driver behavior identified at Site 43.

Discussion of Findings

Field video was reduced to track vehicle paths throughout each study site. Although data were captured on both through and exiting traffic, the focus of this review was on the exiting vehicles. Because exiting vehicles served as the focus of the analysis, in many cases, sample sizes were relatively small and data coding was very time consuming, particularly when combining several camera views. As such, the findings below highlight major safety or mobility issues captured in the field video and not a complete traffic analysis of each interchange.

One common finding across sites is that exiting traffic was found to most commonly use the EXIT ONLY lane rather than the option lane. The cluster analysis found only one site where the most common group of driver behavior did not include moving to the EXIT ONLY lane before, or near, the first guide sign with a yellow EXIT ONLY panel. The analysis also identified examples in which drivers would use the EXIT ONLY lane and, ultimately, make a lane change on the C-D roadway. These instances include signing that does not indicate the presence of an option lane (e.g., Site 11 eastbound) or groups several destinations on a single sign (e.g., Site 43-1). Conversely, the cluster analysis did not find any instances of common behavior in which drivers exited using the option lane, and then changed lanes on the C-D roadway.

The site with the highest option lane utilization was found to be Washington State Site 43-2. At this site, nearly half of the drivers were observed using the option lane. This site also included a guide sign configuration not seen in any of the other sites. On the Advance Guide sign, the option lane is signed with the standard

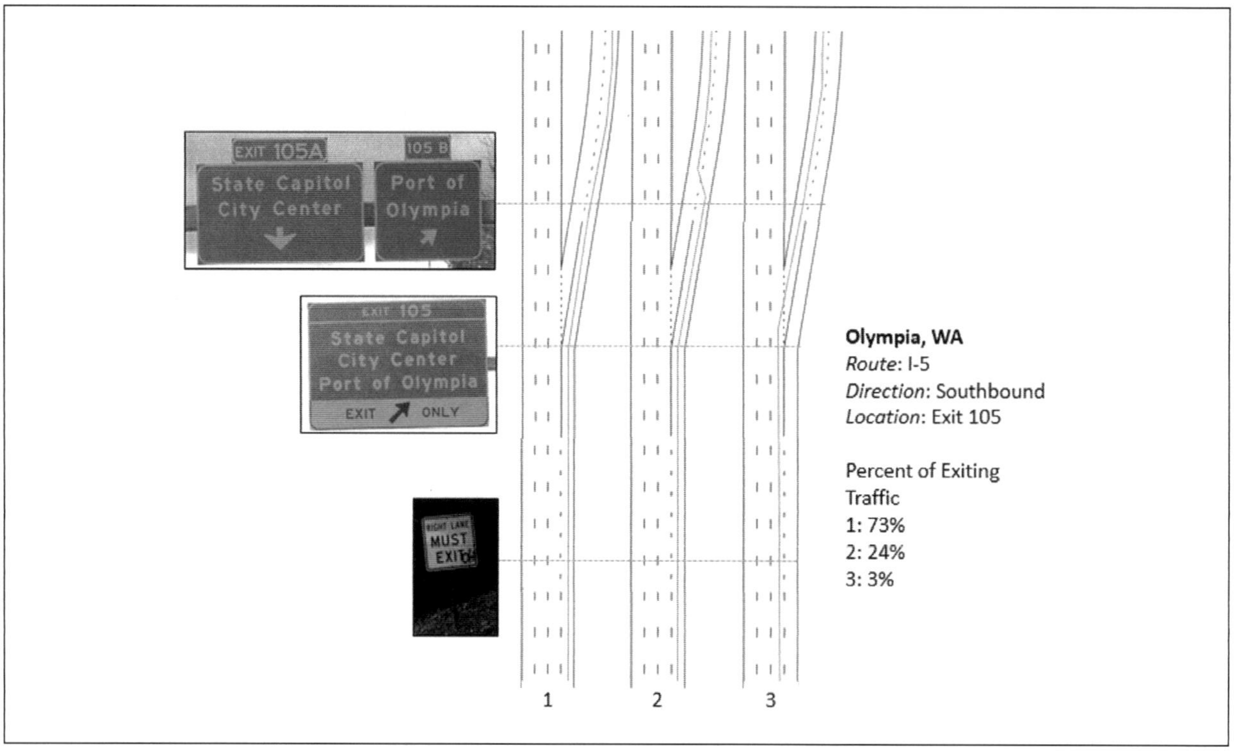

Figure 9. Common exiting behaviors at Site 43-1, Exit 105 ($n = 97$).

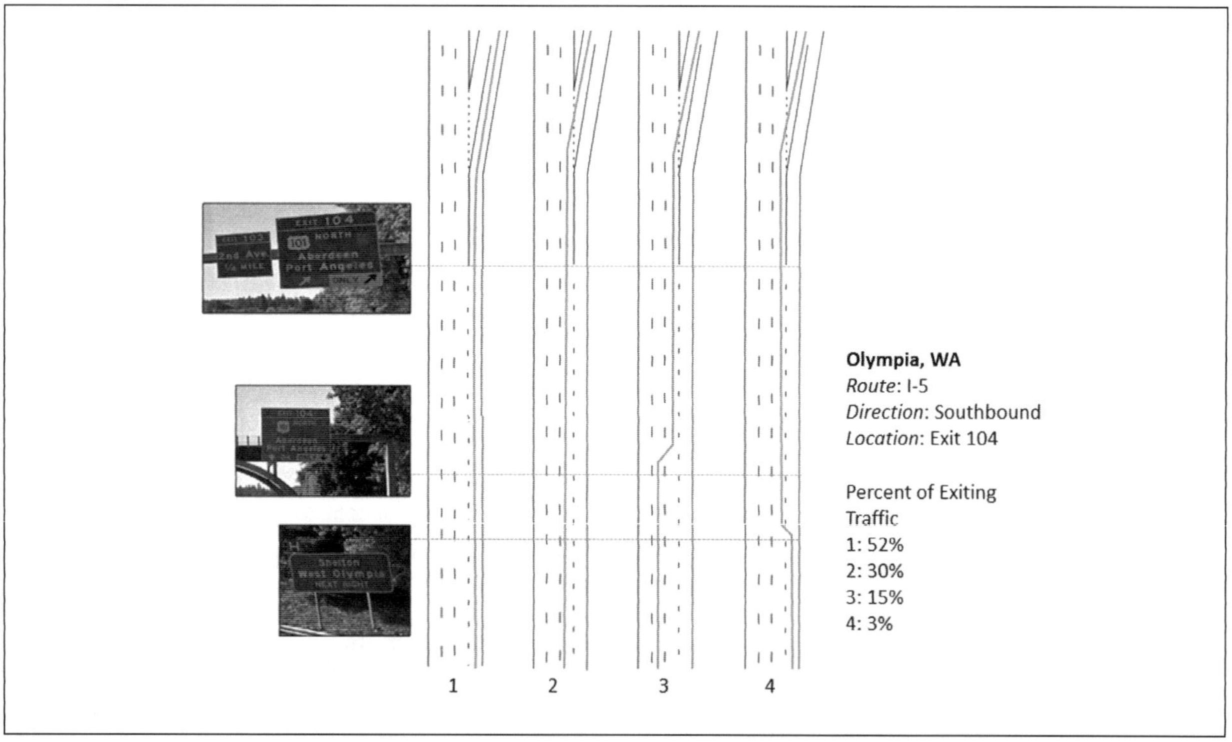

Figure 10. Common exiting behaviors at Site 43-2, Exit 104 ($n = 141$).

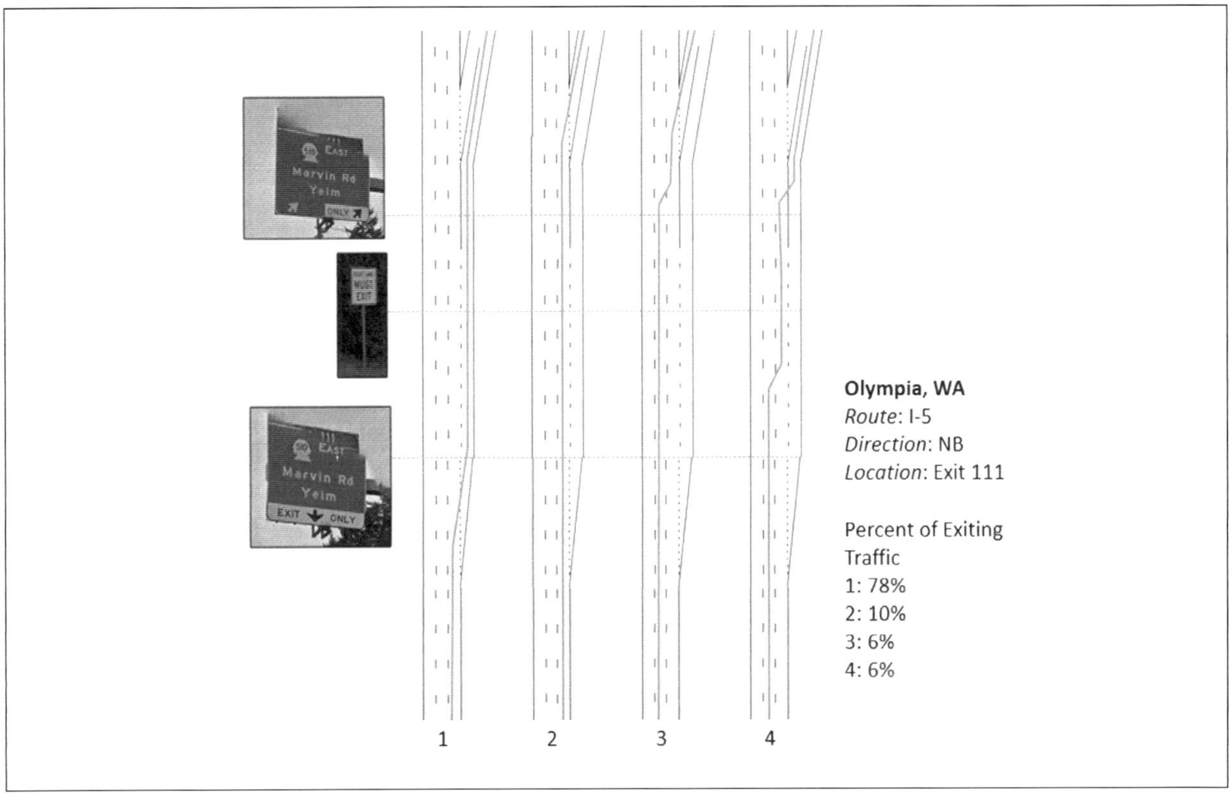

Figure 11. Common exiting behaviors at Site 43-3, Exit 111 ($n = 84$).

downward white arrow on a green background; however, the guide sign also includes the letters "OK" adjacent to the downward arrow.

Few common behaviors identified through the field video show last-minute lane changes. Behaviors with last-minute lane changes were found in the Atlanta, Georgia, site (both exits 51A and 51B), as well as Washington State Site 43-3. In all cases, the percentage of vehicles completing these lane changes was small. Further, for the case of Exit 51B in Atlanta, these lane changes may be influenced by interchange geometry as well as (or in lieu of) signing because exits 51A and 51B are separated by slightly more than 0.25 mi.

There was one instance of a common behavior found that could be interpreted as a misunderstanding of the guide signs. In the Atlanta, Georgia, site, about 12% of the drivers exiting at the second exit (51B) were classified in a cluster of driving behaviors that showed the drivers traveling in the EXIT ONLY lane for the first exit (51A) before changing out of the lane near the exit ramp for Exit 51A to proceed to Exit 51B. Drivers who followed this behavior passed two sign gantries where the sign designated the lane as an EXIT ONLY lane for Exit 51A, with the first gantry also having a separate Advanced Guide sign for Exit 51B.

Findings from this field study support the notion that each interchange is unique, and it is often an interaction of different factors that contributes to complexity. However, using common practices across interchanges that help to satisfy driver expectations should result in less driver confusion.

Acknowledgments

The authors thank William Perez of Leidos for his management oversight on the project. This project was funded by the Federal Highway Administration and Jim Shurbutt of the US Department of Transportation served as the Technical Manager on this project.

Author Contributions

The authors confirm to the paper as follows: study conception and design: BK, SK, JS; data collection: SK, NK; analysis and interpretation of results: SK, NK; draft manuscript preparation: BK, NK. All authors reviewed the results and approved the final version of the manuscript.

References

1. Fitzpatrick, K., S. T. Chrysler, M. A. Brewer, A. Nelson, and V. Iragavarapu. *Simulator Study of Signs for a*

Complex Interchange and Complex Interchange Spreadsheet Tool. Report No. FHWA-HRT-13-047. Federal Highway Administration, Washington, D.C., 2013.
2. Richard, C. M., and M. G. Lichty. *Driver Expectations When Navigating Complex Interchanges*. Report No. FHWA-HRT-13-048. Federal Highway Administration, McLean, Va., 2013.
3. Lichty, M., P. Bacon, and C. Richard. *Collecting and Analyzing Stakeholder Feedback for Signing at Complex Interchanges*. Report No. FHWA-HRT-14-069. Federal Highway Administration, McLean, Va., 2014.
4. Upchurch, J., D. L. Fisher, and B. Waraich. Guide Signing for Two-Lane Exits with an Option Lane: Evaluation of Human Factors. *Transportation Research Record: Journal of the Transportation Research Board*, 2005. 1918: 35–45.
5. Roberts, K.M., and A.G. Klipple (1976). "Driver Expectations at Freeway Lane Drops." Public Roads, 40(1), 32–35.
6. Fitzpatrick, K., M. Lance, and T. Lienau. Effects of Pavement Markings on Driver Behavior at Freeway Lane Drop Exits. *Transportation Research Record: Journal of the Transportation Research Board*, 1995. 1495: 17–27.
7. Borowsky, A., D. Shinar, and Y. Parmet. The Relation Between Driving Experience and Recognition of Road Signs Relative to their Locations. *Human Factors*, Vol. 50, No. 2, 2008, pp. 173–182.
8. Doctor, M., G. Merritt, and S. Moler. Designing Complex Interchanges. *Public Roads*, Vol. 73, No. 3, 2009, pp. 3–11.
9. Ray, B. L., J. Schoen, P. Jenior, and J. Knudsen. *NCHRP Report 687: Guidelines for Ramp and Interchange Spacing*. Transportation Research Board of the National Academies, Washington, D.C., 2011.
10. Bared, J. *Safety Assessment of Interchange Spacing on Urban Freeways*. TechBrief Publication No. FHWA-HRT-07-031. Federal Highway Administration, Washington, D.C., 2007.
11. Ismail, A., A. Aram, R. Aminzadeh, and R. A. Rahmat. Crashes and Effective Safety Factors Within Interchanges and Ramps on Urban Freeways and Highways. *Australian Journal of Basic and Applied Sciences*, Vol. 5, No. 7, 2011, pp. 397–404.
12. Glad, R. W., J. C. Milton, and D. K. Olson. *Weave Analysis and Performance: The Washington State Case Study*. Report No. WA-RD 515.1. Washington State Department of Transportation, Olympia, Wash., 2001.
13. Hall, M., E. Frank, J. Holmes, B. Pfahringer, P. Reutemann, and I. Witten. The WEKA Data Mining Software: An Update. *SIGKDD Explorations*, Vol. 11, No. 1, 2009, pp. 10–18.

The Users Performance Section (AND00) peer-reviewed this paper (18-05065).

Article

TRR
JOURNAL OF THE TRANSPORTATION RESEARCH BOARD

Transportation Research Record
2018, Vol. 2672(37) 31–44

DOI: 10.1177/0361198118798713
journals.sagepub.com/home/trr
SAGE

Variations in Driver Behavior: An Analysis of Car-Following Behavior Heterogeneity as a Function of Road Type and Traffic Condition

Andrew L. Berthaume[1], Rachel M. James[2,3], Britton E. Hammit[3,4], Christina Foreman[1], and Christopher L. Melson[3]

Abstract

Microsimulation modeling is a tool used by practitioners and researchers to predict and evaluate the flow of traffic on real transportation networks. These models are used in practice to inform decisions and thus must reflect a high level of accuracy. Microsimulation models are comprised of sub-models, which control individual vehicle movements throughout the simulated network. These sub-models must be calibrated to accurately capture realistic driving behavior. This research utilizes data collected by the FHWA Living Laboratory instrumented research vehicle to produce evidence of global trends in car-following behavior. Unlike similar studies, this analysis focuses on the physical action taken by the driver—the acceleration—rather than the outcome of that action—speed selection or temporal/spatial gap. This approach enables better interpretation and comparison between car-following behavior in varying "driving environments," that is, on different roadway functional classifications (freeway v. interstate), operational conditions (work zone v. non-work zone), and traffic conditions (congested v. uncongested). This analysis produced conclusive evidence that intra-driver car-following behavior is heterogeneous and is a function of the driving environment. Trends in acceleration behavior were examined on an aggregated psychophysical plane, which accounted for inter-driver heterogeneity, and a statistical analysis identified regions of significantly different acceleration behavior. Lastly, heterogeneity in car-following acceleration behavior in work zones and non-work zones was also verified.

With the enactment of Moving Ahead for Progress in the 21st Century (MAP-21), there has been an increase in emphasis on performance-based planning and scenario development for federally funded transportation investments (*1*). Thus, the accuracy and robustness of modeling tools is of interest to transportation agencies, consultants, and researchers alike. Microsimulation software can be leveraged as a powerful tool in scenario analyses, as the outputs are an explicit representation of different characteristics of an individual driver. However, increased resolution of model output requires the modeler to ensure simulation accuracy in producing reasonable approximations of individual driver behavior. Outputs of these models are used to inform multimillion dollar investments in transportation, and thus the realism of the inputs and algorithms housed within a model are of upmost importance.

Several studies have shown that both intra-driver heterogeneity and inter-driver heterogeneity exist within microscopic data (*2, 3*). Intra-driver heterogeneity describes the phenomenon in which a driver exhibits inconsistent driving behavior when exposed to different driving environments. On the other hand, inter-driver heterogeneity describes the notion that different drivers exposed to the same driving environment may behave differently. Multiple studies have identified intra-driver heterogeneity when comparing driver behavior among different functional classifications (e.g., interstate v. freeway), operational conditions (e.g., work zone v. non-work zone segments), and traffic conditions (e.g., congested v. uncongested conditions); the combination of these conditions is defined as the *driving environment.*

Research into driver heterogeneity has focused on the consequence of a driver's action (e.g., headway or speed)

[1]John A. Volpe National Transportation Systems Center, Cambridge, MA
[2]University of Texas at Austin, Austin, TX
[3]Turner-Fairbank Highway Research Center, McLean, VA
[4]University of Wyoming, Laramie, WY

Corresponding Author:
Address correspondence to Rachel M. James: Rachel.James@dot.gov

instead of the action itself—the acceleration. Hence, many common methods used to evaluate driving behavior were not designed to highlight the nuances that exist in car-following behavior, which involves both human perception and relative behavior between a pair of vehicles. Therefore, this study utilizes the psychophysical car-following plane, which is the foundation of perception–reaction (action-point) car-following models, as a more appropriate method to analyze car-following behavior.

Using instrumented research vehicle (IRV) data from FHWA's Living Laboratory (*4*), this study provides qualitative and quantitative evidence that intra-driver differences in car-following acceleration behavior exists both (a) among distinctively different traffic conditions and (b) between different functional classifications or operational conditions holding traffic conditions constant. This evidence is illustrated and described using an innovative procedure for merging trajectories of multiple vehicles to identify global behavioral trends while accounting for inter-driver heterogeneity.

Due to the nature of the FHWA Living Laboratory IRV data, this study highlights key differences in driving behavior in work zone (WZ) and non-work zone (non-WZ) interstate segments. The insights of this research set the stage for building and calibrating new car-following models that will improve the accuracy and realism of the models, whose outputs inform important decisions and reporting metrics.

The remainder of this paper is organized as follows. The following section provides a brief literature review. We then provide details of the data collection and post-processing methodologies. Next we highlight key results from the analysis and, finally, summarize conclusions and insights gleaned from the data.

Literature Review

There are three key areas of prior research that were used to build this analysis. The section "Intra-Driver Heterogeneity" highlights the state of practice for intra-driver heterogeneity research. Future work will explore insights into inter-driver heterogeneity, however, for the purposes of this study, inter-driver heterogeneity is only accounted for in the aggregated psychophysical framework discussed in the section "Data Analysis." The section "Psychophysical Car-Following Framework" discusses the psychophysical car-following framework and the state phase (relative velocity [*dV*] – relative distance [*dX*]) plane of car-following behavior.

Intra-Driver Heterogeneity

It has long been established that roadway functional classification (e.g., interrupted flow v. uninterrupted flow), operational conditions (including the presence of a WZ), and the onset of traffic congestion significantly impact operational performance (*5, 6*). The last 20 years of research have been marked by efforts to better understand the presence and impacts of intra-driver differences attributable to varying conditions in the driving environment.

Al-Kaisey and Hall examined loop detector data for six WZs, compiling over 72.5 h of traffic data (*6*). They concluded that driver familiarity with a segment of roadway is correlated with following headway selection, and this strongly affects the capacity and performance of the facility. Similar conclusions were made by Maze et al. (*7*) and Benekohal et al. (*8*).

In later research, Lochrane et al. quantified the differences in driving behavior between WZs/non-WZs and congested/uncongested conditions via microscopic IRV data (*9*). This study used mean time gap as the indirect measure of driver behavior. Results showed that 75% of participants increased their time gap while driving in WZs compared with non-WZs under congested conditions.

Distributions of headway have also been used to quantify the differences in driving behavior in WZs/non-WZs and congested/uncongested conditions. Dijker, Bovy, and Vermjis examined the differences in headway—segmented by vehicle type—on two Dutch freeways (*10*). Holding speed fixed, they found significant differences in headways between congested and uncongested conditions. The existence of different headways in congested and uncongested conditions is corroborated by Yin et al. (*11*). This research concluded that vehicle headways are best described by a lognormal distribution in uncongested conditions and a log-logistic distribution in congested conditions. This analysis was confirmed to be statistically significant using Kolmogorov–Smirnov tests. Brackstone et al. found conflicting results; they concluded that the localized flow of traffic (congestion level) has negligible impact on headways (*12*).

Though analyses of indirect measures of driver behavior are heavily represented in literature, results are occasionally conflicting. Hence, analyzing driver behavior directly using acceleration—the driver's reaction—as presented in this paper, is hypothesized to be a more robust and reliable approach.

Psychophysical Car-Following Framework

Driving behavior has been analyzed using a variety of methods, including time-series trajectory analyses (e.g., vehicle speed, acceleration, steering angle) and behavioral profiles (e.g., probability distributions of vehicle speed, acceleration). Although many of these mechanisms for evaluating driving behavior have been used to analyze

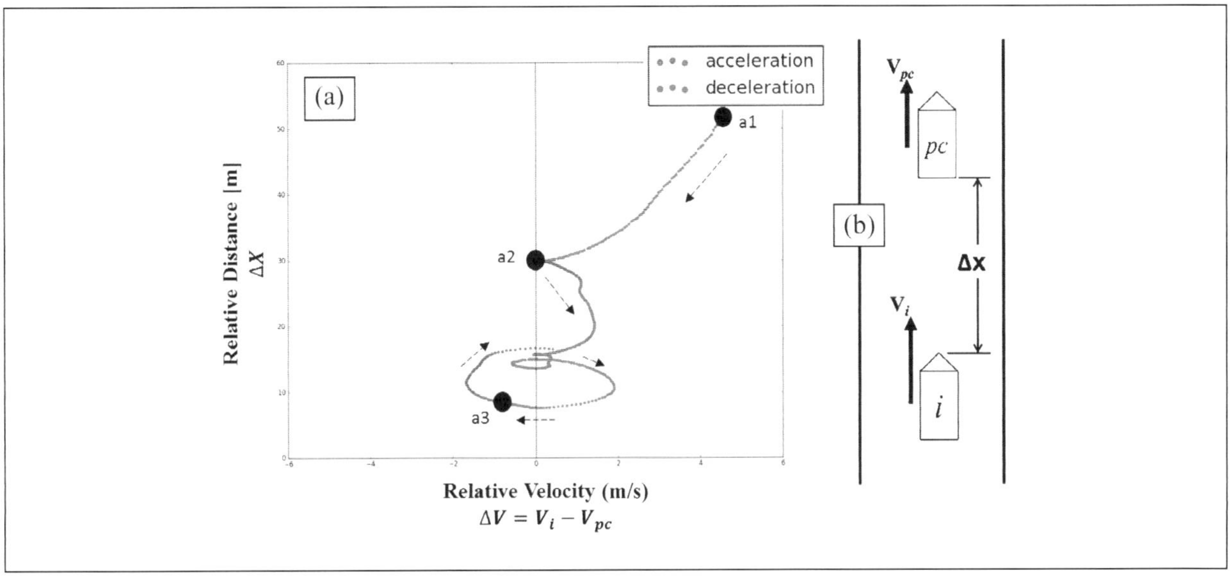

Figure 1. Spiral trend observed in a car-following event. (*a*) Shows a car-following spiral and provides indication of vehicle acceleration (green) and deceleration (red). Starting from "a1," progressing through the event, the following vehicle is traveling faster than the lead vehicle; thus, deceleration behavior is observed until the driver begins traveling at the same speed. At "a2" the following vehicle begins to accelerate—once again gaining on the lead vehicle—thus, quickly begins decelerating until reaching "a3" where the following vehicle is traveling slower than the lead vehicle and accelerates to catch up. (*b*) Defines the physical relationship between the relative distance and relative velocity.

car-following behavior, these studies measure the driver behavior differences using consequences of the driver's action (headway or speed) instead of the direct behavior or action itself (acceleration or deceleration). Hence, they were not explicitly derived for capturing the nuances in car-following behavior.

In the early 1960s, Barbosa (*13*), Michaels (*14*), and Todosoiev (*15*) laid the foundation for the evaluation of relative driver behavior through the introduction of a psychophysical framework, which identifies driver actions and reactions relative to the perceived and detectable behavior of a lead vehicle. The psychophysical car-following framework seeks to explain the natural oscillations in car-following behavior attributed to the limits of human perception. The cornerstone of psychophysical models can be summarized in a car-following phase plane (in the $dV - dX$ plane). The segmentation of this plane into regimes of reaction and no reaction is a function of driver "action" points, defined as points in which a driver makes a change in their behavior because of perceived changes in behavior of the leading vehicle. Drivers can react to changes in relative distance (the spatial gap) and relative velocity once perception thresholds are reached. Each of the regimes within a psychophysical framework are defined under different psychological driving assumptions and represent distinctly different driver behavior (e.g., approaching, separating, no reaction) (*16*). Figure 1 describes subconscious car-following behavior on a psychophysical plane with relative velocity on the *x*-axis and relative distance on the *y*-axis. Subconscious car-following behavior produces a spiral in this plane, in which a driver continuously perceives their relative velocity and following distance with respect to the lead vehicle, and reacts accordingly by accelerating or decelerating (see Figure 1). Psychophysical representations of car-following trends enable the detailed analysis and evaluation of high resolution trajectory-level data, such as the FHWA Living Laboratory IRV data. According to Brackstone and Mcdonald, of all car-following paradigms, psychophysical models are "the most coherent and best able to describe most of the features that we see in everyday driving behavior" (*17*, p. 191). For this reason, the psychophysical car-following framework was determined to be the most appropriate tool to explore the acceleration trends in the car-following data.

Methodology

This research was completed using the FHWA Living Laboratory IRV data, which provides a unique opportunity to explore driving behavior in various driving environments (*4*). This section briefly describes the data collection system, the data processing protocol, the procedure used to extract car-following events and match them with other car-following events occurring in similar driving environments, and the analytical approach used to evaluate groupings of heterogeneous driving behavior.

Data Collection: Living Laboratory

Many data collection strategies aiming to quantify intra-driver behavior heterogeneity relative to variations in the driving environment have been developed. Fitzpatrick identified existing methods of data collection and detailed each of their strengths and weaknesses (*18*). Detailed data is typically collected using either traditional fixed sensors (e.g., pneumatic tubes with automated traffic recorders, RADAR/LiDAR speed guns, and inductive loops) or probe sensors via vehicle instrumentation. The latter are also referred to as IRVs.

This study makes use of the FHWA Living Laboratory IRV data. IRV studies enable the collection of microscopic, trajectory-level information that can better describe driver actions and reactions to their environment—as opposed to the consequence of their actions. IRV studies allow researchers to collect information pertaining to the driving environment (e.g., radar, video) as well as high resolution vehicle data (e.g., acceleration, speed, steering angle).

In 2013, a sport utility vehicle was equipped with the necessary on-board equipment for trajectory-level data collection: two universal medium range radars for the collection of relative velocity and position of adjacent vehicles at 40 Hz, a GPS-based speed sensor to collect and record the IRV's speed at 10 Hz, a 30 frames per second video recording system, and a computer for data acquisition. For details on equipment specifications, please see Lochrane et al. (*4*).

As part of this study, a total of 66 participants drove along a pre-defined route, approximately 50 mi in length, during peak hour and clear weather conditions on a weekday. Four interstate WZs were located along the route: two with a lane closure and two without. Of the 66 participants, trip files from 62 drivers were used in this analysis.

Data Processing

The 62 trip files were processed to extract car-following events and define the conditions under which each car-following event occurred. Car-following events were then sorted and aggregated by similarities in the driving environment to create frameworks.

Car-Following Events. Car-following events were manually identified, classified, and extracted using video and radar data from the 62 trip files. To ensure all extracted car-following events were long enough to illustrate continuity in car-following behavior, a minimum length of continuous car-following was defined to be 10 s. In addition, 80 m was selected as the maximum relative distance between a lead and following vehicle to consider the actions relevant for car following. After car-following event start and stop timestamps were recorded, the radar and GPS data were compiled and filtered to produce 62 processed driver books; the processed driver books summarize each car-following event experienced during a trip. Additional post-processing was required to remove inherent noise from the radar data.

Compiling Car-Following Events into Frameworks. In addition to identifying the start and stop timestamps for car-following events, video observers recorded details related to the driving environment for each car-following event: functional classification (freeway v. interstate), operational condition (WZ v. non-WZ), work zone type (lane closure v. shoulder closure), and traffic condition (uncongested v. congested conditions). Congested and uncongested conditions were defined as a speed reduction caused by surrounding vehicles: below 35 mph in non-WZs or 25 mph in WZs. The local congestion levels were verified in manual video observation of each trip. The driving conditions for each framework are shown in Table 1.

Aggregate Representation of Car-Following Behavior. The limitation of many studies investigating trajectory-level car-following behavior is that they often limit evaluation to an individual driver or a small population of drivers, which does not enable the identification of collective behavioral tendencies. Generally, car-following behavior on a psychophysical plane represents a single car-following event, as shown in Figure 2*a*; however, when considering hundreds of car-following events for a single framework, this representation becomes difficult to interpret, as shown in Figure 2*b*.

Data Analysis

This study introduces an approach to account for inter-driver heterogeneity inherent in a large group of drivers, leveraging the psychophysical $dV - dX$ car-following plane. This study partitions the $dV - dX$ space into a grid, creating the individual bins shown in Figure 3*a*. Every car-following point that falls within the bounds of a bin is summarized as the "contents" of that bin; example summary statistics include the frequency of points and the standard deviation of point values. In this analysis, each bin represents the average acceleration of all points within the bin. To ensure regions of the $dV - dX$ plane with limited data or significant outliers do not skew results, minimum frequency requirements (i.e., a minimum number of data points required for each bin) and a

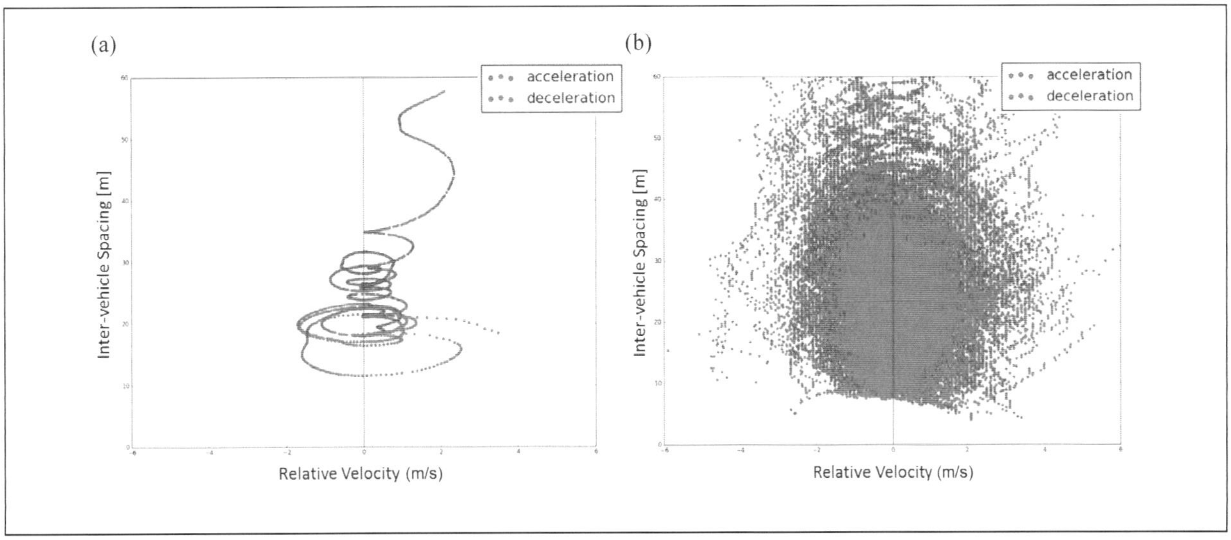

Figure 2. Car-following behavior on a psychophysical plane: (*a*) a single car-following event, in which a distinct spiral of driver reaction can be identified; (*b*) the combination of all car-following events for Framework 2, for which trends cannot be identified.

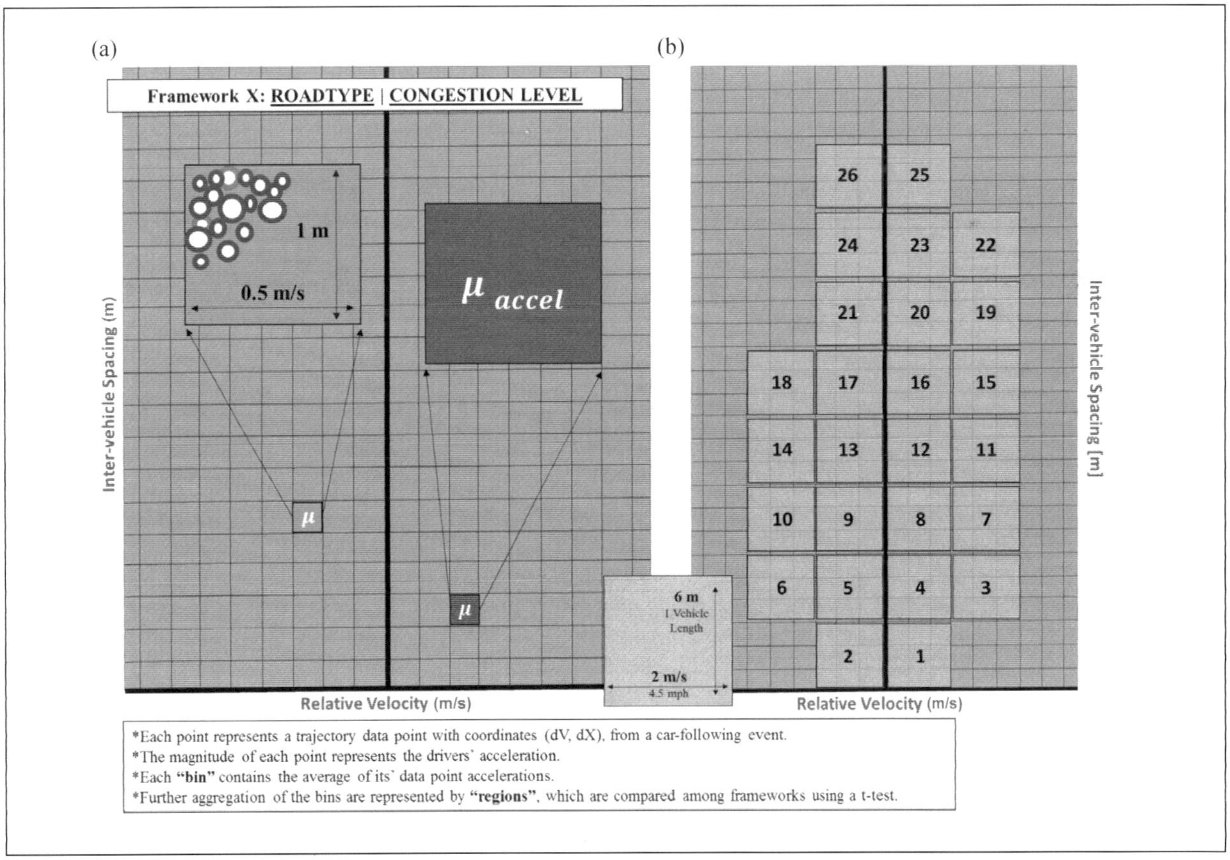

Figure 3. Description of "bins" and "regions" on the psychophysical plane used for data analysis. (*a*) Each square within the grid represents a bin and contains data points with characteristics (i.e., acceleration). The bins are summarized by taking the mean of the acceleration of all included points. Bins highlighted in red indicate a negative average acceleration (deceleration), whereas bins highlighted in green indicate a positive average acceleration. (*b*) Regions represent groups of bins that are used in the statistical analysis to identify significant differences in driver behavior between frameworks.

maximum standard deviation of acceleration points within each bin were established. If a bin failed either criterion, the bin was removed from the analysis and null values were reported.

The data analysis for this study takes both qualitative and quantitative forms. Visual evidence of acceleration behavioral trends is illustrated via aggregated acceleration car-following plots, which are described in Figure 3*a*. Although visual trends produced substantial evidence of acceleration behavioral differences, two-tailed paired *t*-tests were used to compare "regions" of acceleration behavior between framework pairs to determine whether a statistically significant difference (at a 95% confidence level) exists. The regions were selected to represent conceivable areas of car-following behavior: segmented into regions of standard car lengths (6 m) from the lead vehicle for the relative distance increments and 2 m/s (4.5 mph) relative velocity increments, as shown in Figure 3*b*.

Results

An analysis of the observed acceleration behaviors was conducted on two levels: (i) holding level of congestion constant and (ii) holding functional classification/operational condition constant. First, driving behavior observed on each roadway functional classification (i.e., interstate, freeway) was compared in congested conditions. Then, driver behavior on interstates under different operational conditions was compared in congested conditions. Variation in operational conditions, that is, presence of a WZ, was not observed for the freeway functional classification. Lastly, the observed driving behavior on a roadway with a specific functional classification and operational condition was compared under congested and uncongested conditions. The acceleration behavior was compared qualitatively, to visually identify and describe prevalent trends, and quantitatively via two-tailed paired *t*-tests to identify regions in which statistically significant differences in acceleration behavior exist.

Driver Behavior by Road Type

As described in the literature review, many studies have shown intra-driver heterogeneity among different functional classifications and operational conditions. Figure 4 represents the aggregated car-following acceleration behavior for all 62 drivers divided into "frameworks," which are combinations of the driving environment—as described in Table 1—in congested conditions. Positive acceleration trends are denoted on a green color scale and deceleration is represented on a red color scale. Bins with data points of no acceleration—that is, data points with acceleration equal to 0 m/s—are shown in white. Table 2 provides the *t*-test results, as described in the section "Data Analysis," to compare the different functional classifications and operational conditions in congested conditions.

In each framework (FW), acceleration in the upper region of the $dV - dX$ plot—indicative of larger relative distances between vehicles—is of lighter color, which indicates a lower magnitude of acceleration or deceleration. This behavior is expected because as the separation distance between vehicles increases, the perceived necessity to react, due to the lead vehicle's presence, is smaller.

Table 1. Driving Environment Defined for Each Framework

Framework #	Functional classification	Operational condition	Traffic condition
1	Freeway (2 lane divided)	Non-WZ	C
2	Freeway (2 lane divided)	Non-WZ	U
3	Interstate (IS)	Non-WZ	C
4	Interstate (IS)	Non-WZ	U
5	Interstate: Advanced warning (AW)	WZ	C
6	Interstate: Advanced warning (AW)	WZ	U
7	Interstate: Taper zone (TZ)	WZ	C
8	Interstate: Taper zone (TZ)	WZ	U
9	Interstate: Work zone (WZ1) (with lane closure)	WZ	C
10	Interstate: Work zone (WZ1) (with lane closure)	WZ	U
11	Interstate: Work zone (WZ2) (with shoulder closure)	WZ	C
12	Interstate: Work zone (WZ2) (with shoulder closure)	WZ	U

Table 2. *t*-Test Comparison for Different Functional Classifications and Operational Conditions

Region location and number		FW1 v. FW3	FW3 v. FW5	FW3 v. FW7	FW3 v. FW9	FW3 v. FW11	FW5 v. FW7	FW5 v. FW9	FW5 v. FW11	FW7 v. FW9	FW7 v. FW11	FW9 v. FW11
7–8 car-lengths | 0–4.5 mph slower	26				0.935							
7–8 car-lengths | 0–4.5 mph faster	25	0.917			0.047*							
6–7 car-lengths | 4.5–9 mph slower	24	0.003*			0.015*							
6–7 car-lengths | 0–4.5 mph slower	23	0.000*	0.407	0.000*	0.186	0.210	0.001*	0.659	0.368	0.000*	0.000*	0.754
6–7 car-lengths | 0–4.5 mph faster	22	0.702	0.935		0.152	0.010*		0.008*	0.000*			0.002*
5–6 car-lengths | 4.5–9 mph slower	21	0.035*		0.798	0.088					0.041*		
5–6 car-lengths | 0–4.5 mph slower	20	0.001*	0.241	0.722	0.003*	0.266	0.320	0.009*	0.771	0.039*	0.319	0.004*
5–6 car-lengths | 0–4.5 mph faster	19	0.975	0.388	0.003*	0.064	0.016*	0.025*	0.075	0.189	0.094	0.016*	0.027*
4–5 car-lengths | 4.5–9 mph slower	18	0.023*	0.003*	0.097	0.010*	0.229	0.225	0.095	0.111	0.562	0.000*	0.245
4–5 car-lengths | 0–4.5 mph slower	17	0.369	0.066	0.051	0.138	0.001*	0.007*	0.018*	0.000*	0.892	0.170	0.220
4–5 car-lengths | 0–4.5 mph faster	16	0.100	0.821	0.011*	0.054	0.353	0.350	0.117	0.365	0.916	0.431	0.055
4–5 car-lengths | 4.5–9 mph faster	15	0.027*		0.068								
3–4 car-lengths | 4.5–9 mph slower	14	0.114	0.017*	0.001*	0.146	0.132	0.003*	0.018*	0.805	0.096	0.039*	0.116
3–4 car-lengths | 0–4.5 mph slower	13	0.000*	0.331	0.141	0.464	0.001*	0.348	0.187	0.011*	0.100	0.030*	0.039*
3–4 car-lengths | 0–4.5 mph faster	12	0.084	0.868	0.040*	0.394	0.045*	0.091	0.580	0.046*	0.052	0.011*	0.024*
3–4 car-lengths | 4.5–9 mph faster	11	0.003*	0.354									
2–3 car-lengths | 4.5–9 mph slower	10	0.004*	0.243	0.754	0.136	0.329	0.309	0.004*	0.719	0.567	0.432	0.066
2–3 car-lengths | 0–4.5 mph slower	9	0.036*	0.129	0.000*	0.008*	0.004*	0.006*	0.233	0.048*	0.044*	0.410	0.245
2–3 car-lengths | 0–4.5 mph faster	8	0.000*	0.685	0.000*	0.494	0.351	0.000*	0.710	0.275	0.005*	0.000*	0.269
2–3 car-lengths | 4.5–9 mph faster	7	0.313	0.554									
1–2 car-lengths | 4.5–9 mph slower	6	0.039*	0.275	0.889	0.042*	0.918	0.663	0.002*		0.302		
1–2 car-lengths | 0–4.5 mph slower	5	0.307	0.249	0.712	0.729	0.009*	0.880	0.518	0.473	0.856	0.705	0.757
1–2 car-lengths | 0–4.5 mph faster	4	0.016*	0.225	0.558	0.005*	0.023*	0.629	0.000*	0.001*	0.003*	0.006*	0.928
1–2 car-lengths | 4.5–9 mph faster	3	0.113	0.033*									
0–1 car-lengths | 0–4.5 mph slower	2	0.006*	0.042*									
0–1 car-lengths | 0–4.5 mph faster	1	0.052	0.481	0.213	0.774		0.056	0.341	0.001*	0.400	0.630	0.082

Note: FW = framework. # car lengths (6 m) from lead vehicle; # mph speed difference between the lead and following vehicle. Blank cells and regions not included on the table indicate insufficient data to complete the analysis. Results of a two-tailed, paired *t*-test; cells indicate *p*-value at alpha = 0.05 (95% confidence level).

*A significant difference exists between the mean accelerations in that region.

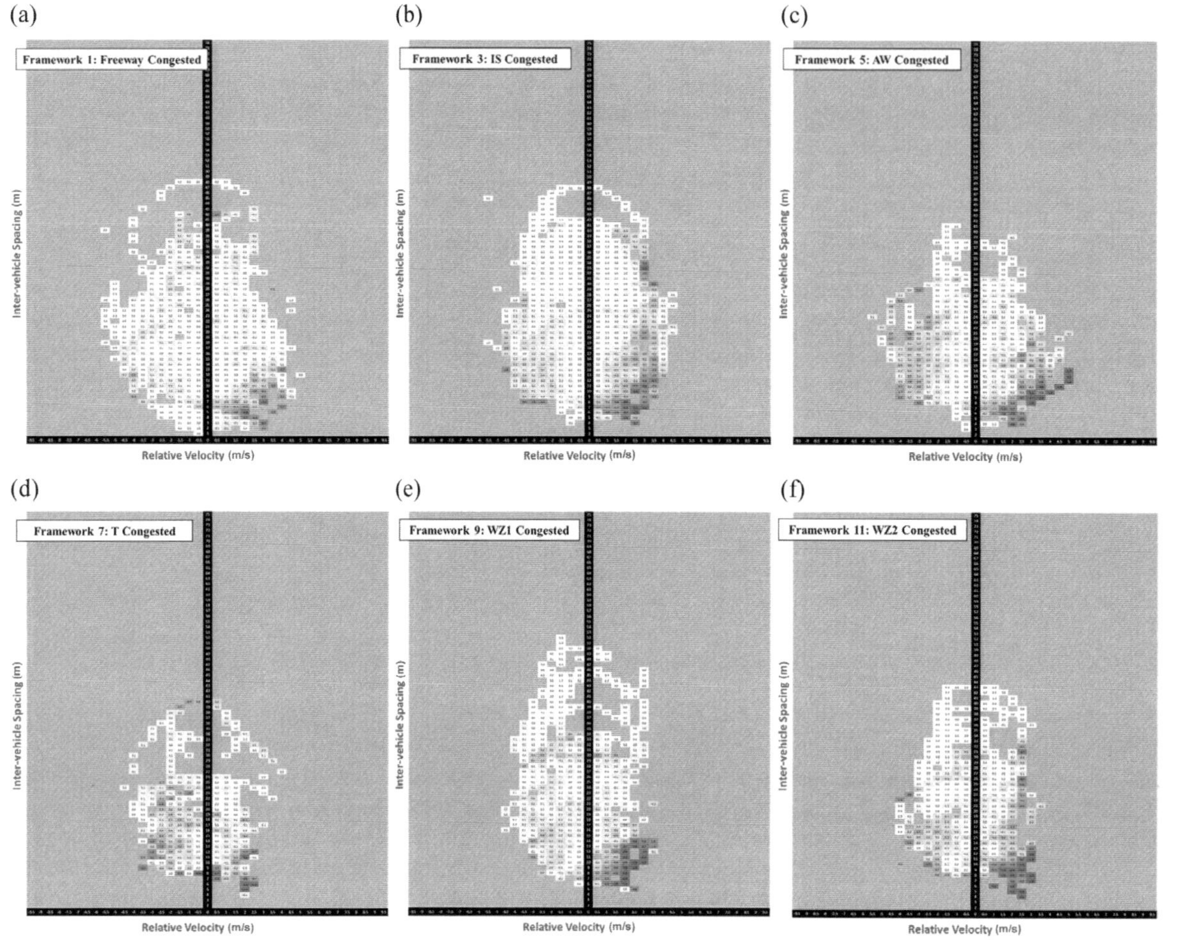

Figure 4. Aggregated car-following acceleration behavior on a psychophysical plane, representing each functional classification and operational condition in congested traffic conditions.

The white, or the "no-reaction" zone, indicates locations on the $dV - dX$ plane where drivers are not reacting to the lead vehicle.

Each framework in Figure 4 shows a clear diagonal originating on the plane where relative distance is small and relative velocity is positive (which indicates that the following vehicle is traveling faster than the lead vehicle), and moves toward a larger relative distance and a more positive relative velocity. The most prominent trend is seen in the congested advanced warning framework (Figure 4*c*), which extends out to a relative velocity of 5.5 m/s at a following distance of 14 m (just over two standard car lengths). This diagonal deceleration trend illustrates the following drivers' recognition of their speed differential and their consequent deceleration action taken to avoid a collision.

The "cleanest" spirals are detected in the interstate framework (Figure 4*b*), where areas of deceleration are seen sweeping from the top (larger values of dX) along the exterior of the observed data in the region of positive relative velocity. The spiral trend then continues from the bottom (smaller values of dX) with heavier acceleration propagating along the exterior of the observed data in the region of negative relative velocity (where the lead vehicle is traveling faster than the following vehicle). Equivalently clean spirals are not detected in freeway conditions. Statistical analysis between FW1 and FW2, as provided in Table 2, supports this observation showing 12 of the 21 regions had significant differences in acceleration behavior. The clearly defined spiral acceleration trend observed in interstate conditions is likely due to the nature of uninterrupted flow facilities.

Another interesting trend is observed in the congested advanced warning framework (Figure 4*c*) and the congested taper zone framework (Figure 4*d*). Strong acceleration patterns are prevalent at greater following distances, indicating drivers of the following vehicles began to accelerate (to close the gap between themselves and their lead vehicle) more often than in other congested interstate operational conditions. This is supported by Table 2 (FW3 v. FW5), which indicates a statistically significant difference in interstate and advanced warning acceleration behavior in regions 13, 14, 17, and 18. This phenomenon can be attributed to drivers' recognition of the WZ advanced warning and their desire to decrease the following distance, likely in anticipation of reduced speeds and in expectation that other drivers may begin to change lanes due to an upstream lane or shoulder closure, or other characteristics of driving in a WZ.

The least populated frameworks are the interstate taper zones (FW7 and FW8). Figure 4*d* (FW7)—interstate taper zone in uncongested conditions—is shown to have a substantially larger acceleration trend, stretching into the space in which relative velocity is positive at smaller following distances. A significantly smaller deceleration trend is also evident. This is supported by Table 2, in which significantly different acceleration behavior is observed in regions 8, 12, and 16 of the $dV - dX$ plane when comparing the interstate taper zone and non-WZ interstate frameworks (FW3 and FW7). This behavior is likely due to previously reduced speeds resulting from advanced warning signage as drivers approach the taper zone; therefore, less deceleration is required and drivers begin to accelerate after passing through the merging/diverging taper zone.

Deceleration behavior within both WZ frameworks (Figure 4, e and *f*) is noticeably stronger at smaller relative velocities, compared with behavior in the other frameworks. This is supported in Table 2, as both WZ frameworks (FW9 and FW11) have statistically significant differences in deceleration rates when compared against non-WZ interstate driving (FW3) in Region 4 of the $dV - dX$ plane. This indicates that drivers react more quickly to perceived velocity "gains" on the lead vehicle while in the WZ.

When comparing WZ1 (Figure 4*d*, i.e., WZ with a lane closure) and WZ2 (Figure 4*e*, i.e., WZ with a shoulder closure), a difference in acceleration magnitude and location is observable. In WZ2, acceleration behavior is stronger on the outer boundary of the acceleration bins, but propagates inward at lower following distances; in comparison, strong acceleration is less frequently detected in WZ1. This reduction in strong acceleration behavior in WZ1 is likely attributable to a decrease in drivers' urgency to maintain their following distance. Practically speaking, when there is a WZ with a lane closure, there is no parallel traffic in a neighboring lane; therefore, no opportunity to pass a slower moving vehicle and a reduced probability of being cut off by an aggressive driver.

A similar comparison of acceleration behavior on different road types in uncongested conditions is not presented here as the trends are much less prevalent, as shown in the following section.

Driver Behavior by Congestion Level

As part of this study, car-following events were split into two traffic condition categories: congested and uncongested. This section identifies trends in driver acceleration behavior between these traffic conditions and presents statistically significant evidence of acceleration differences in congested and uncongested conditions for regions within the $dV - dX$ plane for each roadway functional classification and operational condition. Table 3 provides the *t*-test results, as described in the section "Data Analysis," to compare the same functional classification and operational condition in congested and uncongested conditions.

Figure 5 illustrates the aggregated acceleration behavior for freeway (FW1 and FW2) and interstate (FW3 and FW4) car-following events in congested and uncongested conditions, for each functional classification.

Initial inspection of Figure 5 indicates a distinct difference in trend clarity between freeway and interstate driving. Similar to the finding discussed in the previous section, both congested and uncongested interstate frameworks (Figure 5, *c* and d) are cleaner, illustrating more prevalent spiral trends, than their corresponding frameworks in freeway conditions (Figure 5, *a* and *b*).

Uncongested conditions are substantially more dispersed over the $dV–dX$ plane resulting in the generation of acceleration bins extending to the cut-off following distance of 80 m in both freeway and interstate uncongested conditions (Figure 5, *b* and *d*). Using the video data as corroborating ground truth, this trend can be attributed to the fact that drivers prefer much shorter gaps between themselves and their leading vehicle in congested conditions—likely a byproduct of a reduced gap at lower travel speeds, as well as an attempt to avoid cut-in behavior by more aggressive drivers in adjacent lanes.

The uncongested freeway framework (Figure 5*b*) illustrates substantially more dispersed acceleration behavior, with few clear discernable trends, when compared to the congested freeway framework (Figure 5*a*). This finding is supported by results in Table 3 (FW1 v. FW2), which indicate statistically significant differences in acceleration behavior in regions 8 and 9 of the $dV - dX$ plane. Similarly, the uncongested interstate framework (Figure 5*d*) does not replicate the clear, succinct trends found in the congested interstate framework (Figure 5*c*, FW3 v. FW4).

This decrease in discernable trends is anticipated for uncongested conditions because drivers' desired

Table 3. *t*-Test Comparison of Functional Classifications and Operational Conditions in Congested and Uncongested Conditions

Region location & number		FW1 v. FW2	FW3 v. FW4	FW5 v. FW6	FW7 v. FW8	FW9 v. FW10	FW11 v. FW12
7–8 car-lengths | 0–4.5 mph slower	26		0.937			0.844	
7–8 car-lengths | 0–4.5 mph faster	25	0.985	0.374			0.058	
6–7 car-lengths | 4.5–9 mph slower	24	0.481	0.003*			0.027*	
6–7 car-lengths | 0–4.5 mph slower	23	0.000*	0.067	0.910		0.981	0.236
6–7 car-lengths | 0–4.5 mph faster	22	0.000*	0.010*	0.774		0.130	0.990
5–6 car-lengths | 4.5–9 mph slower	21	0.101	0.117			0.004*	
5–6 car-lengths | 0–4.5 mph slower	20	0.000*	0.325	0.289		0.006*	0.000*
5–6 car-lengths | 0–4.5 mph faster	19	0.351	0.659	0.002*		0.022*	0.250
4–5 car-lengths | 4.5–9 mph slower	18	0.599	0.035*	0.057		0.411	0.243
4–5 car-lengths | 0–4.5 mph slower	17	0.109	0.321	0.557	0.001*	0.263	0.001*
4–5 car-lengths | 0–4.5 mph faster	16	0.117	0.467	0.762		0.462	0.954
4–5 car-lengths | 4.5–9 mph faster	15	0.031*	0.376				
3–4 car-lengths | 4.5–9 mph slower	14	0.054	0.000*			0.801	0.108
3–4 car-lengths | 0–4.5 mph slower	13	0.075	0.000*	0.003*	0.041*	0.033*	0.302
3–4 car-lengths | 0–4.5 mph faster	12	0.027*	0.044*	0.733	0.969	0.588	0.001*
3–4 car-lengths | 4.5–9 mph faster	11	0.051	0.065	0.296			
2–3 car-lengths | 4.5–9 mph slower	10	0.650	0.240			0.020*	
2–3 car-lengths | 0–4.5 mph slower	9	0.000*	0.000*	0.006*	0.003*	0.193	0.000*
2–3 car-lengths | 0–4.5 mph faster	8	0.000*	0.633	0.003*		0.174	0.000*
2–3 car-lengths | 4.5–9 mph faster	7	0.148	0.094				
1–2 car-lengths | 4.5–9 mph slower	6	0.173	0.119				
1–2 car-lengths | 0–4.5 mph slower	5	0.002*	0.001*	0.003*		0.206	0.023*
1–2 car-lengths | 0–4.5 mph faster	4	0.194	0.488	0.062		0.789	0.036*
1–2 car-lengths | 4.5–9 mph faster	3	0.027*	0.005*				
0–1 car-lengths | 0–4.5 mph slower	2		0.197				
0–1 car-lengths | 0–4.5 mph faster	1	0.887	0.061				

Note: FW = framework. # car lengths (6 m) from lead vehicle; # mph speed difference between the lead and following vehicle. Blank cells & regions not included on the table indicate insufficient data to complete analysis. Results of a two-tailed, paired *t*-test; cells indicate *p*-value at alpha = 0.05 (95% confidence level).
*A significant difference exists between the mean accelerations in that region.

acceleration is less likely to be restricted by surrounding traffic and a lead vehicle. When approaching a lead vehicle, drivers are more likely to have an opportunity to overtake the leading vehicle, without making large behavioral adjustments; therefore, car-following spirals are less prominent.

Similarly, Figure 6 illustrates the aggregated acceleration behavior for WZ1 and WZ2 in congested and uncongested conditions. The most apparent distinction between uncongested WZ1 (Figure 6*b*) and uncongested WZ2 (Figure 6*d*) is that no acceleration behavior is observed beyond a 50-m separation distance for WZ1. Thus, uncongested WZs with a lane closure share acceleration behavioral characteristics with different roadway types and operational conditions experiencing congested conditions; this is in direct contrast to all other frameworks in uncongested conditions, which have data reaching the 80-m cut-off. This finding is also shown in Table 3, in which only 32% of regions on the *dV–dX* plane indicate significant differences between WZ1 congested and uncongested conditions. In comparison, 50% of regions indicate significant differences between congested and uncongested conditions in WZ2. This difference can be directly attributed to the lane closure in WZ1, which shifts driver behavioral tendencies to more closely match those in congested conditions; in essence, this is forcing drivers to react in response to their perception of the lead vehicle without the option to overtake the lead vehicle, which results in more consolidated driving behavior and the production of more car-following spirals. Conversely, the uncongested WZ2 behavior closely reflects the behavior seen in uncongested freeway and interstate conditions (Figure 5, *b* and *d*).

Heavy deceleration when relative velocity is negative in both WZ1 and WZ2 uncongested conditions (Figure 6, *b* and *d*) can be attributed to other roadway factors within the WZ that cause drivers to decelerate, even when the lead vehicle is not an imminent threat. A few of these deceleration bins can be identified in congested conditions (Figure 6, *a* and *c*); however, the reduced frequency in their appearance is likely due to the reduced speeds attributed to congested conditions and the consistency of behavioral adjustments relative to the lead vehicle. Statistical evidence of this observation is prevalent when comparing congested and uncongested conditions in WZ2. Results in Table 3 (FW11 v. FW12) indicate that significant differences in acceleration behavior are prevalent for regions 5 and 9.

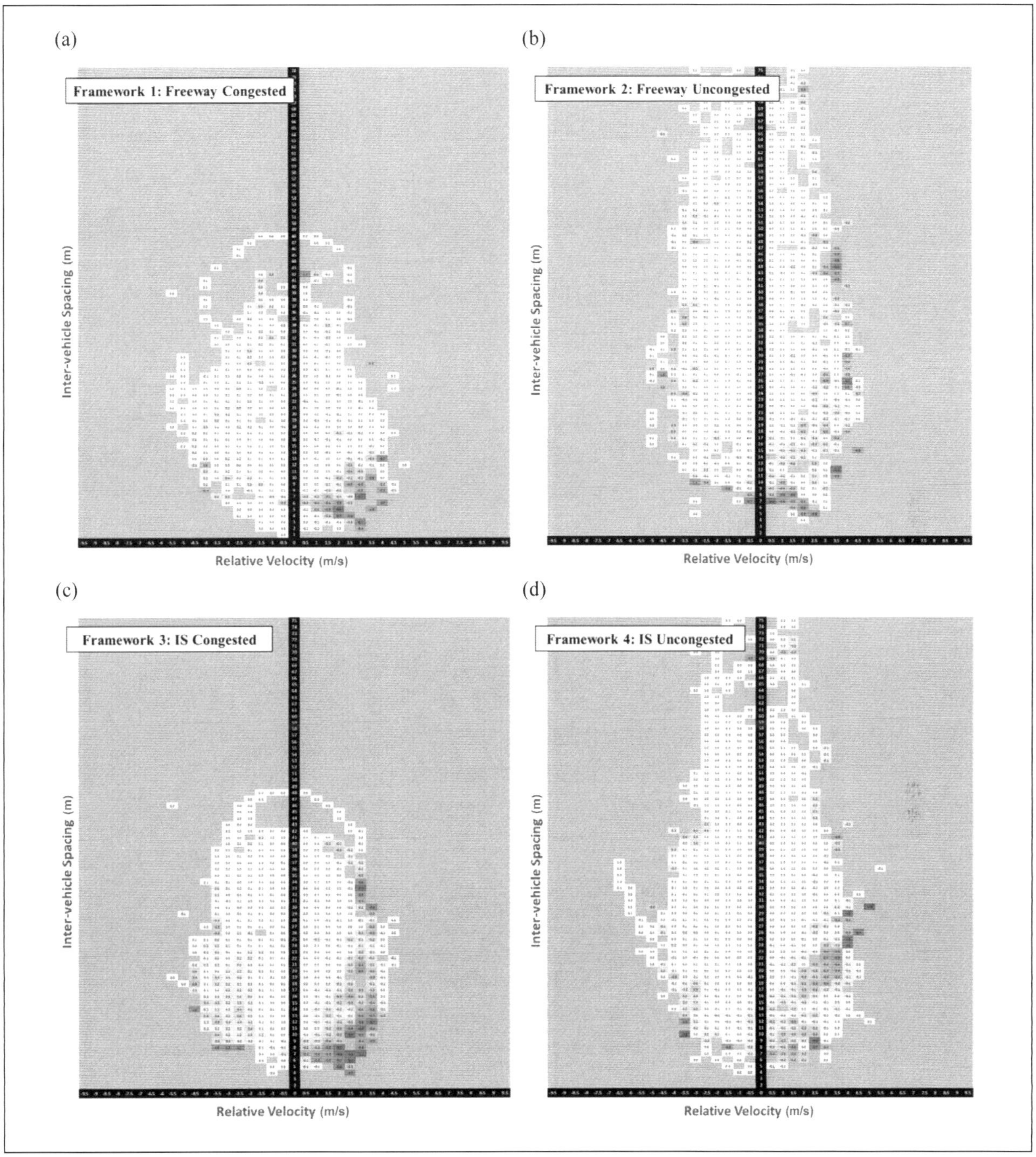

Figure 5. Aggregated car-following acceleration behavior on a psychophysical plane, representing frameworks 1 to 4: freeway and interstate, congested and uncongested conditions.

Acceleration behavior trends in congested and uncongested conditions for advanced warning zones are similar to the previously stated findings: uncongested conditions show significantly fewer trends than those in congested conditions. Although a visual is not provided, this finding is indicated in Table 3 (FW5 and FW6), in which the statistical analysis shows that significant differences in regions 5, 8, 9, 13, and 19 exist at a 99% confidence level. Acceleration trends in the taper zone are less notable in this study, as sufficient data for uncongested taper zones was not available in the data.

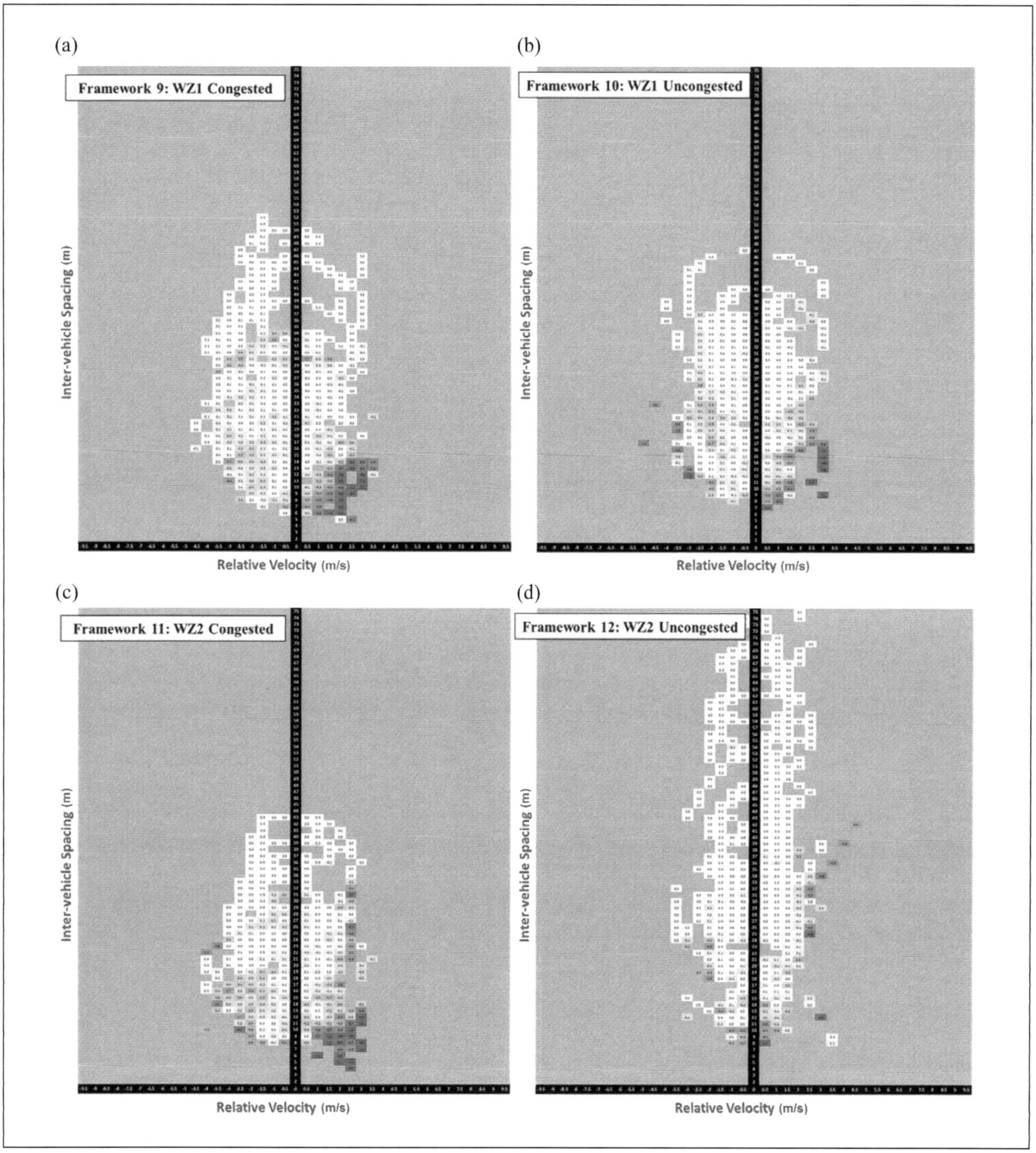

Figure 6. Aggregated car-following acceleration behavior on a psychophysical plane, representing frameworks 9 to 12: Work zone with lane closure and work zone without lane closure, congested and uncongested conditions.

Results Summary

The qualitative and quantitative analysis presented in this section provides evidence of heterogeneity in driving behavior as a function of various driving environments. Car-following behavior for different roadway functional classifications and operational conditions with congested traffic conditions were compared, as congested traffic conditions proved to illustrate the clearest car-following trends. The analysis showed that non-WZ interstate conditions are the cleanest, with the most discernable car-following spirals, whereas acceleration behavior in taper zones is the most sporadic. Strong acceleration trends were identified in WZ frameworks, indicating drivers' motivation to maintain a closer following distance, especially in the advanced warning zone. The results identified regions within the psychophysical plane where driving behavior is significantly different between WZ types and differing levels of congestion. For this analysis, congestion was considered a binary variable, congested or uncongested; however, future analyses will further examine the impact of surrounding traffic conditions at a higher resolution, as studies have shown heterogeneity in desired time gap in differing levels of congestion (*19*).

Conclusions and Recommendations

Given the necessity of microsimulation models to optimally evaluate investment decisions, the accuracy of microsimulation models in depicting traffic conditions is of upmost importance. The precision of a microsimulation model reflects the accuracy of each embedded submodel, including car-following models. To achieve this accuracy, car-following models must be calibrated using appropriate and high-quality data sources.

This research presents conclusive evidence—generated from a robust, trajectory-level dataset—that drivers' car-following actions are different while traveling in different driving environments (intra-driver heterogeneity). This finding is presented in an innovative manner, which aggregates car-following acceleration behavior from 62 drivers participating in the 2013 FHWA Living Laboratory IRV study, to account for inter-driver heterogeneity and enable the identification and calibration of global trends in driver behavior.

The findings serve as the foundation for continued research into the development of predictive car-following and driver behavior models capable of representing heterogeneous intra-driving behavior in various driving environments. Current and future research is focused on developing the global trends portrayed in this paper to calibrate the FHWA Driver Model car-following framework.

Acknowledgments

The authors acknowledge the role of Taylor Lochrane of FHWA, for collecting the FHWA Living Laboratory IRV data, and Stephen Zitzow-Childs, Emily Lawless, Nick Fournier, Deirdre Falling, Nick Fournier, Jesse Davis, Jared Young, and Bianka Mejias of VOLPE for processing that data.

Author Contributions

The authors confirm contribution to the paper as follows: study conception and design: AB; data collection: AB; analysis and interpretation of results: AB, RJ, BH, CF; draft manuscript preparation: AB, RJ, BH, CF, CM. All authors reviewed the results and approved the final version of the manuscript.

References

1. Federal Highway Administration. Performance Measurement. *MAP-21 - Moving Ahead for Progress in the 21st Century Fact Sheet*. https://www.fmcsa.dot.gov/mission/policy/map-21-moving-ahead-progress-21st-century-act.
2. Ossen, S., and S. P. Hoogendoorn. Heterogeneity in Car-Following Behavior: Theory and Empirics. *Transportation Research Part C: Emerging Technologies*, Vol. 19, No. 2, 2011, pp. 182–195.
3. Wang, H., W. Wang, J. Chen, and M. Jing. Using Trajectory Data to Analyze Intradriver Heterogeneity in Car-Following. *Transportation Research Record: Journal of the Transportation Research Board*, 2011. 2188: 85–95.
4. Lochrane, T. W. P., H. Al-Deek, D. J. Dailey, and J. Bared. Using a Living Laboratory to Support Transportation Research for a Freeway Work Zone. *Journal of Transportation Engineering*, Vol. 140, No. 7, 2014, pp. 1–7.
5. Garber, N. J., and L. A. Hoel. *Traffic and Highway Engineering*. Cengage Learning, Toronto, ON, Canada, 2009.
6. Al-Kaisy, A., and F. Hall. Guidelines for Estimating Capacity at Freeway Reconstruction Zones. *Journal of Transportation Engineering*, Vol. 129, No. 5, 2003, pp. 572–577.
7. Maze, T., G. Burchett, J. Hochstein, and S. Initiative. Synthesis of Procedures to Forecast and Monitor Work Zone Safety and Mobility Impacts. *Report*, November, 2005.
8. Benekohal, R., A.-Z. Kaja-Mohideen, and M. Chitturi. Methodology for Estimating Operating Speed and Capacity in Work Zones. *Transportation Research Record: Journal of the Transportation Research Board*, 2004. 1883: 103–111.
9. Lochrane, T. W. P., H. Al-deek, X. Jiang, N. Highway, D. J. Dailey, and J. Shurbutt. Capturing Driver Behavior by Measuring Time Gap in Freeway Work Zones. Presented at 96th Annual Meeting of the Transportation Research Board, Washington, D.C., 2017.
10. Dijker, T., P. H. L. Bovy, and R. G. M. M. Vermijs. Car-Following under Congested Conditions: Empirical Findings. *Transportation Research Record: Journal of the Transportation Research Board*, 1998. 1644: 20–28.
11. Yin, S., Z. Li, Y. Zhang, D. Yao, Y. Su, and L. Li. Headway Distribution Modeling with Regard to Traffic Status. *Proc., IEEE Intelligent Vehicles Symposium*, Xi'an, China, 2009, pp. 1057–1062.
12. Brackstone, M., B. Waterson, and M. McDonald. Determinants of Following Headway in Congested Traffic. *Transportation Research Part F: Traffic Psychology and Behaviour*, Vol. 12, No. 2, 2009, pp. 131–142.
13. Barbosa, L. C. *Studies of Traffic Flow Models*. Columbus, OH, 1961.
14. Michaels, R. M. Perceptual Factors in Car Following. 1963.
15. Todosoiev, E. P. *The Action Point Model of the Driver-Vehicle System*. Columbus, OH, 1963.

16. Olstam, J. J., and A. Tapani. Comparison of Car-Following Models. *Swedish National Road and Transportation Research Institute*, Vol. VTI re, No. 1, 2004, pp. 116–127.
17. Brackstone, M., and M. McDonald. Car-Following: A Historical Review. *Transportation Research Part F: Traffic Psychology and Behaviour*, Vol. 2, No. 4, 1999, pp. 181–196.
18. Fitzpatrick, C. D. *Evaluating the Impacts of Driver Behavior in the Speed Selection Process and the Related Outcomes*. University of Massachusetts, Amherst, MA, 2017.
19. Treiber, M., and D. Helbing. Memory Effects in Microscopic Traffic Models and Wide Scattering in Flow-Density Data. *Physical Review E*, Vol. 68, No. April, 2003, pp. 1–8.

The Standing Committee on Vehicle User Characteristics (AND10) peer-reviewed this paper (18-06005).

TRR
JOURNAL OF THE TRANSPORTATION RESEARCH BOARD

Article

Transportation Research Record
2018, Vol. 2672(37) 45–54

DOI: 10.1177/0361198118794712
journals.sagepub.com/home/trr
SAGE

Impact of Distracted Drivers on Individual and Group Behavior of Following Vehicles: A Networked Multi-Driver Simulator Study

Jing Xu[1] **and Yingzi Lin**[1]

Abstract

As traffic density increases, drivers experience more interaction situations, with a greater need to react and adapt to other road users. Driver distraction as an impairment on road safety and traffic efficiency has been widely discussed. How distracted drivers affect the performance of other drivers and vehicle platoons in traffic flow, however, remains underexplored. This study examines the impact of a low-taskload distraction (cellphone-texting task) not only on the specific distracted driver's behavior, but also how this behavior influences other drivers, individually and collectively as a group, through a networked driving simulation platform. The concept of interactive driving and implications of virtual environments using a networked multi-driver simulation platform were the two major facilitating factors for carrying out this study. The performance indicators from 12 driver groups (three drivers per group) in a car-following situation with a distracted driver involved are analyzed and compared with a no-distraction trial. The results presented show that the impacts of a phone-texting distraction in traffic flow differ when observing drivers individually or as a four-driver platoon. This implementation has the potential to lay a foundation for continued interactive driving experiment design and can serve as an experimental tool for new vehicle technology applications that consider collective vehicle behavior.

The patterns of traffic flow and drivers' behavior mutually determine and influence each other (*1*, *2*). Traffic, especially on highways, is organized and transmitted between a free-flow state and a synchronized flow state (*3*). The ideal free-flow traffic in multi-lane roadways is characterized by high, asynchronous speeds in each lane, modest following distances, and frequent lane changes. However, as the number of interacting vehicles increases, even slight variations in driving behaviors can reduce the stability of traffic flow and result in decreased traffic efficiency (*4*, *5*). The increasing demand for road use, drivers' decision-making, and on–off ramps cause major disruptions to traffic flow. Equally critical are factors related to human distraction, such as in-vehicle use of devices. As reported by the US Centers for Disease Control and Prevention in 2014, "Each day in the United States, more than 9 people are killed, and more than 1,060 people are injured in crashes that are reported to involve a distracted driver" and "nearly 80% of crashes and 65% of near-crashes involved some form of driver inattention within 3 seconds before the event." Distraction is significantly correlated with driver behavior. Studies over the years have concluded that distracted driving is caused by elements like cellphone usage for calls or texting, eating or drinking, smoking, conversations with passengers, reading maps, or any other in-car activities. A significant amount of previous research on distracted driving via phone use has shown that when compared with no phone use, the distracted driver demonstrates general decreases in average speed, response time, information processing, and memory, as well as increases in speed variability, following distance, task duration, number of crashes, and time needed to recover speed (*6–16*). Most of the phone-related distraction studies mentioned above adopted car-following as a primary task in their experimental setup, and focused on studying the performance of the driver who was distracted.

[1]Intelligent Human–Machine Systems Lab, Mechanical and Industrial Engineering Department, College of Engineering, Boston, MA

Corresponding Author:
Address correspondence to Yingzi Lin: yi.lin@northeastern.edu

Despite a large body of literature, the impact of distracted driving on other drivers or on traffic flow has not been widely investigated. Cooper et al. (*4*) studied 36 undergraduate students driving in varying levels of simulated traffic flow with a distracted phone-conservation task, and compared their performance with no-distraction driving. The results showed that both driver distraction and traffic congestion significantly affect lane-change frequency, mean speed, and the likelihood of remaining behind a slower-moving lead vehicle. A later study examined the behavior of teens and young adults driving with various distractions (i.e., cellphone, text, and undistracted) and different driving conditions, such as free flow, stable flow, and oversaturation (*5*). This study revealed that more lane deviations and crashes occurred during texting: Participants exhibited greater fluctuations in speed, significantly fewer lane changes, took longer to complete the scenario, and were overtaken by more vehicles. Both studies provided evidence that distracted driving, particularly texting, may lead to reduced safety and traffic flow efficiency. However, to collect performance data from distracted drivers in traffic flow, both investigations employed a simulated driver model as the lead driver and used simulated vehicles to form traffic. In addition to experiment-based studies, model-based research has also been used to study the influence of distracted driving on groups of drivers or on traffic flow. Xiao and Shi proposed an improved cellular automaton model for simulated traffic flow with distracted drivers based on the Nagel–Schreckenberg model (*17*). The simulation results show that the traffic flow rate and velocity significantly decrease with a greater number of drivers using a phone. Moreover, under low-density conditions, the risk levels of traffic first decrease and then increase as the number of distracted drivers increases. They also found that phone-using distracted driving can reduce traffic flow rates by 5%. A later study from the group proposed a modified optimal velocity model to evaluate the relationship between the distribution of distraction duration times and the influence of distraction on traffic flow (*18*). Their model simulated distraction behavior very well and proves that short-time distractions have a smaller impact on traffic flow than long-time distractions. For driving behavior indicators for traffic flow, the total number of cars the participant driver has passed, the total number of cars that have passed the participant driver, driving speed fluctuations, lane-change frequency, and scenario completion time were indicators mentioned in these studies.

Driving simulators are still widely used as an efficient method for conducting driving behavior studies. Traditional driver simulators can monitor individual drivers. Unlike an experiment that runs in traffic with real drivers, it is difficult to analyze more complex phenomena, such as the interaction between multiple road users, or safety-critical situations, when single-user driving or standalone systems are relied on, as the majority of available simulators do (*19*, *20*). Analysis of the influence of driver distractions, as well as the study of group phenomena in traffic demands a new research approach. Therefore, the topic of driver–driver interaction and multiple-user driving simulation has attracted considerable worldwide attention in recent years. For applications for an interconnected driving simulator platform, components for an advanced driving assistance system, such as lane-merging assistance, hazard warning, and traffic-light assistance, rubbernecking, and collision warnings have been studied (*21–36*). All of these studies show the benefit of connecting multi-driver simulation for the analysis of driving behavior in controlled settings. However, for distracted-driving studies, from our extensive literature review, no previous research has studied multiple drivers simultaneously to investigate the impact of distracted driving on other drivers, or on a platoon of vehicles.

The current research utilizes a networked multi-driver simulation to explore the effect of distractions on driving both on individual and group levels. With this platform, multiple-driver simulators are linked to the same virtual environment: The participants can interact with human drivers instead of simulated driver models. By doing so, assumptions based on driving models are not needed, and more driving behavior indicators can be measured, at both individual and group levels. In this study, in addition to the regular driver behavior indicators (e.g., speed, headway distance [HWD], reaction time, and so forth) (*37*), speed fluctuation, HWD fluctuation, vehicle platoon length, and differences in distance within the platoon (*25*, *38*), as well as speed and HWD profiles, are introduced to describe driver platoon behavior in traffic flow.

Methodology

Participants

A total of 36 participants (13 females, 23 males) ranging from 18 to 56 years old (mean = 26.95, SD = 5.076) were recruited for this study. Participants were required to be experienced drivers with a valid driver's license and active, recent driving experience. Participants were randomly allocated to one of 12 groups. There were three participants per group, who each followed the car in front of them in the virtual environment using a multi-driver simulation platform. Data from one group and three subjects were excluded from analyses because of coordinating issues and experiment failures. This study was approved by the Northeastern University (IRB #17-02-09).

Apparatus

The developed networked multiple-driver simulation system is located in the Intelligent Human–Machine Systems (IHMS) Laboratory at Northeastern University. Four human drivers participated in each simulation, with each driver at the controls of a single driving simulator station, as shown in Figure 1. Each simulator station is composed of a Dell workstation (Minitower 4500) with an NVidia GTX 970 graphics card, a steering system (Logitech G27), and a 24-inch LCD monitor (resolution 1,920 × 1,080 pixels). The driver's input included a steering wheel and gas and brake pedals. Several performance parameters, including speed, acceleration, position, HWD, and more were recorded at 24 Hz.

Participants operated their vehicles on a simulated road from a first-person perspective. A practice trial and two test trials were designed and developed in Unity 3D. The practice trial included a 10 km road in an open area. The two test trials (no distraction and with distraction) consisted of 2.5 km straight four-lane roads with traffic signs and surrounding traffic. The test layout for both trials were the same, but the surrounding trees, buildings, and oncoming traffic varied slightly to provide a sense of difference in environments between the two trials.

Tasks and Procedure

Figure 1 shows the driver layout and setups used for the experiment. The experiment lasted an average of 30 min. The four driving stations were physically separated to reduce potential influences from each driver. After arriving at their station, they were asked to read and sign a consent form, including all the information about the experiment. As further instructions, participants were told to take the time to read a customized instruction manual and the experiment procedure list. During the experiment process, participants did not discuss their instructions and requirements with each other. Next, a car-following practice trial took place to help participants become familiar with the input controls, the actual driving task, and to get to know the basic layout of the test track.

During the test trials, the four vehicles drove in the same virtual environment, through two simulated city blocks, first for the no-distraction trial, and then for the with-distraction trial. The four drivers in each trial are known as the *Lead Car, Follow 1, Follow 2, and Follow 3.* The Lead Car was an experimenter, while Follow 1, 2, and 3 were test participants. The Lead Car was a trained research assistant who was expected to maintain a consistent driving behavior across all trials and subject groups. The Lead Car followed set instructions, while each Follow vehicle was instructed to follow a specific car. Follow 1 followed the Lead Car, Follow 2 followed Follow 1, and Follow 3 followed Follow 2. These rules set the vehicles to drive as a four-vehicle platoon. Subjects were asked to drive as they normally would and stay in the right lane, maintain a safe distance from the front car, try to adhere to the speed limit of 50 mph as best they could, and respond to phone-texting tasks if necessary. It is worth noting that there was no time limit for subjects, and time was not reported until the finish line was reached and a "Finish" sign was displayed. Subjects were instructed, however, to notify the coordinator of any problem or discomfort while driving. After the test, participants were interviewed about the study, and a gift card was given as compensation.

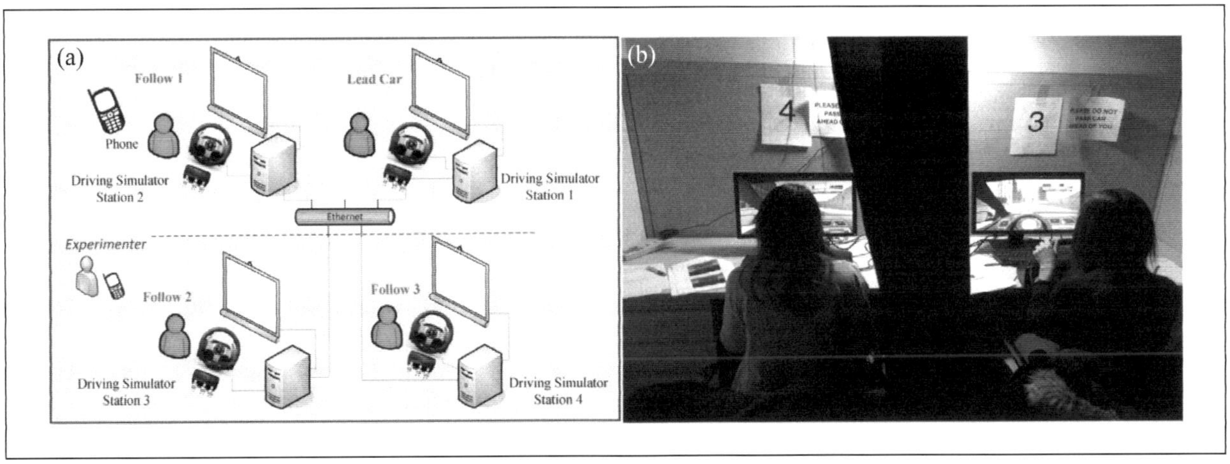

Figure 1. The four-terminal multi-driver simulation system and experiment setups. (*a*) multi-driver simulation system framework; (*b*) the Follow 2 and Follow 3 during experiments.

During each test trial, a "Sudden Brake" command would be displayed on the Lead Car's driving screen, which prompted the Lead Car to create a "Brake to Stop" event. This event took place twice in set locations for each of the two test trials, but in different locations. During the with-distraction trial, only Follow 1 was instructed to send text messages as a distraction task. Follow 1 received instructions and training procedures requiring him/her to save an experiment coordinator's phone number before testing. Throughout the experiment, Follow 1 would receive a "text to Mike" in-vehicle heads-up display (HUD) message at specific locations, 250 m before the Lead Car brake events started. Follow 1 had to use their own phone to send a text message as soon as they received the HUD message. The subject was asked to send a short message, one that they would normally send while driving.

Data Analysis

The impact of distracted driving was assessed at three levels: (1) speed performance; (2) analysis of longitudinal platoon behavior; and (3) safety-related measures. Nine performance measures were examined: (a) average speed of the three follow vehicles along the test trial; (b) standard deviation of average speed, which represents the speed fluctuations; (c) average HWD for each follow driver along the test trial; (d) HWD fluctuation; (e) platoon length, calculated as the sum of the four vehicle lengths and three HWDs; (f) difference in distance among the three follow vehicles, calculated by $HWD_{first\_pair} - HWD_{second\_pair}$, where HWD_{first_pair} represents the distance between Follow 1 and Follow 2, and HWD_{second_pair} represents the distance between Follow 2 and Follow 3 (*25*); (g) reaction time, calculated as the time between the front car braking and subject car beginning to brake; (h) critical gap and safety gap, calculated as the proportion of time to collision (TTC) <1 s and TTC >3 s across the test trial; and (i) speed profile in the event window (50 m before and 200 m after the event). The main analyses comprised a series of analyses of variance (ANOVAs) and pairwise comparisons. The vehicle position in the ANOVAs refers to the individual Follow vehicles, with an emphasis on their order within the platoon. Effect sizes are reported as partial eta squared (η_p^2). A 0.05 significance threshold was tested for and compared in tests.

Results

Speed Performance

A one-way ANOVA was performed to compare the Lead Car's speed performance between distraction and no-distraction trials. No significant differences in average speed ($F_{(1, 17)} = 0.373$, $p = 0.550$) and standard deviation of average speed ($F_{(3, 37)} = 1.852$, $p = 0.156$) were found between the two trials. This supports the idea that the pre-trained experimenter in this experiment maintained a consistent driving speed performance across all experiment trials and groups.

A 3 (vehicle position: Follow 1, Follow 2, Follow 3) × 2 (distraction: no distraction, with distraction) ANOVA was performed on average speed. Distraction ($F_{(1, 51)} = 0.182$, $p = 0.672$) and vehicle position ($F_{(2, 51)} = 2.802$, $p = 0.07$) were not identified as significant main effects. However, the analysis revealed vehicle position as a significant main effect on standard deviation of average speed, $F_{(2, 51)} = 10.199$, $p < 0.001$, $\eta_p^2 = 0.286$. Average speed SD for Follow 1 was significantly higher than for Follow 2 and 3 ($p < 0.001$), and Follow 2 showed a significantly higher speed SD than Follow 3 ($p = 0.04$). No other effects were significant.

Platoon Length

The four-vehicle platoon length from the distraction trial was significantly higher (mean = 145.245, SD = 32.547) than from the no-distraction trial (mean = 111.083, SD = 16.273), $F_{(1, 16)} = 6.557$, $p = 0.021$, $\eta_p^2 = 0.291$.

Average Headway Distance

The 3 (vehicle position: Follow 1, Follow 2, Follow 3) × 2 (distraction: no distraction and with distraction) ANOVA for average HWD did not show distraction ($F_{(1, 51)} = 3.447$, $p = 0.069$) or vehicle position ($F_{(2, 51)} = 1.714$, $p = 0.19$) as main significant effects. The interaction effect between distraction and vehicle position was not significant, $F_{(2, 51)} = 1.335$, $p = 0.272$. However, as Figure 2*c* shows, the three Follow vehicles kept noticeably consistent HWDs to the cars in front during the no-distraction trial. In the distraction trial, the distracted Follow 1 kept a longer HWD, and Follow 2's HWD was also longer than during the no-distraction trial. Follow 3 showed similar HWDs during both conditions. The 3 (vehicle position: Follow 1, Follow 2, Follow 3) × 2 (distraction: no distraction and with distraction) ANOVA for standard deviation of HWD did not show distraction ($F_{(1, 48)} = 0.635$, $p = .429$) or vehicle position ($F_{(2, 48)} = 0.748$, $p = 0.479$) as significant main effects. The interaction effect between distraction and vehicle position was not significant, $F_{(2, 48)} = 1.219$, $p = 0.304$.

The Difference in Distances among the Three Follow Vehicles

The difference in HWD among the three Follow vehicles did not test for significant differences between distraction

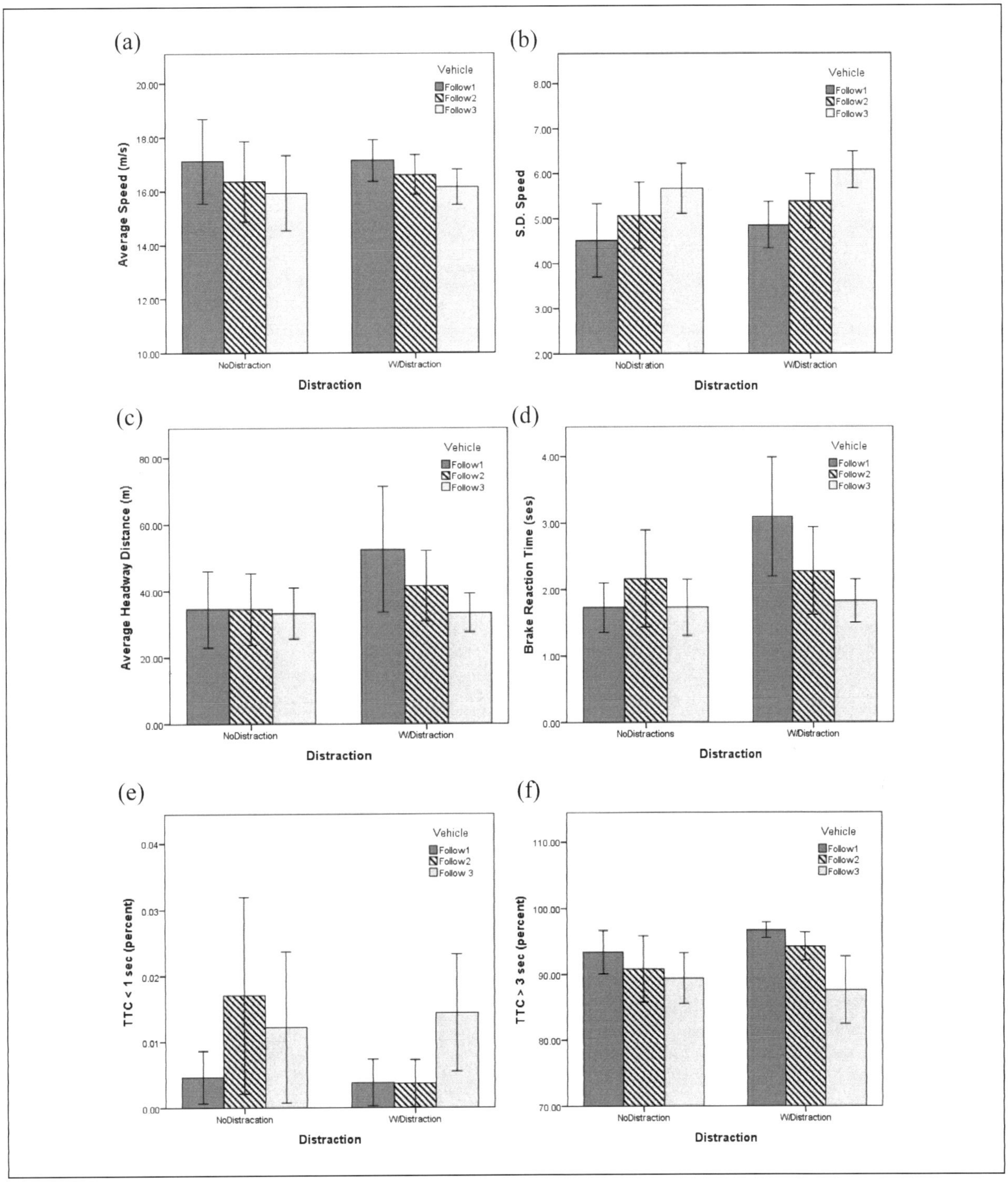

Figure 2. Driving performance measurements for the three Follow vehicles: (*a*) average driving speed; (*b*) standard deviation of average speed; (*c*) average HWD; (*d*) brake reaction time to Lead Car event; (*e*) proportion of critical gap (TTC < 1 s); and (*f*) proportion of safety gap (TTC > 3 s). Error bars represent 95% confidence intervals. Analysis of longitudinal platoon behavior.

and no-distraction trials, $F_{(1,\ 17)} = 0.643$, $p = 0.437$. As Figure 3 shows, the HWD between Follow 1 and Follow 2 was noticeably larger than between Follow 2 and Follow 3 during the distraction trial (mean = 10.950, SD = 24.322). For the no-distraction trial, a marginal deviation from the zero-line indicates a consistent

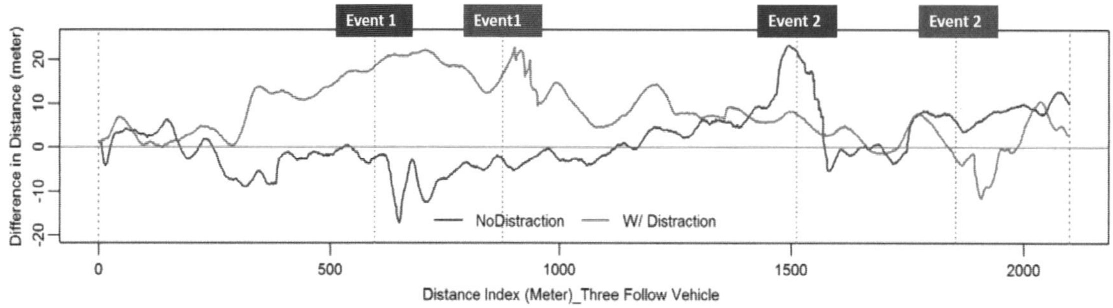

Figure 3. The profile of differences in distance among the three following vehicles during no-distraction and with-distraction trials. Data were sampled every meter. Vertical lines identify Lead Car brake event locations for no-distraction (blue) and with-distraction (red) trials.

longitudinal driving behavior from the vehicle platoon with equal distances between each pair of Follow vehicles (Mean = 0.030, SD = 20.949). However, the resulting non-significant differences between distraction and no-distraction trials were mainly caused by large variances. These large variances can reflect the individual differences among experimental groups and within each test group. On the other hand, inspection of Figure 3 indicates that deviation from the zero-line increased after either distraction were applied, or a lead vehicle event occurred.

Safety-Related Measures

Reaction Time to Lead Vehicle Sudden Brake Event. The reaction times from lead vehicle sudden stops are presented in Figure 2*d*. The 3 (vehicle position: Follow 1, Follow2, Follow 3) × 2 (distraction: no distraction and with distraction) ANOVA showed distraction, $F_{(1, 109)} = 4.515$, $p = 0.036$, $\eta_p^2 = 0.040$, as a main effect. A significant two-way interaction was found between distraction and vehicle position, $F_{(2, 109)} = 2.946$, $p = 0.057$, $\eta_p^2 = 0.051$. Post-hoc analysis tested a significant difference between Follow 1 and Follow 3 ($p = 0.058$). In addition, a pairwise comparison indicated that Follow 1 required significantly longer reaction times while distracted, $F_{(1, 38)} = 7.376$, $p = 0.01$, $\eta_p^2 = 0.163$. However, Follow 2 and Follow 3 did not show significant changes in reaction times between the two conditions.

Critical Gap Measure. The proportion of critical gaps (TTC $\leq$ 1 s) over the entire trial was analyzed by a 3 (vehicle position: Follow 1, Follow 2, Follow 3) × 2 (distraction: no distraction, with distraction) ANOVA. A tested main effect of vehicle position was identified, $F_{(2, 51)} = 3.37$, $p= 0.042$, $\eta_p^2 = 0.117$. Neither distraction as the main effect, nor the distraction × vehicle interaction was significant. Post-hoc tests showed that Follow 1 had a significantly lower critical gap percentage than Follow 3 ($p = 0.029$). In addition, the pairwise comparison indicated that during the distraction trial, Follow 2 had a significantly lower critical gap percentage when compared with no distraction, $F_{(1, 17)} = 5.595$, $p = 0.03$, $\eta_p^2 = 0.248$.

Meanwhile, the analysis of TTCs longer than 3 s revealed a main effect of vehicle position, $F_{(2, 51)} = 8.445$, $p = 0.001$, $\eta_p^2 = 0.249$. No other effects were significantly proven. Post-hoc analysis tested the significant differences between Follow 1 and Follow 3 ($p < 0.001$), and between Follow 2 and Follow 3 ($p = .020$). The pairwise comparison shows that Follow 1 under distraction had a significantly higher percentage of TTCs longer than 3 s, compared with no distraction, $F_{(1, 17)} = 6.183$, $p = 0.024$, $\eta_p^2 = 0.267$. Follow 2 and Follow 3 TTC percentages were not significantly different.

Event Performance Profile. To better observe the influence of distraction on following drivers with the current experimental settings, the average speed values are plotted against distance for front car brake events. Speed was averaged every 10 m over a distance of 50 m before and 150 m after the braking event, as shown in Figure 4. The speed profile shows a clear difference between the two driving conditions. After the phone-texting task, drivers demonstrated higher driving speeds, slower responses, and a slower speed recovery. For example, drivers without a distraction influence started to recover speed at 50 m, while drivers distracted by a phone-texting task began recovery 10–20 m after (at 70 m).

Discussion

This study has found that distracted and undistracted driver groups perform differently as both individual

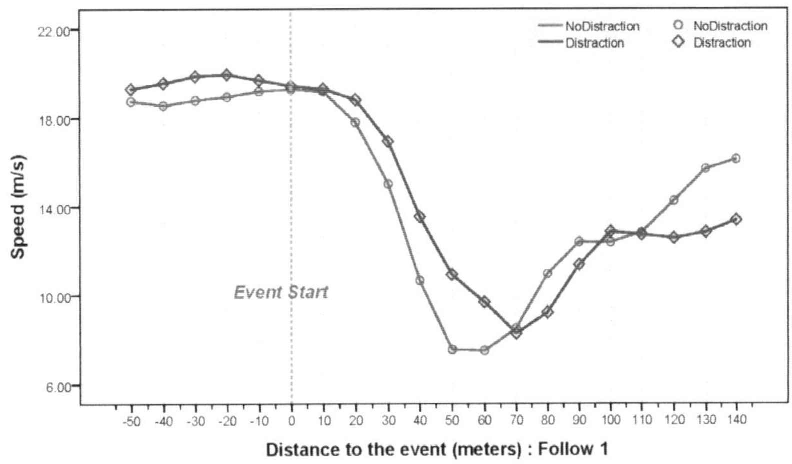

Figure 4. Follow 1 speed performance during the Lead Car brake event.

drivers and as drivers collectively in a group. Evidence from previous studies shows that drivers that are distracted (e.g., talking/texting on a cellphone) perform with slower speeds and greater speed variability (*39, 40*). In this study, distraction trials indicated slightly higher average speeds and lower speed variability compared with no-distraction trials across the driving course. Testing did not identify distraction as a significant main effect either for speed or speed fluctuations, suggesting that no reliable conclusion about the current phone-texting distraction can be made for overall speed behavior. There are also no significant differences among the three following drivers' maintained speeds. In addition, this study found that vehicle position was the main effect on speed fluctuation. Consistent with previous studies, the phone-texting task increased Follow 1's speed variability but did not significantly influence the two vehicles behind.

Treating four drivers as a platoon, the average platoon length was significantly influenced by distracted driver involvement. Distracted trials exhibited reliably higher platoon lengths than those with no distraction. As previous studies have shown, drivers tend to adopt longer HWDs when performing secondary tasks (*41, 42*). This is because drivers distracted by a cell phone task may attempt to compensate for their slower reactions by increasing the separation from the vehicle immediately ahead of them, which has been identified as a compensatory strategy (*6, 15*). The significantly longer vehicle group lengths and inconsistencies in driving patterns may increase the possibility of traffic congestion and impair traffic flow efficiency. To further understand the influence of a distracted driver's behavior on other drivers, the average HWDs for each following vehicle were analyzed; the difference in distances were introduced to describe the arrangement of vehicle positions within the platoons. Though no significant treatment effect of distraction was observed for average headways, the three following vehicles exhibited more stable HWDs with no distracted vehicle in the platoon, as shown in Figures 2*c* and 3. Follow 1 distracted with a phone-texting task demonstrated noticeably longer average HWDs, and Follow 2 also exhibited longer separations with a distracted vehicle in front, yet Follow 3 did not show any influence. The distance between Follow 1 and Follow 2 was noticeably longer than the distance between Follow 2 and Follow 3 in the distracted trial (positive deviations). This distribution of headways may reflect following preferences and a need to respond to the driving behavior in front of them. As we know, the drivers who keep a longer following distance may more easily let others pull into their headway gap, which will further lower road efficiency.

Other than the driver performance indicators above, safety-related driving performance indicators were also analyzed (e.g., brake reaction time to Lead Car braking, critical gap measure). Driver reaction time is an important accident risk indicator. In this study, brake reaction time was measured when the subject vehicles started to respond to Lead Car brake events. Experimental results show that Follow 1's reaction time increased by 44.11% compared with the no-distraction condition. This finding supports previous findings that a driver's ability to detect and react to unexpected road incidents will be impaired by distraction sources (*40, 43, 44*). The two following vehicle reaction times did not change significantly between the two conditions. These results suggest that the phone-texting distraction will significantly increase the distracted driver's reaction time, but will not influence the reaction time of drivers following behind. One interesting observation is that Follow 1 has a lower proportion of TTC $<$ 1 s when distracted, but not enough to be proven significant. However, Follow 2 has significantly lower critical gap proportions. This suggests that Follow 2 might notice the fluctuating behavior of Follow 1 and behave more carefully. In the meantime, Follow 1 exhibited a significantly higher proportion of TTC $>$ 3 s under the distraction condition, which is an indication of safety gaps. This finding, paired with the longer HWD for Follow 1, suggests that the phone-texting distraction's low demand and short duration seems to cause drivers to compensate with cautious behavior.

A related point to consider is the distraction method used in this study. Previous research has employed tasks of different cognitive levels, such as memory tasks (*45*), mental arithmetic tasks (*46*), and phone-conversation tasks (*6*, *15*, *16*). When focusing on phone-texting-related distractions, previous studies chose to send questions such as "What is your favorite television show?" "What do you like to do for fun?" to subject drivers (*5*); or to send predetermined phrases such as "The quick brown fox jumps over the lazy dog" (*14*). Unlike these studies, the current study has *attempted to simulate* a real-life scenario by using simple and short messages commonly used in daily driving as a distraction method. This design may introduce the risk of different distractions experienced among subjects, but one feature that will remain common is the short period of task influences. The messages sent were mostly "hello," "driving," "on the way," or "back soon," with an approximately equal proportion of each. Therefore, the lower significant effects tested in some measures might be a result of the phone-texting task used in this study, which did not have high-demand taskloads or long duration. Continued steps for this study could employ different text messaging tasks to assess impacts of varying levels of task difficulty or complexity on driving safety.

Each platoon consisted of only four vehicles, and the Lead Car served to trigger changes in platoon behavior. It would be interesting to see if congestion levels (traffic density) would have an effect on the difference between distracted and non-distracted driver behavior. Earlier work has found that traffic congestion has no influence on the visual attention of distracted drivers (*47*). Also, no significant differences have been detected between age groups, suggesting that all drivers, regardless of age, may drive in a manner that negatively impacts safety and traffic flow when distracted (*5*). Therefore, both traffic flow density and individual differences among drivers would be relevant topics for further investigation.

It is worth commenting on the car-following task used in this research. Unlike similar research that uses a car-following task with a pre-programmed lead vehicle or driver model, the lead vehicle in this study is driven by an experimenter who was trained to maintain a consistent driving behavior through all experiment trials, which includes following predesigned "brake to stop" instructions to perform sudden braking. As would be expected, a post analysis of the Lead Car's speed profile verified a consistent driving behavior across all groups. In post-experiment interviews, many participants felt the driving behavior of the front car was more natural than computer drivers. Future studies can apply this experimental platform to study how drivers react differently to autonomous vehicles and human drivers in more complex and dynamic road situations.

Conclusion

The influence of distracted driving on traffic safety has been studied for a long time. This study used a networked driving simulation platform to examine the impact of a low-taskload distraction (phone-texting task) not only on the distracted driver's behavior but also how this distracted behavior influences other drivers, both individually and collectively as a group. The concept of interactive driving and implications of the virtual environment using a networked multi-driver simulator were the two major facilitating factors for carrying out this study. This allows active driver–driver interactions in real time, and can also decrease the influences of pre-programmed car models in a simulated environment, which can cause unrealistic driver behavior reactions.

In this study, a four-driver platoon configuration was employed to investigate the influence of distracted driving. Driver interactions in a car-following situation were analyzed for both distraction and no-distraction trials. Results demonstrated that a short phone-texting task influences the driving performance of a single driver as well as a group of drivers; individual drivers and the driver groups exhibited different behavior indicators. In comparison with a no-distraction condition, driver groups with a phone-texting distraction had greater speed variability, longer average following HWDs, significantly slower reaction times and longer distances needed for speed recovery in response to the front car braking events. For the four-vehicle platoon, one distracted driver in the group significantly increased the entire platoon length, and the following drivers showed less consistent HWDs compared with the no-distraction condition. This results in increased gap sizes, which can lead to more frequent traffic flow interruptions, possibly impacting the traffic flow stability. Vehicles behind the distracted driver exhibited more following distance fluctuations, while the second following vehicle seemed not to be influenced by the distracted driving. Within the arrangement of the three following vehicles, a larger separation was found between Follow 2 and the distracted Follow 1 than the rear pair of drivers. The presented work can lay a foundation for continued interactive driving experiment design. This multi-driver simulation scheme can serve as a valuable tool to simultaneously investigate multiple drivers, observe their behaviors, interactions, and mutual influences, and evaluate new vehicle technologies.

Acknowledgments

This work has been financially supported by the National Science Foundation through grants 0954579 and 1333524 awarded to the corresponding author.

Author Contributions

The authors confirm contribution to the paper as follows: study conception and design: YL, JX; data collection: JX, YL; analysis and interpretation of results: JX, YL; draft manuscript preparation: JX, YL. All authors reviewed the results and approved the final version of the manuscript.

References

1. Huang, D. Lane-Changing Behavior on Highways. *Physics Review E*, Vol. 66, No. 2, 2002, p. 026124.
2. Treiber, M., A. Kesting, and D. Helbing. Understanding Widely Scattered Traffic Flows, the Capacity Drop, and Platoons as Effects of Variance-Driven Time Gaps. *Physics Review E*, Vol. 74, No. 1, 2006, p. 016123.
3. Kerner, B. *The Physics of Traffic: Empirical Freeway Pattern Features, Engineering Applications, and Theory*. Springer, Heidelberg, 2004, p. 96.
4. Cooper, J. M. I., N. Vladisavljevic, P. T. Medeiros-Ward, Martin, and D. L. Strayer. An Investigation of Driver Distraction Near the Tipping Point of Traffic Flow Stability. *Human Factors*, Vol. 51, No. 2, 2009, pp. 261–268.
5. Stavrinos, D., J. L. Jones, A. A. Garner, R. Griffin, C. A. Franklin, D. Ball, S. C. Welburn, K. K. Ball, V. P. Sisiopiku, and P. R. Fine. Impact of Distracted Driving on Safety and Traffic Flow. *Accident Analysis & Prevention*, Vol. 61, 2013, pp. 63–70.
6. Strayer, D. L., and F. A. Drews. Profiles in Driver Distraction: Effects of Cell Phone Conversations on Younger and Older Drivers. *Human Factors*, Vol. 46, No. 4, 2004, pp. 640–649.
7. Oviedo-Trespalacios, O., M. M. Haque, M. King, and S. Washington. Understanding the Impacts of Mobile Phone Distraction on Driving Performance: A Systematic Review. *Transportation Research Part C: Emerging Technologies*, Vol. 72, 2016, pp. 360–380. https://doi.org/10.1016/j.trc.2016.10.006.
8. Reimer, B., B. Mehler, J. F. Coughlin, N. Roy, and J. A. Dusek. The Impact of a Naturalistic Hands-Free Cellular Phone Task on Heart Rate and Simulated Driving Performance in Two Age Groups. *Transportation Research Part F: Traffic Psychology and Behaviour*, Vol. 14, No. 1, 2011, pp. 13–25. https://doi.org/10.1016/j.trf.2010.09.002.
9. Collet, C., A. Guillot, and C. Petit. Phoning While Driving I: A Review of Epidemiological, Psychological, Behavioural and Physiological Studies. *Ergonomics*, Vol. 53, No. 5, 2010, pp. 589–601.
10. Caird, J. K., K. A. Johnston, C. R. Willness, and M. Asbridge. The Use of Meta-Analysis or Research Synthesis to Combine Driving Simulation or Naturalistic Study Results on Driver Distraction. Journal of Safety Research, Vol. 49, 2014, pp. 91–6.
11. Brace, C. L., K. L. Young, and M. A. Regan. Analysis of the Literature: The Use of Mobile Phones While Driving. *Analysis*, Vol. 37, No. 3, 2007, pp. 261–265. https://doi.org/10.1016/j.jsr.2014.02.013.
12. Saifuzzaman, M., M. M. Haque, Z. Zheng, and S. Washington. Impact of Mobile Phone Use on Car-Following Behaviour of Young Drivers. *Accident Analysis & Prevention*, Vol. 82, 2015, pp. 10–19. https://doi.org/10.1016/j.aap.2015.05.001.
13. Saifuzzaman, M., and Z. Zheng. Incorporating Human-Factors in Car-Following Models: A Review of Recent Developments and Research Needs. *Transportation Research Part C: Emerging Technologies*, Vol. 48, 2014, pp. 379–403. https://doi.org/10.1016/j.trc.2014.09.008.
14. Drews, F. A., H. Yazdani, C. N. Godfrey, J. M. Cooper, and D. L. Strayer. Text Messaging during Simulated Driving. *Human Factors*, Vol. 51, No. 5, 2009, pp. 762–770. https://doi.org/10.1177/0018720809353319.
15. Strayer, D., F. Drews, and W. Johnston. Cell Phone-Induced Failures of Visual Attention during Simulated Driving. *Journal of Experimental Psychology: Applied*, Vol. 9, No. 1, 2003, pp. 23–32.
16. Strayer, D. L., J. Cooper, and F. A. Drews. *Profiles in Cell Phone-Induced Driver Distraction*. CRC Press, Hoboken, NJ, 2011.
17. Xiao, Y., and J. Shi. Analyzing the Influence of Mobile Phone Use of Drivers on Traffic Flow Based on an Improved Cellular Automaton Model. *Discrete Dynamics in Nature and Society*, 2015. http://dx.doi.org/10.1155/2015/573090.
18. Xiao, Y., and J. Shi. Study on the Influence of Driving Distraction on Traffic Flow Considering the Stochastic Duration Time of Distraction. *Modern Physics Letters B*, Vol. 30, Nos. 32–33, 2016, p. 1650380. https://doi.org/10.1142/S0217984916503802.
19. Xu, J., and Y. Lin. A Networked Multi-Driver Simulation Platform for Interactive Driving Behavior Study. *Proc., 15th Driving Simulation Conference (Europe)*, Paris, France, 2014, pp. 1101–1109.
20. Samareh, A., J. Xu, Y. Lin, and J. Benneyan. *Study the Headway Distance and Physiological Responses of Driver Distraction: An Experiment on Networked Multi-Driver Simulator*. Presented at Human Factors and Ergonomics Society 2015 International Annual Meeting, San Diego, CA, 2015.
21. Abdelgawad, K., J. Gausemeier, M. Grafe, and J. Berssenbrügge. Interest Manager for Networked Driving

Simulation Based on High-Level Architecture. *Designs*, Vol. 1, No.1, 2017, p. 3.
22. Houtenbos, M., J. C. F. de Winter, A. R. Hale, P. A. Wieringa, and M. P. Hagenzieker. Concurrent Audio-Visual Feedback for Supporting Drivers at Intersections: A Study Using Two Linked Driving Simulators. *Applied Ergonomics*, Vol. 60, 2017, pp. 30–42.
23. Ishak, S., O. A. Osman, J. Codjoe, and S. Mousa. Development of a Simulation Test Bed for Connected Vehicles using the LSU Driving Simulator. Gulf Coast Research Center for Evacuation and Transportation Resiliency, Vol. 41, 2016, p. 317.
24. Maag, C., and C. Mark. Assisting Drivers' Merging onto the Motorway: Evaluation of a New Advanced Driver Assistance System (ADAS) Using Multi-Driver Simulation. Presented at 19th ITS World Congress, Vienna, Austria, 2012.
25. Mühlbacher, D., J. Zimmer, F. Fischer, and H. P. Kruger. The Multi-Driver Simulator: A New Concept of Driving Simulation for the Analysis of Interactions between Several Drivers. *Human Centred Automation*, 2011, pp. 147–158.
26. Mühlbacher, D., and H. P. Kruger. The Effect of Car-Following on Lateral Guidance during Cognitive Load: A Study Conducted in the Multi-Driver Simulation. *Proc., 2nd International Conference on Driver Distraction and Inattention*, Goeteborg/Sweden, 2011.
27. Mühlbacher, D., L. Rittger, and C. Maag. Real vs. Simulated Surrounding Traffic: Does it Matter? *Proc., Driving Simulation Conference Europe, Paris, France*, 2014.
28. Mühlbacher, D. Multi-Driver Simulation: The Link Between Driving Simulation and Traffic Simulation. *Proc., mobil. TUM 2015: International Scientific Conference on Mobility and Transport. Munich Technologies, Solutions and Perspectives for Intelligent Transport Systems*, Munich, Germany, 2015.
29. Rittger, L., D. Muehlbacher, C. Maag, and A. Kiesel. Anger and Bother Experience when Driving with a Traffic Light Assistant: A Multi-Driver Simulator Study. *Proc., Human Factors and Ergonomics Society Europe*, 2015, pp. 41–51.
30. Bando, T., and T. Shibata. Networked Driving Simulator Based on SIGVerse and Lane-Change Analysis According to Frequency of Driving. *Proc., 15th International IEEE Conference on Intelligent Transportation Systems (ITSC)*, Anchorage, Alaska, IEEE, New York, 2012.
31. Lin, Y. A Natural Contact Sensor Paradigm for Nonintrusive and Real-time Sensing of Bio-signals in Human–Machine Interactions. *IEEE Sensors Journal, Special Issue on Cognitive Sensor Networks*, Vol. 11, No. 3, 2011, pp. 522–529.
32. Cai, H., and Y. Lin. Coordinating Cognitive Assistances with Cognitive Engagement Control Approaches in Human–Machine Interactions. *IEEE Transactions on Systems, Man and Cybernetics Part A: Humans and Systems*, Vol. 42, No. 2, 2012, pp. 286–294.
33. Lin, Y. i-DRIVE, Are We Ready? *Journal of Automotive Safety and Energy*, Vol. 7, No. 1, 2016, pp. 14–24.
34. Wan, P., C. Wu, Y. Lin, and X. Ma. Optimal Threshold Determination for Discriminating Driving Anger Intensity Based on EEG Wavelet Features and ROC Curve Analysis. *Journal of Information*, Vol. 7, No. 3, 2016, p. 52.
35. Wan, P., C. Wu, Y. Lin, and X. Ma. On-Road Experimental Study on Driving Anger Identification Model Based on Physiological Features by ROC Curve Analysis. *IET Intelligent Transport Systems*, Vol. 11, No. 5, 2017, pp. 290–298.
36. Lin, Y. Towards Intelligent Human–Machine Interactions: Human Assistance Systems (HAS). *Mechanical Engineering*, Vol. 139, No. 6, 2017, pp. 1–5.
37. Brookhuis, K., D. D. Waard, and B. Mulder. Measuring Driving Performance by Car-Following in Traffic. *Ergonomics*, Vol. 37, No. 3, 1994, pp. 427–434. http://dx.doi.org/10.1080/00140139408963661.
38. Oeltze, K., and C. Schießl. Benefits and Challenges of Multi-Driver Simulator Studies. *IET Intelligent Transport Systems*, Vol. 9, No. 6, 2015, pp. 618–625.
39. Papantoniou, P., E. Papadimitriou, and G. Yannis. Review of Driving Performance Parameters Critical for Distracted Driving Research. *Transportation Research Procedia*, Vol. 25, 2017, pp. 1801–1810.
40. Yannis, G., E. Papadimitriou, X. Karekla, and F. Kontodima. Mobile Phone Use by Young Drivers: Effects on Traffic Speed and Headways. *Transportation Planning and Technology*, Vol. 4, No. 33, 2010, pp. 385–394. http://dx.doi.org/10.1080/03081060.2010.494030.
41. Greenberg, J., B. Artz, and L. Cathey. The Effect of Lateral Motion Cues during Simulated Driving. *Proc., Driving Simulation Conference North America*, Dearborn, Mich., 2003.
42. Östlund, J., L. Nilsson, O. Carsten, N. Merat, H. Jamson, S. Jamson, S. Mouta, J. Carvalhais, J. Santos, V. Anttila, H. Sandberg, J. Luoma, D. de Waard, K. Brookhuis, E. Johansson, J. Engström, T. Victor, J. Harbluk, W. Janssen, , and R. Brouwer. *Deliverable 2: HMI and Safety-Related Driver Performance*. Human Machine Interface and the Safety of Traffic in Europe (HASTE) Project, Report No. GRD1/2000/25361 S12.319626. 2004.
43. Brown, R. L., and J. E. Hummer. Determining the Best Method for Measuring No-Passing Zones. *Transportation Research Record: Journal of the Transportation Research Board*, 2000. 1701: 61–67.
44. Ishigami, Y., and R. M. Klein. Is a Hands-Free Phone Safer than a Handheld Phone? *Journal of Safety Research*, Vol. 40, No. 2, 2009, pp. 157–164.
45. Mehler, B., B. Reimer, and J. A. Dusek. *MIT AgeLab Delayed Digit Recall Task (n-back)*. Massachusetts Institute of Technology, Cambridge, Mass., 2011.
46. Janson, J., and N. Rohleder. Distraction Coping Predicts Better Cortisol Recovery after Acute Psychosocial Stress. *Psychoneuroendocrinology*, Vol. 71, 2016. pp. 38–39. https://doi.org/10.1016/j.biopsycho.2017.07.014.
47. Williams, L. A., B. McManus, and D. Stavrinos. Effect of Distraction and Traffic Congestion on Visual Attention. *UAB Inquiro*, Vol. 6, 2012, pp. 21–22.

The Standing Committee on Vehicle User Characteristics (AND10) peer-reviewed this paper (18-03240).

TRR
JOURNAL OF THE TRANSPORTATION RESEARCH BOARD

Article

Transportation Research Record
2018, Vol. 2672(37) 55–63

DOI: 10.1177/0361198118782758
journals.sagepub.com/home/trr
SAGE

Driver's Visual Performance in Rear-End Collision Avoidance Process under the Influence of Cell Phone Use

Xiaomeng Li[1,2], Andry Rakotonirainy[1], Xuedong Yan[2], and Yuting Zhang[2]

Abstract

Rear-end crash is the most common type of on-road traffic crash, and cell phone use contributes to the increase of rear-end crashes. The effects of cell phone use on driving performance have been thoroughly investigated in previous research with various measurements. However, change in driver's visual performance while using a cell phone in situations with high rear-end risk has not yet been fully understood. This driving simulator study investigated drivers' eye movement performance in a rear-end collision avoidance maneuver during cell phone conversation. Eye movement data of 36 participants were collected in a car-following scenario featuring imminent rear-end collision. The whole collision avoidance process was divided into four stages for eye movement data analysis, including normal driving stage, brake response stage, deceleration adjusting stage, and speed recovering stage. Results showed that the average pupil size, fixation duration, and dwell time on the leading vehicle increased significantly during the brake response and deceleration adjusting stages. This indicated that the drivers' cognitive workload increased during these stages. Drivers used blink inhibition and quick saccade as a visual compensation strategy to mitigate the increased workload from cell phone use during the brake response stage. However, in the deceleration adjusting stage, the cell phone use condition led to a lower fixation frequency on the leading vehicle than in the no phone use condition. Professional drivers were found to pay more visual attention to the leading vehicle than non-professional drivers in the normal driving stage.

As information technology is developing at a rapid pace in modern society, the amount of available information increases significantly along with people's demand, which also increases the motivations to multitask (*1*). Driver distraction has been recognized as an increasing contributing factor to road crashes. In the US 100-car naturalistic study, it was found that nearly 80% of the crashes and near-crashes involved driver distraction (*2*). The widespread use of cell phones has made this a main modality of driver distraction while driving. Many studies have shown that both hand-held and hands-free cell phone use led to an increased risk of crash involvement in both simulated situations and real-world situations (*3*). Specifically, driver distraction induced by cell phone use increased the driver's involvement in rear-end collisions (*4*). In an observational study containing over 400 teen drivers, at least 75% of them were observed engaging in a potentially distracting behavior (coded as operating/looking at the phone) before moderate to severe rear-end crashes (*5*).

A growing body of research has shown the negative interference of cell phone use on driving tasks, such as the increase in reaction time (*6*), the deterioration of speed control and lateral control abilities (*7, 8*), and the impairment of allocation of visual attention (*9*), and so on. Meanwhile, other research suggested that the increased workload resulting from cell phone use prompted drivers to compensate and drive more conservatively. The induced compensation actions included a larger following distance gap (*10*), a reduction in driving speed (*11*), and an increase in deceleration rate in risky situations (*12*).

Multitasking drivers may divert their eyes to the secondary tasks and degrade the level of attention required to attend to critical events, which can result in disastrous consequences. Visual information was considered as the main source of information in driving (*13*). Eye movement patterns were predictors of visual attention, and visual attention was crucial to maintain driving safety (*14*). According to a recent 100-car naturalistic driving study, if the drivers could

[1]Centre for Accident Research and Road Safety-Queensland (CARRS-Q), Institute of Health and Biomedical Innovation (IHBI), Queensland University of Technology (QUT), Kelvin Grove, Queensland, Australia
[2]MOE Key Laboratory for Urban Transportation Complex System Theory and Technology, School of Traffic and Transportation, Beijing Jiaotong University, Beijing, China

Corresponding Author:
Address correspondence to Xuedong Yan: xdyan@bjtu.edu.cn

balance visual attention on and off the road effectively in a near-crash situation, they would have more opportunities to avoid the crash (*15*). Driving simulation research showed that receiving and particularly sending text messages decreased the amount of time drivers spent looking at the forward roadway by up to 29% (*16*). Sodhi and Reimer found that during driving with a secondary cognitive task, drivers' standard deviations of fixation displacements were shorter than those observed during normal driving (*17*). In addition to the distraction interference, drivers' demographic factors (such as driving experience and gender) were deemed as potential factors in a traffic crash (*18*, *19*). Underwood suggested that one of the fundamental changes in skill that marked the transition from novice to advanced driver was the efficiency of visual search strategies (*20*). White and Caird classified drivers' looked-but-failed-to-see errors accurately by discriminant function analysis using passenger conversation and driver gender as predictors (*21*).

Up to now, considerable research has investigated the impact of secondary tasks on drivers' visual performance. Most of it focused on the visually based secondary tasks such as texting (*16*) or using In-Vehicle Information Systems (*22*). Some other research investigated the changes of drivers' eye movements during a secondary cognitive task. For example, an increase in blink frequency was observed in research by Tsai et al. when drivers performed auditory and driving combined tasks, and the mean pupil diameter and horizontal vergence also increased when subjects had good performance in the auditory task (*23*). Research by Niezgoda et al. indicated that change in pupil size was the most sensitive measure of change in cognitive demand, whereas the fixation location and blink rate were less sensitive measures (*24*). In most of the research, the effects of secondary tasks on the drivers' eye movements and visual performance were tested in normal driving situations, such as freeway driving or drivers approaching an intersection or negotiating a curve. However, the high-demand and high-risk driving circumstances, such as a car-following situation with high rear-end collision risk in which driver distraction may play an important role in safety, were barely considered in the research. It remains largely unclear how drivers' visual search scanning strategies change and how the visual attention is affected by secondary cognitive tasks under such circumstances.

Therefore, by using a driving simulation experiment combined with advanced eye-tracking technology that provided noninvasive eye movement measures, this study investigated drivers' visual performance in a rear-end collision avoidance process with the interference of hands-free and hand-held cell phone conversation. Various variables that can measure visual demand and cognitive workload, such as pupil size, blink frequency, blink duration, saccade frequency, saccade duration, fixation frequency, fixation duration, and dwell time were collected during the experiment.

The objectives of the study were: (1) to investigate drivers' eye movement pattern during the rear-end collision avoidance process; (2) to identify the effects of cell phone conversation (including both hands-free and hand-held), driving experience, and gender on visual performance during the collision avoidance; and (3) to test whether there were visual compensation strategies adopted by distracted drivers.

Methodology

Participants

Forty-five participants in total were recruited for this research. A balance of driver gender (male and female drivers) and driving experience (professional and non-professional drivers) was considered in the participant recruitment. However, because of motion sickness and problems in eye movement calibration, only 36 drivers, including 17 male (eight professional and nine non-professional) and 19 female (nine professional and 10 non-professional), completed the experiment. The average age of participants was 35. Age ranged from 31 to 40 years old, with a standard deviation of 3.2 years. Each participant held a valid driver's license with at least two years of driving experience. The professional drivers were full-time taxi drivers with an average mileage of 71.2 thousand kilometers per year and an average self-reported crash record of three per million kilometers. The non-professional drivers used their vehicles only for daily travel, and their average mileage and self-reported accident record were 20.5 thousand kilometers per year and 11 per million kilometers, respectively.

Apparatus

This study used the BJTU (Beijing Jiaotong University) driving simulator for the experiment and data collection, as shown in Figure 1*a*. The BJTU driving simulator is a high-performance, high-fidelity driving simulator with a linear motion base capable of operating with 1 degree of freedom. It is an automatic control vehicle composed of a full-size cabin (Ford Focus) with a real operational interface, environmental noise and vibration system, digital video recording system, and the central console. The simulated environment is projected with a front/peripheral field of view of 300 degrees at a resolution of 1400 × 1050 pixels. The simulator lab also provides the software (Simvista and Simcreator) for driving scenario design, virtual traffic environment simulation, and virtual road modeling.

A head-mounted eye-tracking system (SensoMotoric Instruments Eye Tracking Glasses [SMI ETG]) was used to collect the participants' eye movement data (see Figure 1*b*). The eye-tracking system recorded the front view and eye movements of participants. The specific data about eye movement activities (e.g., fixation, saccade, blink), such as time, duration, and counts, were extracted with the SMI software.

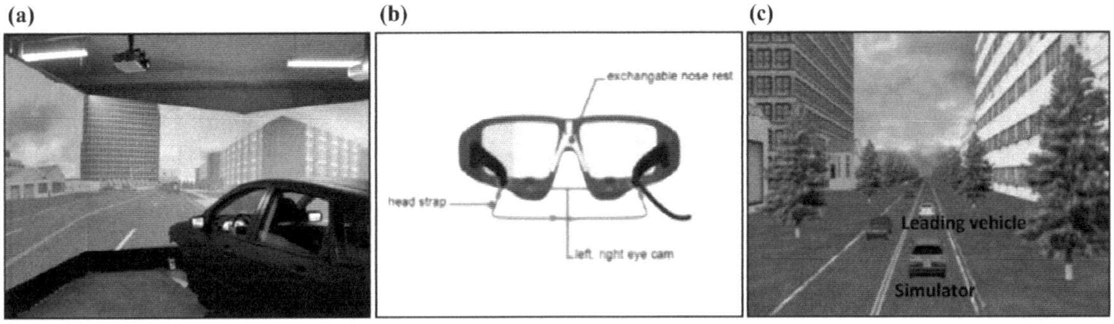

Figure 1. (*a*) BJTU driving simulator, (*b*) eye-tracking system, and (c) car-following scenario.

Scenario Design

The driving route was a 4-km, two-way, two-lane road with a speed limit of 80 km/h. A high density vehicle flow (600~700 vehicles per hour per lane) was scripted to create a car-following situation for participants and prevent overtaking maneuvers (Figure 1*c*). As the traffic volume was close to capacity, the operating speed of the vehicles on the road was about 40–50 km/h, which was nearly down to half of the posted speed limit. At the end of the car-following event, each participant would encounter a potential rear-end collision situation by decelerating the leading vehicle (LV) from 40 km/h to 4 km/h with a deceleration rate of 6 m/s^2. The speed of the following vehicle, the distance, and the time headway between the leading and following vehicles were all non-pre-set for drivers when the potential rear-end collision was triggered. The scenario was created to mimic a common real-life driving situation in which the drivers followed the LV in an individually preferred way.

Each participant drove the same route three times under three types of cell phone use conditions. The three conditions were no phone conversation, hands-free cell phone conversation, and hand-held cell phone conversation. Each participant was subjected to only one type of cell phone conversation per drive. To eliminate the experiment order effect, the three sessions of drive were administered in a random sequence by using the "RANDBETWEEN" function in Excel. Some other measures were also taken to reduce the expectancy effects from multiple exposures to the same event. For example, several intersections and curve segments were arranged on the route before the car-following road segments to enrich the drivers' driving experience; ambient vehicles were also arranged to travel in front of the simulator and drive away at intersections before drivers entered the test segments.

The cell phone conversation task in the experiment was a series of unit digit arithmetic problems that the drivers were requested to answer. In the hand-held situation, the participants were asked to hold the cell phone to their ear with one hand, whereas in the hands-free task the participants could use both their hands to control the steering wheel.

Experiment Procedure and Data Collection

On arrival, each participant was briefed on the requirements of the experiment and asked to read and sign an informed consent form. The participants were instructed to drive as they normally do in real-life situations. They started with a practice drive of at least 5 min to become familiar with the driving simulator. Before an actual recording of driver's eye movement data, a "3 points" method was applied to calibrate each driver's fixation points. After wearing the head-mounted eye-tracking glasses, the driver was instructed to gaze at three different points one by one in the front screen to complete the fixation calibration process. During the experiment, participants were also notified that they could quit driving at any time in case of motion sickness or discomfort. Between each session the participants had 10 min rest to mitigate motion sickness and learning effects.

During the experiment, the raw eye movement data were sampled at 30 Hz. Key eye movement variables extracted from the raw data included (1) pupil size, that is, average pupil diameter (mm); (2) blink frequency (number of blinks per second) and average blink duration (ms); (3) saccade frequency (number of saccades per second) and average saccade duration (ms); (4) fixation frequency (number of fixations per second) and average fixation duration (ms), the LV being selected as the area of interest (AOI) of the simulator drivers, on which the fixation details were collected; and (5) dwell time on LV (%), representing the percentage of the sum of durations from all fixations and saccades that hit the AOI.

Drivers' Rear-End Collision Avoidance Process

The driving scenario forced the driver to perform an emergency brake to avoid a rear-end collision in a car-following situation. It should be noted that as the drivers could choose

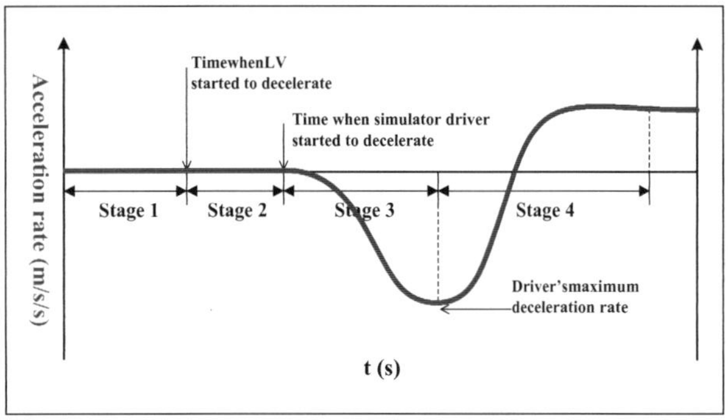

Figure 2. Four stages of rear-end collision avoidance process.

their own speed, distance, and time headway in the car-following situation, the urgency level of the potential rear-end collision might be different for different drivers. Therefore, the driving style that the driver adopted in the car-following stage might influence the subsequent rear-end collision avoidance performance and the collision result. According to the findings of a previous study, no matter which kind of car-following strategy the drivers applied, the rear-end collision avoidance process could be divided into several main stages (*25*). A maximum value of deceleration rate existed during the braking process, after which the drivers started to release the brake pedal. Thus the collision avoidance process was divided into four stages which were normal driving stage, brake response stage, deceleration adjusting stage, and speed recovering stage (see Figure 2). Recent research suggested that drivers' brake actions in emergent situations came in response to the developing situation kinematics, probably mediated by visual looming, rather than some pre-defined "hazard onset" (e.g., LV brake light onset) (*26*). However, because of the current limitation in data collection, it was difficult to determine at what point the visual looming became a hazard for drivers that they had to respond to. Thus in this study, it was assumed that the driver taking evasive response was aware of the LV brake light onset, and the brake response stage started from when the LV's brake light was activated and ended when the driver initiated brake action. The following stage definitions and consequent analysis sectioning were based on these assumptions.

Stage 1: normal driving stage. This stage represented a normal car-following situation before the LV started to decelerate. Both the leading and following vehicles kept a steady speed and the participant had no perception of the LV's intention to decelerate. In the study, the duration of this stage for analysis was set to 2s.

Stage 2: brake response stage. The stage started when the brake light of the LV turned red and ended as the participant started to brake. In this stage the drivers kept their initial speeds but perceived the LV's deceleration. The duration time of the stage varied with each individual.

Stage 3: deceleration adjusting stage. The stage started from when the drivers began to decelerate (hit the brake pedal) to the moment when their deceleration rates reached the maximum. The duration of the stage varied from individual to individual.

Stage 4: speed recovering stage. In this stage, the drivers loosened the pressure on the brake pedal and the deceleration rate started to decrease, indicating the end of rear-end collision avoidance and drivers starting to resume their speeds to a normal (non-emergency) level.

Independent and Dependent Variables

The independent variables in the study included cell phone use conditions, driver gender, and driving experience. It should be noted that in addition to the three cell phone use conditions, the study also considered whether drivers used a cell phone or not (binary variable) as a potential factor that might impose a significant effect on the eye movement variables. In such cases, the hands-free and hand-held cell phone use conditions were deemed as one condition to compare with the condition under no phone use. The dependent variables included pupil size, blink frequency, average blink duration, saccade frequency, average saccade duration, fixation frequency, average fixation duration, and dwell time on LV.

Based on the above four-stage division, basic statistical information on the eye movement variables in four stages was extracted. Specially, the frequency variables (i.e., blink frequency, saccade frequency and fixation frequency) represented the number of blinks, saccades and fixations per second within each stage duration, and they were calculated by the numbers within a certain stage divided by the duration of this stage. The average duration variables (i.e., average blink duration, average saccade duration, and average fixation duration) represented the average duration of all blinks, saccades, or fixations during each stage, and they were calculated by the sum duration of all blinks, saccades, or fixations within a certain stage divided by the numbers of blinks, saccades, or fixations in this stage, respectively. Because the duration of stages varied from individual to individual, the frequency variable and its corresponding duration variable were mutually independent and irreplaceable.

In the following data analysis, a repeated-measures analysis of variance (ANOVA) test (Type III) by stage was conducted to investigate significant changes of the eye movement variables across stages. At different stages, one-way ANOVA tests by cell phone use conditions, driver gender, and driving experience were conducted to examine the effects of these factors on drivers' eye movement performance. Because the rear-end crash risk terminated before the last stage when drivers started to recover their speeds, the ANOVA tests would mainly focus on the first three stages.

Table 1. Descriptive Statistics and Repeated Analysis of Variance for Drivers' Eye Movement Variables

		Collision avoidance stages				Stages		
Variables	Parameters	Stage 1	Stage 2	Stage 3	Stage 4	d.f.	*F*	Sig
Average pupil size (mm)	Mean	4.42	4.50	4.62	4.67	3	26.760	.000
	SD	0.69	0.71	0.79	0.77			
Blink frequency (/s)	Mean	0.90	1.32	1.47	0.94	3	2.573	.065
	SD	0.61	0.74	0.92	0.58			
Average blink duration (ms)	Mean	384.54	361.26	305.27	385.19	3	1.156	.336
	SD	166.00	187.64	172.66	155.20			
Saccade frequency (/s)	Mean	3.71	3.71	3.79	3.72	3	.108	.956
	SD	2.21	2.19	2.24	2.35			
Average saccade duration (ms)	Mean	75.59	74.08	74.27	74.65	3	1.257	.289
	SD	7.87	9.30	9.41	7.35			
Fixation frequency (/s)	Mean	2.19	2.39	2.46	2.15	3	1.049	.371
	SD	0.98	1.10	1.16	0.97			
Average fixation duration (ms)	Mean	249.33	320.26	278.95	200.42	3	4.133	.007
	SD	237.11	436.94	301.18	99.85			
Dwell time on LV (%)	Mean	62.37	68.75	69.20	57.53	3	4.667	.003
	SD	23.38	27.40	28.22	21.68			

Note: d.f. = degrees of freedom; *F* = *F* value; Sig = p value; SD = standard deviation.

Experiment Results

Eye Movement Descriptions in Four Stages

The basic measurements of all the eye movement variables in four collision avoidance stages are listed in Table 1. It shows that drivers' average pupil size dilated after the normal driving stage. The blink frequency increased and the average blink duration decreased during the brake response and deceleration adjusting stage. There was no significant change in saccade frequency but the average saccade duration was shortened after the first stage. The drivers' fixation frequency, average fixation duration, and dwell time on LV increased during the brake response and deceleration adjusting stages. The repeated-measures (within stages) ANOVA test (see Table 1) indicated that changes in drivers' average pupil size, average fixation duration, and dwell time were significant among different stages.

Stage 1: Normal Driving Stage

In the normal driving stage, drivers drove at steady speeds and followed the LV without any hint of an impending critical event. ANOVA results showed that driving experience had a significant effect on driver's fixation frequency ([F] = 4.33, [p] < 0.05), but no significant effects of cell phone use or gender on the eye movement variables were observed. During the car-following process, professional drivers (mean [M] = 2.01, standard deviation [SD] = 0.93) had a higher fixation frequency on the LV than non-professional drivers (M = 2.40, SD = 1.00).

Stage 2: Brake Response Stage

In the brake response stage when drivers perceived the LV's deceleration, significant effects of cell phone use on a driver's eye movement activities were observed. ANOVA results showed that the different cell phone use conditions significantly influenced driver's average saccade duration (F = 3.53, $p < 0.05$). Whether drivers used a cell phone or not also had a significant effect on the blink frequency ($F = 4.65$, $p < 0.05$). The average saccade duration was smaller when the driver used a cell phone (hands-free: M = 73.27, SD = 7.32; hand-held: M = 71.65, SD = 8.78) compared with the no phone use condition (M = 77.21, SD = 10.73). The blink frequency was also smaller in the cell phone use condition (M = 1.19, SD = 0.64) compared with the no phone use condition (M = 1.77, SD = 0.92). No significant effects of driver gender or driving experience were observed.

Stage 3: Deceleration Adjusting Stage

The deceleration adjusting stage was a stage when drivers performed collision avoidance actions to avoid a rear-end crash. Whether drivers used a cell phone or not had a significant effect on the fixation frequency ($F = 5.84$, $p < 0.05$). No significant effects of driver gender or driving experience were observed on the eye movements in this stage. When drivers were adjusting deceleration rate in response to the LV's deceleration, their fixation frequencies on the hazard source were lower in cell phone use conditions (M = 2.87, SD = 1.29) than in the no phone use condition (M = 2.27, SD = 1.04).

Discussion

The study investigated drivers' eye movement pattern during a rear-end collision avoidance process. Several visual measurements that can indicate changes of driver's cognitive workload were effectively identified. The study also tested the effects of cell phone use and drivers' characteristics (gender and driving experience) on the eye movement activities during different collision avoidance stages.

In the study, the changes of pupil size, fixation duration, and dwell time demonstrated the change of drivers' cognitive workload across stages. From the normal driving stage to the speed recovering stage, drivers' average pupil size increased by 5.7%. Pupil size is one of the most commonly recognized workload indicators that has been applied in different kinds of cognitive tasks (*23*, *24*). Early research by Hess noted the famous bi-directional effect of emotion on pupil change, reporting that the pupil constricted when people viewed unpleasant pictures but dilated when they viewed pleasant pictures (*27*). More recent research suggested that pupil dilation was determined by emotional arousal, independent of whether pictures were pleasant or unpleasant (*28*). The present study further supported the more recent discovery that pupil diameter could also increase when people processed unpleasant visual stimuli.

In addition to the pupil size, change of driver's cognitive workload during the collision avoidance was also reflected by their fixation duration and dwell time on the hazard source. Both the fixation duration and dwell time on LV were comparatively higher in the brake response stage and deceleration adjusting stage than in the normal driving stage and speed recovering stage. In the brake response stage, drivers perceived the LV's deceleration and the consequent rear-end collision risk, and so a preliminary collision avoidance strategy was formed accordingly. During the deceleration adjusting stage, the collision avoidance strategy was implemented by drivers through controlling the deceleration time as well as the deceleration rate. The increase of fixation duration and dwell time on the LV indicated that drivers' visual demand increased with the increase of cognitive workload, and thus the visual attention on the hazard source was enhanced so that the drivers could better prepare for the imminent rear-end risk.

The effects of cell phone use on drivers' eye movements were found in both the brake response stage (blink and saccade activities) and deceleration adjusting stage (fixation activity), and, interestingly, the effects in these two stages seemed to be contrary. In the brake response stage, efficient eye movements were conducted by drivers as a visual compensation strategy to mitigate the increased cognitive workload imposed by cell phone use, whereas in the deceleration adjusting stage the effects of cell phone use on eye movements seemed to be negative.

In the brake response stage, drivers with cell phone use had lower blink frequency than drivers without cell phone use. It has been commonly agreed in past research that eyeblink, especially endogenous eyeblink, can provide information about the process involved in attention and response programming and execution (*29*, *30*). Many studies have reported that blink frequency decreased significantly when subjects were performing tasks associated with heightened visual attention demands (*31*, *32*), and the effects were not restricted to visual task modality only. Some other research also found a decrease in the blink frequency of subjects when they were performing nonvisual tasks such as mental arithmetic, digit span, and auditory tracking (*33*, *34*). In this study, the significant decrease in blink frequency in the brake response stage further demonstrated that cognitive demands were important in determining blink occurrence and a correlation might exist between inhibition of blink and high cognitive demand. It should be noted that this finding of change in drivers' blink frequency in a secondary auditory task conflicted with the finding of Tsai et al. (*23*). Instead of a decrease in blink frequency, Tsai et al.'s study indicated that the blink frequency increased when subjects performed an unchanging visual driving task and a taxing auditory task concurrently. The authors explained the result that the auditory task reduced subjects' concentration on the visual driving task, and therefore the number of blinks increased because of the decrease of visual attention. However, previous research has reached a consensus that the inhibition of blinks was associated with increased visual attention, though it is still not clear whether the decrease of visual attention would lead to an increase in blinks or not.

Along with blink frequency, a significant effect of cell phone use was also observed in drivers' saccade duration in the brake response stage. Drivers in the hand-held condition had the smallest saccade duration, followed by the hands-free condition and no phone use condition. Similarly, in the research of Fogarty and Stern, a time-locked relationship of blinks and saccadic eye movements was observed (*35*). The act of blinking caused a temporary loss of visual information as the eyelid created a barrier between the eye and the external environment, and long saccadic eye movements, which meant slow fixation transfer, caused a delay in the perception of important information. Therefore, it was suggested that the simultaneous occurrence of blink inhibition and quick saccadic eye movements was the most efficient strategy for drivers using cell phone to minimize visual information missing in the brake response stage.

In the deceleration adjusting stage, drivers using a cell phone were found to have lower fixation frequency on LV than no-phone-use drivers. Many previous studies showed that secondary in-vehicle visual-demand tasks (such as texting, using a navigation system or other In-Vehicle Information Systems) resulted in degraded visual attention, impaired driving performance, and increased crash risk (*36*, *37*). The present study further demonstrated that non-visual-demand task distraction could also interrupt drivers' visual attention to the hazard source and this was probably because of the overlapped multi-tasks at this stage, which included both manual operation (adjusting the strength exerted on the brake pedal) and cognitive activities (calculating arithmetic problems). Finally, it should be pointed out that even though the study tested driver distraction from both hands-free and hand-held cell phone use, little difference between the two types of cell phone use with regards to their effects on the drivers' eye movements was observed. From another point of view, however, this result reflected a notion that drivers' conversation through hands-free phone or hand-held phone affected their visual search strategies in much the same way.

In regard to the drivers' demographics, no difference of eye movements between male and female drivers was observed during the collision avoidance process. However, drivers' profession significantly affected the fixation frequency on the LV in the normal driving stage and professional drivers had a higher fixation frequency than non-professional drivers. The professional taxi drivers paid more visual attention to the front LV in the car-following process, probably because their work required them to drive more carefully and their accumulated experience made it easier for them to focus more on the potential hazards in complex traffic situations. Similarly, in the research of Konstantopoulos et al., driving instructors and learner drivers were organised to drive under day, night, and rainy conditions in a simulator, and results showed that driving instructors were able to collect more information about the scene by employing more fixations across all three visibility conditions (*38*). The strategy of deploying frequent short fixations was considered crucial for experienced drivers in hazardous situations to maintain awareness and anticipate potential sources of hazard on the road.

Some limitations of the study should be pointed out. First, the study was conducted by using a driving simulator and the experiment scenario created was an emergent situation with high rear-end collision risk. Differences in drivers' eye movements, risk assessment, and collision evasive response may exist between simulation and naturalistic on-road driving. However, it is not ethically possible to validate the present results against a real-world driving experiment as participants cannot be put in a critical situation with high collision risk. Second, an emergent pre-crash scenario was a small-likelihood event in reality, and drivers were unlikely to encounter one emergent conflict after another in a short span of time. Participants' multiple exposure to safety-critical events may lead to potential confounding (*39*). Therefore, a stricter experiment control method (e.g., a long interval time between formal experiments) should be applied in future study.

Conclusion

Previous research investigated the effects of cognitive secondary tasks on drivers' visual performance by using various eye movement measurements. However, only a few of them considered their effects in emergent situations, such as rear-end collisions, in which distraction tasks occur frequently. Therefore, the study investigated drivers' eye movement pattern in a rear-end collision avoidance process under the influence of cell-phone-induced driver distraction. Specifically, a four-stage rear-end collision avoidance process was discovered in the study for eye movement pattern analysis. Significant changes were observed in several eye movement variables that can indicate the status of a driver's cognitive workload, including the average pupil size, average fixation duration, and dwell time on the LV. The effects of cell phone conversation on visual performance were found to be contrary in the brake response stage and deceleration adjusting stage. For the former stage, visual compensation strategies were conducted by cell phone use drivers through reducing the blink frequency and saccade duration. However, in the latter stage, driver's cell phone use behavior led to a low fixation frequency on the LV, which might explain the high near-collision risk of cell phone use drivers from the view of visual performance. In regard to the drivers' characteristic factors, professional drivers were found to pay more visual attention to the LV than non-professional drivers in a normal car-following situation whereas no significant gender effect was observed during the whole collision avoidance process. Future study will test drivers' eye movement activities in other collision scenarios (right-angle collision or head-on collision, for example) and investigate the correlation mechanism between eye movements and behavior actions to provide effective predictions of collision risk by applying a comprehensive measurement of collision avoidance performance.

Acknowledgments

This research was supported by the National Natural Science Foundation of China (71621001, 71771014) and Australian Research Council (ARC DP140102895).

Author Contributions

The authors confirm contribution to the paper as follows: study conception and design: Xuedong Yan; data collection: Xiaomeng Li and Yuting Zhang; analysis and interpretation of results: Xiaomeng Li and Andry Rakotonirainy; draft manuscript preparation: Xiaomeng Li, Andry Rakotonirainy, Xuedong Yan, Yuting Zhang. All authors reviewed the results and approved the final version of the manuscript.

References

1. Just, M. A., T. A. Keller, and J. Cynkar. A Decrease in Brain Activation Associated with Driving when Listening to Someone Speak. *Brain Research*, Vol. 1205, 2008, pp. 70–80.
2. Klauer, S. G., T. A. Dingus, V. L. Neale, J. D. Sudweeks, and D. J. Ramsey. *The Impact of Driver Inattention on Near-Crash/Crash Risk: An Analysis Using the 100-Car Naturalistic Driving Study Data*. Report No. DOT HS 810 594. National Highway Traffic Safety Administration, Washington, D.C., 2006.

3. Oviedo-Trespalacios, O., M. M. Haque, M. King, and S. Washington. Understanding the Impacts of Mobile Phone Distraction on Driving Performance: A Systematic Review. *Transportation Research Part C: Emerging Technologies*, Vol. 72, 2016, pp. 360–380.
4. Neyens, D. M., and L. N. Boyle. The Effect of Distractions on the Crash Types of Teenage Drivers. *Accident Analysis and Prevention*, Vol. 39, No. 1, 2007, pp. 206–212.
5. Carney, C., K. K. Harland, and D. V. McGehee. Using Event-Triggered Naturalistic Data to Examine the Prevalence of Teen Driver Distractions in Rear-End Crashes. *Journal of Safety Research*, Vol. 57, 2016, pp. 47–52.
6. Strayer, D. L., and F. A. Drews. Profiles in Driver Distraction: Effects of Cell Phone Conversations on Younger and Older Drivers. *Human Factors*, Vol. 46, No. 4, 2004, pp. 640–649.
7. Reimer, B., B. Mehler, J. Coughlin, N. Roy, and J. Dusek. The Impact of Anaturalistic Hands-Free Cellular Phone Task on Heart Rate and Simulated Driving Performance in Two Age Groups. *Transportation Research Part F: Traffic Psychology and Behaviour*, Vol. 14, No. 1, 2011, pp. 13–25.
8. Haque, M. M., and S. Washington. The Impact of Mobile Phone Distraction on the Braking Behaviour of Young Drivers: A Hazard-Based Duration Model. *Transportation Research Part C: Emerging Technologies*, Vol. 50, 2015, pp. 13–27.
9. Reimer, B., B. Mehler, Y. Wang, and J. F. Coughlin. A Field Study on the Impact of Variations in Short-Term Memory Demands on Drivers' Visual Attention and Driving Performance across Three Age Groups. *Human Factors*, Vol. 54, No. 3, 2012, pp. 454–468.
10. Ranney, T. A., J. L. Harbluk, and Y. I. Noy. Effects of Voice Technology on Test Track Driving Performance: Implications for Driver Distraction. *Human Factors*, Vol. 47, No. 2, 2005, pp. 439–454.
11. Rakauskas, M. E., L. J. Gugerty, and N. J. Ward. Effects of Naturalistic Cell Phone Conversations on Driving Performance. *Journal of Safety Research*, Vol. 35, No. 4, 2004, pp. 453–464.
12. Benedetto, A., A. Calvi, and F. D'Amico. Effects of Mobile Telephone Tasks on Driving Performance: A Driving Simulator Study. *Advances in Transportation Studies*, Vol. 26, No. 26, 2012, pp. 29–44.
13. Recarte, M., and L. Nunes. Effects of Verbal and Spatial-Imagery Task on Eye Fixations while Driving. *Journal of Experimental Psychology: Applied*, Vol. 6, No. 1, 2000, pp. 31–43.
14. Zhang, Y., X. Yan, X. Li, and Q. Xue. Drivers' Eye Movements as a Function of Collision Avoidance Warning Conditions in Red Light Running Scenarios. *Accident Analysis and Prevention*, Vol. 96, 2016, pp. 185–197.
15. Seppelt, B. D., S. Seaman, J. Lee, L. S. Angell, B. Mehler, and B. Reimer. Glass Half-Full: On-Road Glance Metrics Differentiate Crashes from Near-Crashes in the 100-Car Data. *Accident Analysis and Prevention*, Vol. 107, 2017, pp. 48–62.
16. Young, K. L., C. M. Rudin-Brown, C. Patten, R. Ceci, and M. G. Lenné. Effects of Phone Type on Driving and Eye Glance Behaviour while Text-Messaging. *Safety Science*, Vol. 68, 2014, pp. 47–54.
17. Sodhi, M., and B. Reimer. Glance Analysis of Driver Eye Movements to Evaluate Distraction. *Behavior Research Methods, Instruments, & Computers*, Vol. 34, No. 4, 2002, pp. 529–538.
18. Amarasingha, N., and S. Dissanayake. Gender Differences of Young Drivers on Injury Severity Outcome of Highway Crashes. *Journal of Safety Research*, Vol. 49, 2014, pp. 113–120.
19. Machado-León, J. L., J. de Oña, R. de Oña, L. Eboli, and G. Mazzulla. Socio-Economic and Driving Experience Factors Affecting Drivers' Perceptions of Traffic Crash Risk. *Transportation Research Part F: Traffic Psychology and Behaviour*, Vol. 37, 2016, pp. 41–51.
20. Underwood, G. Visual Attention and the Transition from Novice to Advanced Driver. *Ergonomics*, Vol. 50, No. 8, 2007, pp. 1235–1249.
21. White, C. B., and J. K. Caird. The Blind Date: The Effects of Change Blindness, Passenger Conversation and Gender on Looked-but-Failed-to-See (LBFTS) Errors. *Accident Analysis and Prevention*, Vol. 42, No. 6, 2010, pp. 1822–1830.
22. Benedetto, S., M. Pedrotti, L. Minin, T. Baccino, A. Re, and R. Montanari. Driver Workload and Eye Blink Duration. *Transportation Research Part F: Traffic Psychology and Behavior*, Vol. 14, No. 3, 2011, pp. 199–208.
23. Tsai, Y., E. Viirre, C. Strychacz, B. Chase, and T. Jung. Task Performance and Eye Activity: Predicting Behavior Relating to Cognitive Workload. *Aviation, Space, and Environmental Medicine*, Vol. 78, No. 5, 2007, pp. B176–B185.
24. Niezgoda, M., A. Tarnowski, M. Kruszewski, and T. Kamiński. Towards Testing Auditory–Vocal Interfaces and Detecting Distraction while Driving: a Comparison of Eye-Movement Measures in the Assessment of Cognitive Workload. *Transportation Research Part F: Traffic Psychology and Behaviour*, Vol. 32, 2015, pp. 23–24.
25. Li, X., X. Yan, J. Wu, E. Radwan, and Y. Zhang. A Rear-End Collision Risk Assessment Model Based on Drivers' Collision Avoidance Process under Influences of Cell Phone Use and Gender—a Driving Simulator based Study. *Accident Analysis and Prevention*, Vol. 97, 2016, pp. 1–18.
26. Markkula, G., J. Engström, J. Lodin, J. Bärgman, and T. Victor. A Farewell to Brake Reaction Times? Kinematics-Dependent Brake Response in Naturalistic Rear-End Emergencies. *Accident Analysis and Prevention*, Vol. 95, 2016, pp. 209–226.
27. Hess, E. H., and J. M. Polt. Pupil Size as Related to Interest Value of Visual Stimuli. *Science*, Vol. 132, No. 3423, 1960, pp. 349–350.
28. Bradley, M. M., L. Miccoli, M. A. Escrig, and P. J. Lang. The Pupil as a Measure of Emotional Arousal and Autonomic Activation. *Psychophysiology*, Vol. 45, No. 4, 2008, pp. 602–607.
29. Recarte, M., E. Perez, A. Conchillo, and L. Nunes. Mental Workload and Visual Impairment: Differences between Pupil, Blink, and Subjective Rating. *The Spanish Journal of Psychology*, Vol. 11, No. 2, 2008, pp. 374–385.
30. Ahlstrom, U., and F. Friedman-Berg. Using Eye Movement Activity as a Correlate of Cognitive Workload. *International Journal of Industrial Ergonomics*, Vol. 36, No. 7, 2006, pp. 623–636.
31. Stern, J. A., and J. J. Skelly. The Eye Blink and Workload Consideration. *Proc., Human Factors Society 28th Annual Meeting* (Human Factors Society, Santa Monica, CA), Vol. 28, No. 11, 1984, pp. 942–944.
32. Veltman, J. A., and A. W. Gaillard. Physiological Indices of Workload in a Simulated Flight Task. *Biological Psychology*, Vol. 42, No. 3, 1996, pp. 323–342.

33. Holland, M. K., and G. Tarlow. Blinking and Mental Load. *Psychological Reports*, Vol. 31, No. 1, 1972, pp. 119–127.
34. Gregory, R. L. Variations in Blink Rate during Nonvisual Tasks. *Quarterly Journal of Experimental Psychology*, Vol. 4, No. 4, 1952, pp. 165–169.
35. Fogarty, C., and J. A. Stern. Eye Movements and Blinks: their Relationship to Higher Cognitive Processes. *International Journal of Psychophysiology*, Vol. 8, No. 1, 1989, pp. 35–42.
36. Drews, F. A., H. Yazdani, C. N. Godfrey, J. M. Cooper, and D. L. Strayer. Text Messaging during Simulated Driving. *Human Factors*, Vol. 51, No. 5, 2009, pp. 762–770.
37. Caird, J. K., C. R. Willness, P. Steel, and C. Scialfa. A Meta-Analysis of Cell Phone Use on Driver Performance. *Accident Analysis and Prevention*, Vol. 40, No. 4, 2008, pp. 1282–1293.
38. Konstantopoulos, P., P. Chapman, and D. Crundall. Driver's Visual Attention as a Function of Driving Experience and Visibility. Using a Driving Simulator to Explore Drivers' Eye Movements in Day, Night and Rain Driving. *Accident Analysis and Prevention*, Vol. 42, No. 3, 2010, pp. 827–834.
39Sawyer, B. D., A. A. Calvo, and V. S. Finomore. Serendipity in Simulation: Building Environmentally Valid Driving Distraction Evaluations of Google Glass™ and an Android™ Smartphone. Presented at 19th Triennial Congress of the International Ergonomics Association, Melbourne, 2015.

The Standing Committee on Vehicle User Characteristics (AND10) peer-reviewed this paper (18-03392).

Article

TRR
JOURNAL OF THE TRANSPORTATION RESEARCH BOARD

Transportation Research Record
2018, Vol. 2672(37) 64–73

DOI: 10.1177/0361198118791666
journals.sagepub.com/home/trr
SAGE

Electrophysiological Measure of Impaired Information Processing in Drivers with Hematological Malignancy

David E. Anderson[1], Vijaya R. Bhatt[2], Kendra Schmid[3], Matthew Lunning[2], Sarah A. Holstein[2], and Matthew Rizzo[1]

Abstract

The broad goal of this study is to measure remote effects of cancer on brain physiology and behaviors that underpin instrumental activities of daily living such as automobile driving. Studies of hematological malignancies (HM) have demonstrated impairments in multiple brain functions shown to be critical for safe automobile driving. In the current pilot study, brain physiology during driving simulation was examined in 14 HM patients and 13 healthy comparison drivers. Electroencephalography was used to measure the eye fixation-related potential (EFRP)—a positive amplitude deflection evoked approximately 100 milliseconds after eye movement termination. Previous studies have demonstrated sensitivity of EFRP activity to information-processing capacity. All drivers completed visual search tasks to evaluate the relationship between driving-related changes in performance and EFRP activity. Results showed smaller EFRP amplitudes in drivers who had: (1) greater driving-related changes in visual search performance (p = 0.03, Cohen's d = 0.91); and (2) HM diagnosis (p = 0.18, Cohen's d = 0.54). Extending previous studies, these results provide neural evidence of reduced information-processing capacity associated with cancer diagnosis. Future large-scale studies are needed to confirm these results, given the high level of uncertainty and small sample size. This study provides a novel platform for linking changes in brain physiology and safety-critical driving behaviors.

Cancer, the "emperor of all maladies," comprises a range of conditions that affects nearly 0.5% of the U.S. population each year (*1*). Advances in treatment aim to reduce cancer mortality, and survivors commonly experience long-term quality-of-life impairments (*2–4*). Multiple studies demonstrate cognitive impairment before treatment (*5–11*) and potential worsening after chemotherapy (*6–9, 12–16*). Neuroimaging studies show associated changes to brain structure and function (*12, 14, 16–19*). These results suggest cognitive impairment in cancer survivors is related to a combination of cancer- and treatment-related toxicities.

The overarching goal of the current work is to measure how cancer-related toxicity affects brain physiology and behaviors that underpin instrumental activities of daily living such as automobile driving. Insights into potential links between cancer-related toxicity and driving performance come from investigations of medical disorders and aging. For example, aging (*20, 21*) and age-related disorders such as dementia (*22–25*) and glaucoma (*26, 27*) are associated with poor driving performance and increased risk for motor vehicle crashes. Impaired visual attention is thought to play a key role in at-risk driving in these cohorts (*28*).

Visual attention is commonly measured using eye-tracking methods and visual psychophysics to understand its role in driving performance (*29–35*). Saccadic behaviors, including fixation frequency, duration, and variability, are recorded to estimate the locus and control of overt visual attention. When viewing high-density traffic scenes, drivers exhibit greater fixation variability and shorter fixation duration (*31*). Multiple studies have demonstrated experience-dependent differences in saccadic behaviors (*30–32*). More experienced drivers show

[1]Department of Neurological Sciences, University of Nebraska Medical Center, Nebraska Medical Center, Omaha, NE
[2]Division of Oncology & Hematology, Department of Internal Medicine, Nebraska Medical Center, Omaha, NE
[3]Department of Biostatistics, University of Nebraska Medical Center, Nebraska Medical Center, Omaha, NE

Corresponding Author:
Address correspondence to David E. Anderson: david.anderson@unmc.edu

shorter fixation durations (*31*) and greater fixation variability (*30*) than novice drivers while viewing complex traffic scenes. Saccadic behaviors also depend on concurrent cognitive load (*29, 31, 33–35*). Drivers show lower fixation frequency, longer fixation durations, and lower fixation variability when viewing dangerous scenes (*31*) or completing a secondary attention task (*29*). Together, these studies suggest inexperience and cognitive load are associated with similar changes to overt attention while driving.

Eye-tracking studies are essential to studying the relationship between saccadic eye movement patterns and driving performance. Critically, however, the spatial loci of fixation and attention may differ (*36–38*), as is the case of "looking but not seeing." Electroencephalography (EEG) can provide information on brain activity associated with attention that complements saccadic eye movement patterns. Eye fixation-related potentials (EFRP) are evoked in EEG signals immediately following saccade termination (*39–42*). The P1 response is a prominent positive deflection in EFRP amplitude approximately 100 milliseconds after saccade termination and has been localized to visual cortex (*42, 43*). P1 amplitude is associated with accuracy during a proof-reading task (*44*) and task engagement (*45*), suggesting EFRPs reflect information-processing mechanisms in sensory and attentional networks.

Few studies have examined EFRP activity in the context of driving (*46–48*). Takeda et al. measured EFRPs in a simulated environment while drivers performed a verbal or spatial working memory task (*46*). Increasing cognitive load in the verbal and spatial conditions led to a reduction in saccade frequency, consistent with previous studies (*34*). Only spatial cognitive load affected P1 amplitudes, suggesting fewer resources were available for visual information processing under a concurrent visual load. Kamienkowski et al. demonstrated comparable spatiotemporal EEG profiles of stimulus-evoked potentials and EFRPs (*42*), but found that early evoked activity in extrastriate cortex distinguished between target and distractor items only in EFRPs. Similar spatiotemporal EEG profiles were found in a driving simulation study (*49*), suggesting EFRPs can provide a window on driver visual attention when eye fixation and attention loci differ.

The current study used state-of-the-art driving simulation, eye movement, and EEG measures to test the hypothesis that visual attention is impaired in cancer patients diagnosed with hematological malignancy (HM), including myelodysplastic syndrome, non-Hodgkin lymphoma, and multiple myeloma. A few EEG studies have demonstrated chemotherapy-related reductions in stimulus-evoked measure of information processing up to 6 years post-treatment in breast cancer patients (*50–53*). It is currently unknown whether similar EEG measures are affected in HM patients with putative attention impairments (*54–56*). Consequently, the study measured visual search performance, eye movement behavior, and EFRP activity to investigate visual attention in a group of HM patients who received either chemotherapy or best supportive care. It is hypothesized that information-processing mechanisms of visual attention are impaired in HM relative to comparisons without cancer, as indexed by reduction in EFRP activity.

Methods

Subjects

Participants included cancer patients diagnosed with a hematological malignancy (HM; $n = 30$) and a comparison group without cancer (HC; $n = 15$). HM patients were recruited from University of Nebraska Medical Center (UNMC) Fred & Pamela Buffett Cancer Center. Demographically matched HC participants were recruited from an existing research registry in the UNMC Department of Neurological Sciences. Inclusion criteria for both groups were: (1) age 19–80 years, and (2) normal or corrected-to-normal vision. Exclusion criteria for both groups were: (1) previous cancer diagnosis, and (2) previous radiotherapy or chemotherapy. All participants provided informed consent approved by the UNMC Institutional Review Board and Scientific Review Committee.

Driving Simulator

SENSEI (Simulator for Ergonomics, Neuroscience, Safety Engineering and Innovation; Figure 1*a*), a DriveSafety (Salt Lake City, UT) RS-600 Research Simulator, provided the driving environment. SENSEI is constructed around a full-sized automotive cab (2004 Ford Focus). SENSEI's cab is centered within 7 Ultra-HD (3840 × 2160) curved LED displays that provide a 290-degree panoramic field of view. Additional side and center rear-view LCD monitors provide a full 360-degree simulated environment. SENSEI is fully integrated with instrumentation for recording vehicle and driver performance measures.

Driving Scenario

Scenarios were created in HyperDrive Authoring Suite (DriveSafety). Drivers completed a baseline drive on a rural road (5.6 km; 88.5 km/h speed limit) as part of a longer driving simulation study. Drivers were instructed to drive as they would normally. No other task demands were imposed during the baseline drive. Simulator adaptation was implemented prior to driving simulation to prevent simulator sickness; simulator adaptation

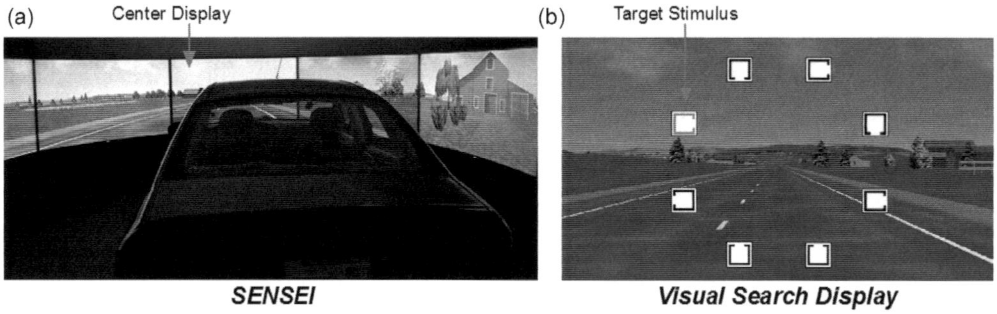

Figure 1. Driving Simulator and Visual Search Task. (*a*) Participants completed simulated driving scenarios in SENSEI (Simulator for Ergonomic, Neuroscience, Safety, Engineering, and Innovation). (*b*) Participants completed visual search tasks by responding with the direction of a red target stimulus presented among black distractors. Visual search displays were presented on the center display in SENSEI. During the driving (shown in *b*) condition, visual search stimuli were presented against the simulated background. During the baseline condition, stimuli were presented against a black background.

involved driving through a suburban roadway (1.95 km; 40.2 km/h speed limit) and stopping at a total of four 4-way intersections.

Overall, drivers completed three secondary tasks in addition to baseline driving. First, drivers completed a baseline visual search task (Task 1; see *Visual Search Task and Procedures*). Drivers were then presented with the suburban roadway and completed simulator adaptation. Following simulator adaptation, drivers transitioned onto a rural road. As vehicle speed approached the speed limit (88.5 km/h), drivers completed the visual search task while maintaining vehicle speed and lane position (Task 2). Upon completing the visual search task, drivers completed the baseline drive. Finally, drivers completed a car-following task that required maintaining a two car-length distance behind a lead vehicle whose speed pseudo-randomly fluctuated around 88.5 km/h (Task 3).

Visual Search Task and Procedures

The visual search task required drivers to respond with the identity of a red target Landolt square stimulus presented among black distractor Landolt square stimuli. Stimulus configurations are depicted in Figure 1*b*. Eight square placeholders (diameter = 2.0° visual angle) were rendered in white and presented around an imaginary concentric circle (radius = 7.3°) centered over a 0.8° white fixation dot. Each placeholder location was separated by 45°, starting at 22.5° and ending at 337.5°. Landolt square stimuli (diameter = 1.7°) were rendered with line thickness of 0.2° and a 0.6° gap centered on one side. Each stimulus display comprised 7 black distractor Landolt squares and 1 red target Landolt square. Distractor squares were rendered in black with the gap on any one side of a given square, and target squares were rendered in red with the gap on either the left or right side of the square. Driver head position was approximately 2.74 m from the central stimulus display.

Each trial started with the presentation of the square placeholders to cue trial initiation. Following a 50-ms cueing period, the stimulus display was presented until a response was made. Drivers were instructed to indicate whether the gap was on the left or right side of the target square by pressing the left or right steering wheel button, respectively. Following each response, placeholder and stimulus displays were removed for an inter-stimulus interval period randomly sampled with replacement from values ranging between 1 and 3 seconds in increments of 1/60th of a second. Target position and target direction were counterbalanced across a total of 48 trials.

Drivers completed the visual search task under two conditions: stationary (low load) and driving (high load). In the stationary condition, visual search displays were presented against a black background and central fixation was visible; in the driving condition, visual search displays were presented against the naturalistic background (Figure 1*b*) while drivers traveled along a straight rural road.

EEG Recordings

EEG was recorded using a NeuroScan NuAmps (Compumedics) digital amplifier and silver-chloride electrodes distributed in a fitted elastic cap according to the International 10–20 System. EEG was measured from 8 scalp locations (O1, O2, PO1, PO2, P3, P4, P7, P8) with a ground reference electrode at the AFz site. Electrooculography (EOG) was recorded for both horizontal (HEOG) and vertical (VEOG) eye movements;

HEOG was recorded from electrodes placed 1 cm outside the lateral canthus of each eye, and VEOG was recorded from electrodes placed above and below the left eye. Impedances of all electrodes were maintained below 10 kΩ. EEG signals were digitized at a sampling rate of 1000 Hz and re-referenced offline to mathematically averaged left and right mastoids (A1–A2).

Data Analysis

Visual Search Performance. Visual search performance was estimated as the mean response time across all analyzed trials. Behavioral costs associated with an additional driving load were estimated as the difference between visual search performance in baseline and driving conditions. Incorrect trials and response times >4 s were excluded from analysis.

EFRP. The primary goal of the current work was to measure differences between HM patients and HC comparisons in a neural correlate of information processing during naturalistic viewing conditions. To this end, EFRPs were measured during driving simulation with unconstrained eye movements. The P1 EFRP component, a positive amplitude deflection approximately 100 ms after saccade offset, was studied. EFRP waveforms were measured from EEG recordings during baseline driving using the following methods.

EEG was analyzed using EEGLab (*57*) and custom MATLAB scripts. EEG recordings were high (0.01 Hz)- and low (30 Hz)-pass filtered using a 2nd order Butterworth filter. EEG recordings were then extracted 3–240 s after the driving visual search task (Task 2) was completed. HEOG recordings were used to measure eye movements. Saccade offset events were created for timepoints meeting all of the following criteria: (1) for 30 ms of HEOG activity preceding time *t*: R^2 and β parameter of best fitted regression line were >0.9 and 2.0 uV/ms, respectively; (2) for 150 ms of HEOG activity after time *t*: root mean squared of residuals and β parameter of best fitted regression line were <20 and 2.0 uV/ms, respectively; (3) difference between β parameters for both regression lines were >2.0 uV/ms.

EFRP epochs were extracted 0–400 ms after each saccade offset event with a 100-ms baseline period. Saccade amplitude was estimated as the absolute difference between HEOG activity observed −150 to −100 ms prior to and 0–50 ms following saccade offset events (*47*). Individual waveforms were analyzed only for epochs with saccade amplitudes within the range of 100–400 μV (*47*) and EEG peak-to-peak amplitudes <160 μV. P1 amplitudes were measured from O1 and O2 electrodes by averaging EFRP activity ± 50 ms relative to peak amplitude observed 80–120 ms after saccade offset.

Statistical Analysis. Statistical analyses were performed with SAS Studio 3.6 (SAS Institute Inc.) using PROC FREQ to analyze categorical data and PROC TTEST to analyze numerical data. Numeric data were descriptively summarized using means and standard deviations, and categorical data were descriptively summarized using frequencies and percentages. Between-group effects on categorical data were assessed using chi-square statistics, between-group effects on numerical data were assessed using independent sample *t*-tests, and within-group effects on numerical visual search data were assessed using dependent sample *t*-tests. Odds ratios were used to assess the odds of HM patients showing worse outcomes (based on median-split) than HC comparisons in visual search and EFRP measures. Cohen's *d* was used to estimate standardized effect sizes.

Results

Study Enrollment

Forty-five study participants were enrolled in the study at the time of analysis. Of enrolled participants, 17 participants did not complete study procedures for the following reasons: simulator discomfort ($n = 4$), dropped out prior to initiating study procedures ($n = 2$), acute leukemia patients who could not be transported to the driving simulator ($n = 4$), technical difficulties with EEG ($n = 1$), research assistant error ($n = 1$), patient not wearing EEG during driving simulation ($n = 1$), or ongoing study participation ($n = 4$).

Of the 17 participants who did not perform the current study, 88% were HM patients ($n = 15$) and 12% were HC participants ($n = 2$). Of the 15 HM patients who did not perform the current study, 60% were receiving chemotherapy ($n = 9$) and 40% were receiving best supportive care ($n = 6$).

Study Patient Characteristics

One HM patient was excluded from analysis because of anisometropia, resulting in 14 HM patients and 13 HC comparisons that were included in the current study. HM patients were diagnosed with myelodysplastic syndrome ($n = 4$), non-Hodgkin lymphoma ($n = 9$), or multiple myeloma ($n = 1$). Of the 14 HM patients that completed task procedures, six were receiving chemotherapy and eight were receiving best supportive care. HM and HC groups did not differ along demographic dimensions of gender (HM: 50% female; HC: 46% female; $\chi^2 = 0.04$, $p = 0.84$), age (HM: 61.6 ± 14.0; HC: 58.0 ± 17.0; $t_{25} = 0.61$, $p = 0.55$), or years of education (HM: 13.8 ± 1.8; HC: 14.4 ± 1.7; $t_{25} = 0.90$, $p = 0.38$).

Task Performance

Visual search performance was faster during baseline (989 ± 302 ms) compared with driving (1027 ± 316 ms) conditions ($t_{26} = 1.79$, $p = 0.08$). HM and HC groups demonstrated comparable visual search performances in both baseline (HM = 1055 ± 380, HC = 917 ± 173; $t_{25} = 1.20$, $p = 0.24$) and driving (HM = 1105 ± 386, HC = 945 ± 56; $t_{25} = 1.33$, $p = 0.20$) conditions. Driving load-dependent differences in response times did not differ between HM (49 ± 131 ms) and HC (28 ± 94 ms) groups ($t_{25} = 0.48$, $p = 0.64$). Although between-group differences failed to reach statistical significance, odds of HM patients to show larger load-dependent differences in response times were 2.9 (CI 0.60–13.7) times the odds of HC patients.

Saccade Behavior

Previous EFRP studies have shown P1 amplitudes are more reliable following larger saccades (*47*). Therefore, saccade events were submitted for analysis only if they fell within the amplitude range of 100–400 μV. HM and HC groups did not differ from each other in either total number of saccade events (HM: 148.4 ± 97.1; HC: 158.1 ± 111.4; $t_{25} = 0.24$, $p = 0.81$) or average saccade amplitude (HM: 179.0 ± 16.8 μV; HC: 183.5 ± 22.0 μV; $t_{25} = 0.61$, $p = 0.55$). These results are suggestive of similar saccadic behaviors between HM and HC groups.

EFRP

Grand-averaged EFRP waveforms for HM and HC groups are depicted in Figure 2*a*. Both groups demonstrated a clear positive deflection in EFRP amplitude in O1/O2 electrodes approximately 101 ms after saccade offset consistent with P1 component. Average P1 amplitudes were larger for HC (2.51 ± 1.60 μV) compared with HM (1.43 ± 2.37 μV) groups ($t_{25} = 1.37$, $p = 0.18$; Cohen's $d = 0.54$; Figure 2*c*). Using a median-split of P1 amplitudes to partition individual drivers into high or low-amplitude groups, odds of HM patients being in the low-amplitude group were 4.1 (CI 0.81–20.2) times to the odds of HC participants being in the low-amplitude group. These results suggest the capacity of information processing following saccadic eye movements is reduced in HM patients compared with HC participants.

Slower response times during visual search tasks have important implications for detecting hazardous events. Next, the relationship between post-saccadic information processing and behavioral costs sustained in visual search performance completed during driving relative to baseline performance was examined. Using a median-split of response time differences, different EFRP patterns were noted between drivers showing low and high costs (Figure 2*b*). During the P1 window, larger amplitudes were observed in the drivers sustaining smaller behavioral costs ($r = -0.44$, $p = 0.022$). P1 amplitudes were larger for the low-cost group (2.85 ± 1.95 μV) compared with the high-cost group (1.11 ± 1.87 μV; $t_{25} = 2.37$, $p = 0.03$; Cohen's $d = 0.91$). Indeed, odds of low-cost drivers showing larger P1 amplitudes were 20.2 (CI 2.80–144.93; $\chi^2 = 10.8$, $p = 0.001$) times the odds of high-cost drivers, suggesting lower P1 amplitudes may be associated with greater sensitivity to cognitive load on visual search behavior.

Next, the relationship between behavioral costs and P1 amplitudes was examined separately for HM and HC groups. It was found that the odds of low-cost drivers showing larger P1 amplitudes were 31.9 (OR 1.6–666.7; $\chi^2 = 6.64$, $p = 0.01$) and 10.5 (OR 0.67–163.9; $\chi^2 = 3.26$, $p = 0.07$) times the odds of high-cost drivers in HM and HC groups, respectively. These results demonstrate post-saccade information-processing capacity is similarly reduced across groups for drivers slower to detect targets within visual search displays.

Discussion

HM patients are at risk for impaired processing speed and attention (*54–56*), and impaired visual attention contributes to increased risk for vehicle crashes (*28*). Consequently, this pilot study used state-of-the-art driving simulation and EEG to test the hypothesis that visual attention is impaired in cancer patients with HM. EFRP activity, an electrophysiological measure of information processing (*39–41*, *43*), showed promising evidence for reduced information processing in HM patients within 150 milliseconds of saccade termination. Amplitudes of the P1 EFRP component were reduced in HM compared with HC drivers. In contrast, HM and HC groups did not differ in saccade frequency or saccade magnitude. Thus, HM patients differed from HC comparisons in neural information processing and not in saccade measures.

A relationship was also found between driving load-dependent declines in visual search performance and P1 amplitudes (Figure 2, *b* and *d*). Previous studies show that older drivers are slower to respond and use more eye movements during visual search within traffic scenes (*58*). Patients with glaucoma, an age-related eye disease characterized by peripheral visual field loss, are slower to respond and deploy less efficient eye movements during visual search (*59*). These behavioral patterns may reflect recruited auxiliary mechanisms to compensate for reduced information-processing capacity in posterior cortex (*60*). In the context of studies demonstrating age- and disease-related changes in visual search performance (*58*, *59*), increased saccade frequency may be an auxiliary

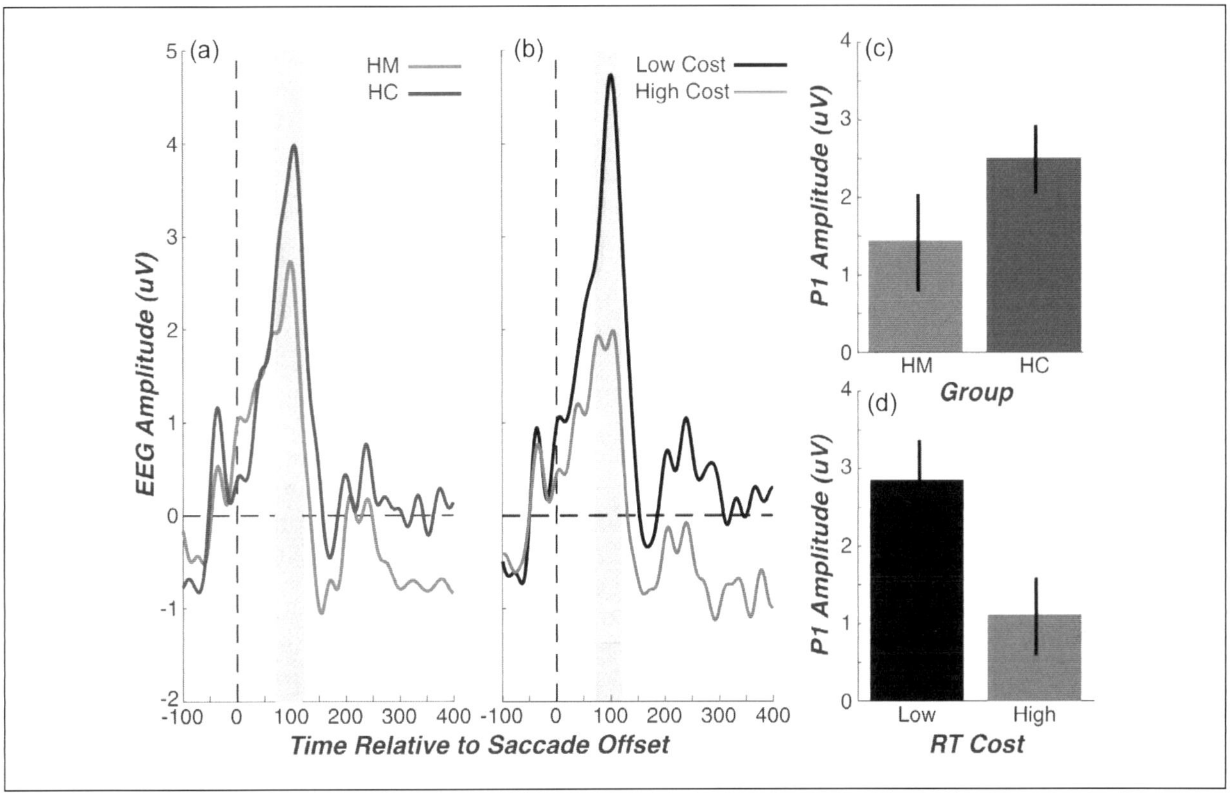

Figure 2. EFRP results. (*a*, *b*) EFRP activity was measured by estimating mean EEG amplitude as a function of time relative to saccade offset (time = 0). P1 activity (gray boxes) emerged approximately 100 milliseconds after saccade offset. (*a*) EFRP activity is plotted separately for healthy control (HC; blue line) drivers and drivers diagnosed with a hematological malignancy (HM; red line). (*b*) Drivers were split into two groups based on differences in visual search response time (RT) between driving and baseline conditions. EFRP activity is plotted separately for drivers showing smaller (low cost; black line) and larger (high cost; gray line) RT differences. (*c*, *d*) P1 amplitude was estimated as mean EFRP activity during the P1 time window (gray boxes in A, B). (*c*) P1 amplitudes were larger for HC (blue) drivers compared with HM drivers (red; $p = 0.18$). (*d*) P1 amplitudes were larger for drivers with low RT costs (black) compared with drivers with high RT costs (gray; $p = 0.03$).

mechanism for overcoming reduced neural information-processing capacity by sampling driving scenes more frequently and consequently boosting signal-to-noise ratio. The current work shows reduced neural information-processing capacity (i.e., P1 amplitudes) in drivers showing greater load-dependent declines in visual search performance, which was more prevalent in HM drivers. These findings promote further investigations to elucidate the relationship between reduced neural information-processing capacity and saccade behavior in HM patients during more complex visual search scenes.

Spatial loci of fixation and attentional resources were found to be separable. This recalls the situation of inattentional blindness, (*61*, *62*) in which observers fail to detect an unexpected object within the locus of fixation (*61*, *63*). These findings militate against exclusive reliance on eye-tracking methods to study visual attention. Further investigation is needed on relationships between disease-related changes in saccade behavior and information-processing metrics in visual cortex. Combining eye-tracking and EFRP methods (*64*, *65*) promises to advance the understanding of how disease-related changes in attention translate to failures in hazard detection while driving.

No difference was found between HM and HC groups in visual search performance. One possible explanation is that the stimulus display promoted highly efficient visual search performance. Visual search efficiency has been shown to be affected by target and distractor similarity (*66*). Displays comprising low target–distractor similarity (e.g., red target, black distractors) and high distractor–distractor similarity (e.g., all black distractors) promote efficient search, such that competing distractors fail to impose response time costs. Conversely, displays increasing in target–distractor similarity and decreasing in distractor–distractor similarity result in inefficient search, such that competing distractors impose response time costs that increase with distractor number. Thus, more complex visual search displays demand focused attention to discriminate between target and distractor search elements (*67, 68*). Here, the results indicate that performance is indistinguishable between cancer patients and healthy comparisons during efficient visual search. Future studies are needed to determine whether performance during visual search within complex displays is impaired in cancer patients.

EFRP measures can help probe how the brain processes information before, during, and after eye saccades. First, pre-saccadic shifts of attention enhance perceptual processing of locations and features present in the saccadic targets (*69, 70*). Second, saccadic suppression inhibits neural activity in low-level visual regions during eye movements (*71–73*). Third, information processing is enhanced immediately following saccade offset (*74, 75*). Importantly, most of what we know about the neural mechanisms of visual attention during saccades has been gained through studies of non-human primates. EFRP studies in human subjects engaged in real-world relevant tasks may complement and advance our current understanding of visual attention mechanisms.

Limitations

First, this study is limited by a small sample size. Although study participation is ongoing for some patients ($n = 4$), these data will not provide sufficient power to overcome this limitation. Future large-scale studies will need to compensate for subject dropout for a variety of reasons, such as simulator discomfort (9%) and patient mobility restrictions (e.g., acute leukemia inpatients; 9%). Nevertheless, these pilot data provide promising initial evidence for reduced neural information processing in HM patients. Second, it is assumed that cancer biology and chemotherapy toxicity both contributed to reduced EFRP activity in HM patients. This study was designed to prospectively evaluate changes in performance during driving simulation, which included a pretreatment assessment for patients scheduled to undergo chemotherapy. Given the current number of HM patients who completed driving simulation, the authors are currently unable to dissociate contributions of cancer-related factors from chemotherapy-related factors. Third, EFRP activity while drivers completed the visual search task could not be reliably measured. EEG recorded during visual search included both stimulus-evoked activity and EFRP activity. It is not possible to dissociate these two patterns of activity with current methods. Thus, the study was unable to evaluate load-dependent changes in P1 amplitudes. Future studies will incorporate eye-tracking methods and non-visual attention tasks to reliably measure cognitive load-dependent changes in EFRP activity while driving.

Conclusion

Driving simulation, eye movement, and EEG measures were used to study visual attention and driving in cancer patients with HM. Lower EFRP activity following saccade offset was observed in drivers showing greater impairment in visual search performance while driving, consistent with previous studies showing EFRP activity is associated with task performance. Similarly reduced EFRP activity was found in drivers diagnosed with a HM; because of the high level of uncertainty estimated for this effect, further replication of reduced EFRP activity in HM drivers is necessary. In contrast, measures of saccade behavior such as frequency and magnitude were unrelated to both cognitive function and disease status. These results support a functional dissociation between saccade behavior and post-saccade information processing during driving and may serve as a neural measure of impaired information processing in diseased and cognitively impaired populations.

Acknowledgments

We would like to thank Andrew Alexander for his assistance with EEG analysis scripting, Doug Evans and Ken Melnrick (DriveSafey) for simulator support, and Drs. Yuji Takeda and Satoshi Kitazaki (National Institute of Advanced Industrial Science and Technology) for technical assistance with EFRP analysis. This study was supported by the Fred & Pamela Buffett Cancer Center Support Grant from the National Cancer Institute (P30 CA036727), National Institute on Aging R01 AG017177, Research Support Funds from the Clinical Research Center at the University of Nebraska Medical Center, and the Department of Neurological Sciences at the University of Nebraska Medical Center.

Author Contributions

The authors confirm contribution to the paper as follows: study conception and design: D.E.A., V.R.B., M.R.; data collection: D.E.A.; analysis and interpretation of results: D.E.A., K.S., M.R.; draft manuscript preparation: D.E.A. All authors reviewed the results and approved the final version of the manuscript.

References

1. U.S. Cancer Statistics Working Group. *United States Cancer Statistics: 1999–2014 Incidence and Mortality Web-Based Report*. U.S. Department of Health and Human Services, Centers for Disease Control and Prevention and National Cancer Institute, Atlanta, 2017.
2. Berger, A. M., S. A. Mitchell, P. B. Jacobsen, and W. F. Pirl. Screening, Evaluation, and Management of Cancer-Related Fatigue: Ready for Implementation to Practice? *CA: Cancer Journal for Clinicians*, Vol. 65, No. 3, 2015, pp. 190–211.

3. Berger, A. M., C. Visovsky, M. Hertzog, S. Holtz, and F. R. Loberiza, Jr. Usual and Worst Symptom Severity and Interference with Function in Breast Cancer Survivors. *The Journal of Supportive Oncology*, Vol. 10, No. 3, 2012, pp. 112–118.
4. Berger, A. M., J. L. Grem, C. Visovsky, H. A. Marunda, and J. M. Yurkovich. Fatigue and Other Variables During Adjuvant Chemotherapy for Colon and Rectal Cancer. *Oncology Nursing Forum*, Vol. 37, No. 6, 2010, pp. 359–369.
5. Hedayati, E., H. Alinaghizadeh, A. Schedin, H. Nyman, and M. Albertsson. Effects of Adjuvant Treatment on Cognitive Function in Women with Early Breast Cancer. *European Journal of Oncology Nursing*, Vol. 16, No. 3, 2012, pp. 315–322.
6. Quesnel, C., J. Savard, and H. Ivers. Cognitive Impairments Associated with Breast Cancer Treatments: Results from a Longitudinal Study. *Breast Cancer Research and Treatment*, Vol. 116, No. 1, 2009, pp. 113–123.
7. Shilling, V., V. Jenkins, R. Morris, G. Deutsch, and D. Bloomfield. The Effects of Adjuvant Chemotherapy on Cognition in Women with Breast Cancer: Preliminary Results of an Observational Longitudinal Study. *The Breast*, Vol. 14, No. 2, 2005, pp. 142–150.
8. Stewart, A., B. Collins, J. Mackenzie, E. Tomiak, S. Verma, and C. Bielajew. The Cognitive Effects of Adjuvant Chemotherapy in Early Stage Breast Cancer: A Prospective Study. *Psychooncology*, Vol. 17, No. 2, 2008, pp. 122–130.
9. Tager, F. A., P. S. McKinley, F. R. Schnabel, M. El-Tamer, Y. K. K. Cheung, Y. Fang, C. R. Golden, M. E. Frosch, U. Habif, M. M. Muligan, and I. S. Chen. The Cognitive Effects of Chemotherapy in Post-Menopausal Breast Cancer Patients: A Controlled Longitudinal Study. *Breast Cancer Research and Treatment*, Vol. 123, No. 1, 2010, pp. 25–34.
10. Askren, M. K., M. Jung, M. G. Berman, M. Zhang, B. Therrien, S. Peltier, L. Ossher, D. F. Hayes, P. A. Reuter-Lorenz, and B. Cimprich. Neuromarkers of Fatigue and Cognitive Complaints Following Chemotherapy for Breast Cancer: A Prospective fMRI Investigation. *Breast Cancer Research and Treatment*, Vol. 147, No. 2, 2014, pp. 445–455.
11. Vardy, J., H. M. Dhillon, G. R. Pond, S. B. Rourke, W. Xu, A. Dodd, C. Renton, A. Park, T. Bekele, J. Ringash, and H. Zhang. Cognitive Function and Fatigue After Diagnosis of Colorectal Cancer. *Annals of Oncology*, Vol. 25, No. 12, 2014, pp. 2404–2412.
12. Amidi, A., M. Agerbæk, L. M. Wu, A. D. Pedersen, M. Mehlsen, C. R. Clausen, D. Demontis, A. D. Børglum, A. Harbøll, and R. Zachariae. Changes in Cognitive Functions and Cerebral Grey Matter and Their Associations with Inflammatory Markers, Endocrine Markers, and APOE Genotypes in Testicular Cancer Patients Undergoing Treatment. *Brain Imaging and Behavior*, Vol. 11, No. 3, 2016, pp. 1–15.
13. Collins, B., J. Mackenzie, A. Stewart, C. Bielajew, and S. Verma. Cognitive Effects of Hormonal Therapy in Early Stage Breast Cancer Patients: A Prospective Study. *Psychooncology*, Vol. 18, No. 8, 2009, pp. 811–821.
14. Lepage, C., A. M. Smith, J. Moreau, E. Barlow-Krelina, N. Wallis, B. Collins, J. MacKenzie, and C. Scherling. A Prospective Study of Grey Matter and Cognitive Function Alterations in Chemotherapy-Treated Breast Cancer Patients. *Springerplus*, Vol. 3, No. 1, 2014, p. 444.
15. Wefel, J. S., D. J. Vidrine, S. K. Marani, R. J. Swartz, T. L. Veramonti, C. A. Meyers, H. J. Hoekstra, J. E. Hoekstra-Weebers, and E. R. Gritz. A Prospective Study of Cognitive Function in Men with Non-Seminomatous Germ Cell Tumors. *Psychooncology*, Vol. 23, No. 6, 2014, pp. 626–633.
16. Deprez, S., F. Amant, A. Smeets, R. Peeters, A. Leemans, W. Van Hecke, J. S. Verhoeven, M. R. Christiaens, J. Vandenberghe, M. Vandenbulcke, and S. Sunaert. Longitudinal Assessment of Chemotherapy-Induced Structural Changes in Cerebral White Matter and Its Correlation with Impaired Cognitive Functioning. *Journal of Clinical Oncology*, Vol. 30, No. 3, 2012, pp. 274–281.
17. Jung, M. S., M. Zhang, M. K. Askren, M. G. Berman, S. Peltier, D. F. Hayes, B. Therrien, P. A. Reuter-Lorenz, and B. Cimprich. Cognitive Dysfunction and Symptom Burden in Women Treated for Breast Cancer: A Prospective Behavioral and fMRI Analysis. *Brain Imaging and Behavior*, Vol. 11, No. 1, 2016, pp. 1–12.
18. McDonald, B. C., S. K. Conroy, T. A. Ahles, J. D. West, and A. J. Saykin. Alterations in Brain Activation During Working Memory Processing Associated with Breast Cancer and Treatment: A Prospective Functional Magnetic Resonance Imaging Study. *Journal of Clinical Oncology*, Vol. 30, No. 20, 2012, pp. 2500–2508.
19. McDonald, B. C., S. K. Conroy, T. A. Ahles, J. D. West, and A. J. Saykin. Gray Matter Reduction Associated with Systemic Chemotherapy for Breast Cancer: A Prospective MRI Study. *Breast Cancer Research and Treatment*, Vol. 123, No. 3, 2010, pp. 819–828.
20. Marottoli, R. A., L. M. Cooney, R. Wagner, J. Doucette, and M. E. Tinetti. Predictors of Automobile Crashes and Moving Violations Among Elderly Drivers. *Annals of Internal Medicine*, Vol. 121, No. 11, 1994, pp. 842–846.
21. Cooper, P. J., K. Tallman, H. Tuokko, and B. L. Beattie. Vehicle Crash Involvement and Cognitive Deficit in Older Drivers. *Journal of Safety Research*, Vol. 24, No. 1, 1993, pp. 9–17.
22. Dubinsky, R. M., A. C. Stein, and K. Lyons. Practice Parameter: Risk of Driving and Alzheimer's Disease (An Evidence-Based Review): Report of the Quality Standards Subcommittee of the American Academy of Neurology. *Neurology*, Vol. 54, No. 12, 2000, pp. 2205–2211.
23. Lundberg, C., K. Johansson, K. Ball, B. Bjerre, C. Blomqvist, A. Braekhus, W. H. Brouwer, F. W. Bylsma, D. B. Carr, L. Englund, and R. P. Friedland. Dementia and Driving: An Attempt at Consensus. *Alzheimer Disease and Associated Disorders*, Vol. 11, No. 1, 1997, pp. 28–37.
24. Rizzo, M., D. V. McGehee, J. D. Dawson, and S. N. Anderson. Simulated Car Crashes at Intersections in Drivers with Alzheimer Disease. *Alzheimer Disease and Associated Disorders*, Vol. 15, No. 1, 2001, pp. 10–20.
25. Uc, E. Y., M. Rizzo, S. W. Anderson, J. Sparks, R. L. Rodnitzky, and J. D. Dawson. Impaired Visual Search in

Drivers with Parkinson's Disease. *Annals of Neurology*, Vol. 60, No. 4, 2006, pp. 407–413.
26. Kwon, M., C. Huisingh, L. A. Rhodes, G. McGwin, J. M. Wood, and C. Owsley. Association Between Glaucoma and At–fault Motor Vehicle Collision Involvement Among Older Drivers. *Ophthalmology*, Vol. 123, No. 1, 2016, pp. 109–116.
27. Owsley, C., G. McGwin, and K. Ball. Vision Impairment, Eye Disease, and Injurious Motor Vehicle Crashes in the Elderly. *Ophthalmic Epidemiology*, Vol. 5, No. 2, 1998, pp. 101–113.
28. Ball, K. Attentional Problems and Older Drivers. *Alzheimer Disease and Associated Disorders*, Vol. 11(Suppl. 1), 1997, pp. 42–47.
29. Bian, Z., R. Pierce, and G. Andersen. Eye Movement Patterns and Driving Performance. *Driving Assessment*, 2011.
30. Crundall, D. E., and G. Underwood. Effects of Experience and Processing Demands on Visual Information Acquisition in Drivers. *Ergonomics*, Vol. 41, No. 4, 1998, pp. 448–458.
31. Chapman, P. R., and G. Underwood. Visual Search of Driving Situations: Danger and Experience. *Perception*, Vol. 27, No. 8, 1998, pp. 951–964.
32. Pradhan, A. K., K. R. Hammel, R. DeRamus, A. Pollatsek, D. A. Noyce, and D. L. Fisher. Using Eye Movements to Evaluate Effects of Driver Age on Risk Perception in a Driving Simulator. *Human Factors*, Vol. 47, No. 4, 2005, pp. 840–852.
33. Recarte, M. A., and L. M. Nunes. Effects of Verbal and Spatial-Imagery Tasks on Eye Fixations While Driving. *Journal of Experimental Psychology: Applied*, Vol. 6, No. 1, 2000, pp. 31–43.
34. Recarte, M. A., and L. M. Nunes. Mental Workload While Driving: Effects on Visual Search, Discrimination, and Decision Making. *Journal of Experimental Psychology: Applied*, Vol. 9, No. 2, 2003, pp. 119–137.
35. Victor, T. W., J. L. Harbluk, and J. A. Engströ. Sensitivity of Eye-Movement Measures to In-Vehicle Task Difficulty. *Transportation Research Part F: Traffic Psychology and Behaviour*, Vol. 8, No. 2, 2005, pp. 167–190.
36. Martínez, A., F. Dirusso, L. Anllo-Vento, M. I. Sereno, R. B. Buxton, and S. A. Hillyard. Putting Spatial Attention on the Map: Timing and Localization of Stimulus Selection Processes in Striate and Extrastriate Visual Areas. *Vision Research*, Vol. 41, 2001, pp. 1437–1457.
37. Luck, S. J., S. A. Hillyard, M. Mouloua, M. G. Woldorff, V. P. Clark, and H. L. Hawkins. Effects of Spatial Cuing on Luminance Detectability: Psychophysical and Electrophysiological Evidence for Early Selection. *Journal of Experimental Psychology: Human Perception and Performance*, Vol. 20, No. 4, 1994, pp. 887–904.
38. Vogel, E. K., S. J. Luck, and K. L. Shapiro. Electrophysiological Evidence for a Postperceptual Locus of Suppression During the Attentional Blink. *Journal of Experimental Psychology: Human Perception and Performance*, Vol. 24, No. 6, 1998, pp. 1656–1674.
39. Yagi, A. Lambda Waves Associated with Offset of Saccades: A Subject with Large Lambda Waves. *Biological Psychology*, Vol. 8, No. 3, 1979, pp. 235–238.
40. Yagi, A. Visual Signal Detection and Lambda Responses. *Electroencephalography and Clinical Neurophysiology*, Vol. 52, No. 6, 1981, pp. 604–610.
41. Yagi, A., S. Imanishi, H. Konishi, Y. Akashi, and S. Kanaya. Brain Potentials Associated with Eye Fixations During Visual Tasks Under Different Lighting Systems. *Ergonomics*, Vol. 41, No. 5, 1998, pp. 670–677.
42. Kamienkowski, J. E., M. J. Ison, R. Q. Quiroga, and M. Sigman. Fixation-Related Potentials in Visual Search: A Combined EEG and Eye Tracking Study. *Journal of Vision*, Vol. 12, No. 7, 2012, p. 4.
43. Kazai, K., and A. Yagi. Comparison Between the Lambda Response of Eye-Fixation-Related Potentials and the P100 Component of Pattern-Reversal Visual Evoked Potentials. *Cognitive, Affective, & Behavioral Neuroscience*, Vol. 3, No. 1, 2003, pp. 46–56.
44. Takeda, Y., M. Sugai, and A. Yagi. Eye Fixation Related Potentials in a Proof Reading Task. *International Journal of Psychophysiology*, Vol. 40, No. 3, 2001, pp. 181–186.
45. Takeda, Y., T. Okuma, M. Kimura, T. Kurata, T. Takenaka, and S. Iwaki. Electrophysiological Measurement of Interest During Walking in a Simulated Environment. *International Journal of Psychophysiology*, Vol. 93, No. 3, 2014, pp. 363–370.
46. Takeda, Y., N. Yoshitsugu, K. Itoh, and N. Kanamori. Assessment of Attentional Workload while Driving by Eye-Fixation-Related Potentials. *Kansei Engineering International Journal*, Vol. 11, No. 3, 2012, pp. 121–126.
47. Takeda, Y., T. Sato, K. Kimura, H. Komine, M. Akamatsu, and J. Sato. Electrophysiological Evaluation of Attention in Drivers and Passengers: Toward an Understanding of Drivers' Attentional State in Autonomous Vehicles. *Transportation Research Part F: Traffic Psychology and Behaviour*, Vol. 42, 2016, pp. 140–150.
48. Wiberg, H., E. Nilsson, P. Lindén, B. Svanberg, and L. Poom. Physiological Responses Related to Moderate Mental Load During Car Driving in Field Conditions. *Biological Psychology*, Vol. 108, 2015, pp. 115–125.
49. Renold, H., R. Chavarriaga, L. Gheorghe, and J. D. R. Millan. EEG Correlates of Active Visual Search During Simulated Driving: An Exploratory Study. *Proc., 2014 IEEE International Conference on Systems, Man, and Cybernetics (SMC)*. IEEE, 2014, pp. 2815–2820.
50. Kam, J. W. Y., C. A. Brenner, T. C. Handy, L. A. Boyd, T. Liu-Ambrose, H. J. Lim, S. Hayden, and K. L. Campbell. Sustained Attention Abnormalities in Breast Cancer Survivors with Cognitive Deficits Post Chemotherapy: An Electrophysiological Study. *Clinical Neurophysiology*, Vol. 127, No. 1, 2016, pp. 369–378.
51. Kreukels, B. P. C., S. B. Schagen, K. R. Ridderinkhof, W. Boogerd, H. L. Hamburger, and F. S. A. M. van Dam. Electrophysiological Correlates of Information Processing in Breast-Cancer Patients Treated with Adjuvant Chemotherapy. *Breast Cancer Research and Treatment*, Vol. 94, No. 1, 2005, pp. 53–61.
52. Kreukels, B. P. C., F. S. A. M. van Dam, K. R. Ridderinkhof, W. Boogerd, and S. B. Schagen. Persistent Neurocognitive Problems After Adjuvant Chemotherapy for Breast

Cancer. *Clinical Breast Cancer*, Vol. 8, No. 1, 2008, pp. 80–87.

53. Kreukels, B. P. C., S. B. Schagen, K. R. Ridderinkhof, W. Boogerd, H. L. Hamburger, M. J. Muller, and F. S. van Dam. Effects of High-Dose and Conventional-Dose Adjuvant Chemotherapy on Long-Term Cognitive Sequelae in Patients with Breast Cancer: An Electrophysiologic Study. *Clinical Breast Cancer*, Vol. 7, No. 1, 2006, pp. 67–78.
54. Scherwath, A., L. Schirmer, M. Kruse, G. Ernst, M. Eder, A. Dinkel, S. Kunze, F. Balck, M. Bornhäuser, G. Ehninger, and K. Dolan. Cognitive Functioning in Allogeneic Hematopoietic Stem Cell Transplantation Recipients and Its Medical Correlates: A Prospective Multicenter Study. *Psychooncology*, Vol. 22, No. 7, 2013, pp. 1509–1516.
55. Zimmer, P., A. Mierau, W. Bloch, H. K. Strüder, T. Hülsdünker, A. Schenk, L. Fiebig, F. T. Baumann, M. Hahn, N. Reinart, and M. Hallek. Post-Chemotherapy cognitive impairment in patients with B-cell non-Hodgkin lymphoma: a first comprehensive approach to determine cognitive impairments after treatment with rituximab, cyclophosphamide, doxorubicin, vincristine and prednisone or rituximab and bendamustine. *Leukemia & Lymphoma*, Vol. 56, No. 2, 2015, pp. 347–352.
56. Meyers, C. A., M. Albitar, and E. Estey. Cognitive Impairment, Fatigue, and Cytokine Levels in Patients with Acute Myelogenous Leukemia or Myelodysplastic Syndrome. *Cancer*, Vol. 104, No. 4, 2005, pp. 788–793.
57. Delorme, A., and S. Makeig. EEGLAB: An Open Source Toolbox for Analysis of Single-Trial EEG Dynamics Including Independent Component Analysis. *Journal of Neuroscience Methods*, Vol. 134, 2004, pp. 9–21.
58. Maltz, M., and D. Shinar. Eye Movements of Younger and Older Drivers. *Human Factors*, Vol. 41, No. 1, 1999, pp. 15–25.
59. Crabb, D. P., N. D. Smith, F. G. Rauscher, C. M. Chisholm, J. L. Barbur, D. F. Edgar, and D. F. Garway-Heath. Exploring Eye Movements in Patients with Glaucoma When Viewing a Driving Scene. *PLoS One*, Vol. 5, No. 3, 2010, p. e9710.
60. Reuter-Lorenz, P. A., and K. A. Cappell. Neurocognitive Aging and the Compensation Hypothesis. *Current Directions in Psychological Science*, Vol. 17, No. 3, 2008, pp. 177–182.
61. Simons, D. J., and C. F. Chabris. Gorillas in Our Midst: Sustained Inattentional Blindness for Dynamic Events. *Perception*, Vol. 28, No. 9, 1999, pp. 1059–1074.
62. Mack, A. Inattentional Blindness. *Current Directions in Psychological Science*, Vol. 12, No. 5, 2003, pp. 180–184.
63. Simons, D. J. Attentional Capture and Inattentional Blindness. *Trends in Cognitive Sciences*, Vol. 4, No. 4, 2000, pp. 147–155.
64. Nikolaev, A. R., S. Pannasch, J. Ito, and A. V. Belopolsky. Eye Movement-Related Brain Activity During Perceptual and Cognitive Processing. *Frontiers in Systems Neuroscience*, Vol. 8, 2014, p. 62.
65. Wenzel, M. A., J. E. Golenia, and B. Blankertz. Classification of Eye Fixation Related Potentials for Variable Stimulus Saliency. *Frontiers in Neuroscience*, Vol. 10, 2016, p. 23.
66. Duncan, J., and G. W. Humphreys. Visual Search and Stimulus Similarity. *Psychological Reviews*, Vol. 96, No. 3, 1989, pp. 433–458.
67. Treisman, A. M., and G. Gelade. A Feature-Integration Theory of Attention. *Cognitive Psychology*, Vol. 12, No. 1, 1980, pp. 97–136.
68. Wolfe, J. M. Guided Search 2.0 A Revised Model of Visual Search. *Psychonomic Bulletin & Review*, Vol. 1, No. 2, 1994, pp. 202–238.
69. Bichot, N. P., A. F. Rossi, and R. Desimone. Parallel and Serial Neural Mechanisms for Visual Search in Macaque Area V4. *Science*, Vol. 308, No. 5721, pp. 529–534.
70. Schall, J. D., and N. P. Bichot. Effects of Similarity and History on Neural Mechanisms of Visual Selection. *Nature Neuroscience*, Vol. 2, No. 6, 1999, pp. 549–554.
71. Bremmer, F., M. Kubischik, K. P. Hoffmann, and B. Krekelberg. Neural Dynamics of Saccadic Suppression. *Journal of Neuroscience*, Vol. 29, No. 40, 2009, pp. 12374–12383.
72. Burr, D. C., M. C. Morrone, and J. Ross J. Selective Suppression of the Magnocellular Visual Pathway During Saccadic Eye Movements. *Nature*, Vol. 371, No. 6497, 1994, pp. 511–513.
73. Reppas, J. B., W. M. Usrey, and R. C. Reid. Saccadic Eye Movements Modulate Visual Responses in the Lateral Geniculate Nucleus. *Neuron*, Vol. 35, No. 5, 2002, pp. 961–974.
74. Johns, M., K. Crowley, R. Chapman, A. Tucker, and C. Hocking. The Effect of Blinks and Saccadic Eye Movements on Visual Reaction Times. *Attention, Perception, and Psychophysics*, Vol. 71, No. 4, 2009, pp. 783–788.
75. Ibbotson, M., N. Price, N. Crowder, S. Ono, and M. Mustari. Enhanced Motion Sensitivity Follows Saccadic Suppression in the Superior Temporal Sulcus of the Macaque Cortex. *Cerebral Cortex*, Vol. 17, No. 5, 2006, pp. 1129–1138.

The Standing Committee on Vehicle User Characteristics (AND10) peer-reviewed this paper (18-06350).

Article

TRR
JOURNAL OF THE TRANSPORTATION RESEARCH BOARD

Transportation Research Record
2018, Vol. 2672(37) 74–83

DOI: 10.1177/0361198118787631
journals.sagepub.com/home/trr
SAGE

Comparison of Self-Declared Mobile Use While Driving in Canada, the United States, and Europe: Results from the European Survey of Road Users' Safety Attitudes

Heather Woods-Fry[1], Ward G. M. Vanlaar[1], Robyn D. Robertson[1], Katrien Torfs[2], Woon Kim[3], Wouter Van den Berghe[2], and Uta Meesmann[2]

Abstract

Existing literature on hand-held mobile use while driving includes little research about the differences in prevalence worldwide. The current research aims to increase the available knowledge by comparing rates of self-reported hand-held mobile use behaviors while driving from three different regions (Canada, the United States, and Europe). Self-declared mobile use (talking on a hand-held mobile, sending a text message or email), personal acceptability, and attitudes toward mobile use while driving were measured as part of the E-Survey of Road Users' Attitudes (ESRA 1) conducted in 25 countries during 2015–2016. Survey data was collected uniformly, allowing for full comparability among regions. The descriptive analysis compared rates of drivers' mobile use behaviors, opinions, and attitudes by region. Two multivariate models predicting self-declared talking on a hand-held phone while driving, and self-declared sending of a text message or email while driving were also estimated. Regional comparisons of the descriptive results demonstrate that both self-declared behaviors were the lowest in Canada. Multivariate models show that U.S. drivers' personal acceptability, attitudes, and support for zero-tolerance measures had a significant effect on their self-declared rates of hand-held mobile use while driving. Furthermore, when examining self-declared talking on a hand-held mobile device while driving across all regions, a common social phenomenon known as the bandwagon effect was found, as those who agreed that almost all car drivers occasionally talk on a hand-held mobile while driving were 2.41 times more likely to report doing so themselves. Implications and suggestions for further research are discussed.

Distracted driving has become one of the most significant road safety concerns worldwide, with distraction from mobile devices and other in-vehicle technology being at the forefront of this discussion. In North America, distraction is estimated to be a factor in about 20% to 30% of motor vehicle collisions (*1*).

Distracted driving is defined as the act of driving while performing another activity that takes one's attention away from the road. This includes activities such as personal grooming, eating while driving, using a GPS navigation system, and using a mobile phone. Distraction can be classified as visual, manual, and cognitive. Texting while driving incorporates all three types of distractions and is therefore particularly dangerous. On average it takes 4.6 s to read a text message, and when driving at 55 mph (88.5 km/h), the car will travel the length of a football field in the time span taken to read a text message (*2*). Consequently, studies have shown that drivers who text are approximately 2 to 3 times more likely to be involved in a crash or near crash (*3*, *4*). Talking or dialing on a hand-held mobile device increased a driver's risk of experiencing a crash or near crash by 2 to 4 times (*4–7*). In comparison, young and novice drivers were 7 to 8 times more likely to experience a crash or near crash when reaching for, or dialing, a phone and were almost four times more likely to experience such safety-critical events when texting (*4*). Novice teenage drivers (15 to 19 years old) constitute the largest proportion of drivers who were distracted at the time of the fatal crashes (*8*). Furthermore, older

[1]Traffic Injury Research Foundation, Ottawa, ON, Canada
[2]Vias institute, Brussels, Belgium
[3]AAA Foundation for Traffic Safety, Washington, D.C.

Corresponding Author:
Address correspondence to Heather Woods-Fry: heatherw@tirf.ca

drivers are also negatively affected by distraction while driving, however, they tend to engage in distracting tasks much less than other age groups (*9*).

Although studies clearly demonstrate that distracted driving increases crash risk, many drivers can still be observed using a mobile device while driving. In Canada, an observational study showed that 4.6% of Canadian drivers were found to be using a hand-held mobile at any given moment while driving. The type of usage was also recorded: 2.3% of drivers were observed talking on their hand-held device and 1.7% of drivers were observed typing on their hand-held device (*7*). A similar U.S. study observed 3.8% of drivers talking on their hand-held device. Using this figure along with the estimated number of drivers on the road at any given daylight moment, the authors calculated that approximately 6.9% of all drivers were using some type of mobile device (hand-held or hands-free) at any given daylight moment in 2015. During this study, they also observed 2.2% of drivers visibly manipulating their mobile device. Mobile use was more frequently observed in females, and drivers aged 16 to 24 continued to be the age group with the highest proportion of observed mobile usage (*8*). In Europe, the proportion of drivers observed using a mobile device at any given time is approximately 1% to 7% (*9*).

While a majority of drivers feel that it is unacceptable to use a mobile device while driving, some drivers hold permissive attitudes toward this and feel that it is an acceptable behavior. The 2016 Traffic Safety Culture Index by the AAA Foundation for Traffic Safety (*10*) demonstrated that although 70% of licensed drivers reported that it was unacceptable for a driver to talk on a hand-held mobile device while driving, 29% found this behavior to be acceptable. Attitudes toward the use of hands-free devices was also measured, and 33% of respondents felt that it was unacceptable to talk on a hands-free device while driving, whereas 66% felt it was acceptable to talk on a hands-free device while driving.

Drivers also self-report concerning rates of mobile use while driving. A Canadian public opinion poll demonstrated that 36.3% of respondents indicated they had used their cell phone while driving during the last 7 days (*11*). In the United States, over 68% of drivers report talking on their mobile device while driving during the past 30 days, and approximately 32% of drivers report doing so regularly (*10*). Self-reported text messaging behaviors were also measured. Thirty one percent of drivers admit to writing or sending a text/email while driving in the past 30 days, and 40.2% admit to reading a text message while driving in the past month. Similar percentages were reported across Europe, as demonstrated by the 2017 European Responsible Driving Barometer. On average, 32% of Europeans report talking on their hand-held device while driving (ranging from 54% in Greece to 14% in the United Kingdom). Other behaviors such as sending or reading a text message/email were reported by 24% of Europeans, on average (ranging from 32% in Italy to 16% in the Netherlands) (*12*).

Although there have been various individual studies measuring the self-reported attitudes and mobile usage behaviors of drivers across Canada, the United States and Europe, research comparing this issue across these regions has not been conducted to the best of our knowledge. Therefore, the primary objective of this research paper is to compare the attitudes and behaviors of self-reported mobile use in Canada, the United States, and Europe to determine what quantitative differences exist between these three regions. Observed differences may begin to help researchers understand the qualitative differences influencing mobile use while driving in each country such as how laws, enforcement techniques, and prevention strategies affect the prevalence of distracted driving.

Methodology

The ESRA project (E-Survey of Road Users' Attitudes) is a joint initiative of research organizations and road safety institutes all over the world. The aim is to collect comparable national data on road users' opinions, attitudes, and behavior with respect to traffic and road safety and to provide policy makers with data to inform policy measures. The first ESRA survey (ESRA1) was conducted in the following 25 countries: Australia, Austria, Belgium, Canada, Czech Republic, Denmark, Finland, France, Germany, Greece, Hungary, Ireland, Israel, Italy, the Republic of Korea, the Netherlands, Norway, Poland, Portugal, Slovenia, Spain, Sweden, Switzerland, the United Kingdom and the United States.

The ESRA1 survey is a web survey using access panels. Two Belgian market research agencies (iVOX and GfK) organized the fieldwork under the supervision of the Belgian Road Safety Institute. The target population was the adult population (≥18 years) of each country. The targeted number of respondents was 1,000 in each country. At least 600 respondents should be regular car drivers (i.e., having a valid driving license and driven at least 1,500 km with a car or a van in the last 6 months). If needed, the sample of 1,000 respondents could be extended in order to fulfill both aforementioned requirements. The market research agencies used a quota for age and gender (interlaced), and monitored geographical distribution. The total sample size consisted of 26,856 road users from 25 countries.

In view of comparability, not just between the countries but also with past surveys, almost all questions of the ESRA1 survey were based on, and often identical to, questions previously used and validated in Belgian (*13*, *14*), European (*15*) and American (*16*) surveys. The survey covered diverse themes in varying detail due to survey duration constraints (all information on the questionnaire is available on www.esranet.eu). The questionnaire was produced in 30 different country-language versions. Translated versions were reviewed by the ESRA team to ensure consistency and concept validity within each language. Data collection took place in two phases: the first phase of field work took place

Table 1. Sample Distribution per Region

		Region					
		Canada		United States		Europe	
Male	18 to 34 years	133	(13%)	155	(14%)	2,770	(13%)
	35 to 54 years	217	(20%)	207	(19%)	4,527	(21%)
	55+	162	(15%)	151	(14%)	3,316	(16%)
Female	18 to 34 years	150	(14%)	163	(15%)	3,424	(16%)
	35 to 54 years	224	(21%)	224	(21%)	4,262	(20%)
	55+	173	(16%)	175	(16%)	2,885	(14%)
N_{total}		1,059		1,075		21,184	

Note: Interlaced gender by three age groups; unweighted sample distribution.

in June and July 2015 (17 countries) and the second phase from September until December of 2016 (8 countries). Having a uniform sampling method, an identical questionnaire, and uniform programming of this questionnaire allows ESRA1 results to be fully comparable among the participating countries. Questionnaires were hosted by various national panel providers, and the panel provider and the respondent had pre-established consent. For this kind of online panel survey, no additional internal review board or ethics board approval was necessary. The present analysis focuses on a comparison of self-declared mobile use while driving and related potential behavioral determinants, in three regions: Canada, the United States, and Europe. Table 1 shows the distribution of the sample per region.

In the first part of our analysis, we report descriptive results of the three regions for self-declared mobile use while driving, personal acceptability, and attitudes toward mobile phone use while driving. In the second part, we carried out multivariate analyses, investigating factors predicting self-declared talking on a hand-held mobile device and sending a text message or email while driving. Binary logistic regression models were used to investigate the association between several explanatory variables and self-declared behaviors. Note that a weighting of the data was applied to both descriptive and multivariate analyses (i.e., binary logistic regression). This weighting took into account small corrections with respect to the national representativeness of the sample based on gender and three age groups: <35; 35 to 54 years; and ≥55 years, based on population statistics from United Nations data (*17*). For Europe, the weighting also took into account the population size of each country in the total set of 20 countries. SPSS 22.0 was used for all analyses (*18*).

Results

Descriptive Analysis

As part of the ESRA1 questionnaire, respondents in each region were asked questions about hand-held mobile use while driving. Results are grouped into three sections: self-declared behaviors in traffic, the personal acceptability of unsafe traffic behaviors, and attitudes toward unsafe traffic behaviors.

Self-Declared Behaviors in Traffic. Rates of self-declared hand-held mobile use and sending or reading a text message while driving at least once within the last 12 months are reported by region in Table 2. The proportions of those who report that they have talked on a hand-held mobile device while driving were significantly different between all three regions, with Canadians reporting the lowest incidence of this behavior (25% for Canada, 49.7% for the United States, and 37.8% for Europe) (Pearson's chi-squared = 102.59; *df* = 2; *p*-value = .000). A significantly higher proportion of Americans indicated that they have sent a text message or email while driving compared with Canadians or Europeans (35.3% for the United States, versus 24% for Canada, and 27.2% for Europe) (Pearson's chi-squared = 30.45; *df* = 2; *p*-value = .000). The proportions of those who report that they have read a text message or email while driving were also significantly different between all three regions (29.4% for Canada, 41.6% for the United States, and 36.4% for Europe) (Pearson's chi-squared = 25.91; *df* = 2; *p*-value = .000), with Canadians reporting the lowest incidence of this behavior.

Personal Acceptability of Unsafe Traffic Behaviors. Rates of personal acceptability of unsafe traffic behaviors are located in Table 3. Personal acceptability was measured on a five-point Likert scale ranging from unacceptable to acceptable. For the purpose of the analysis, values 1 to 3 were coded as unacceptable, and values 4 and 5 were coded as acceptable. When asked whether they felt it to be personally acceptable for a driver to talk on a hands-free mobile while driving, Americans reported significantly higher rates of acceptance compared with Canadians or Europeans (12.6% for the United States, versus 6.0% for Canada, and 7.2% for Europeans) (Pearson's chi-squared = 43.17; *df* = 2; *p*-value = .000). Personal acceptance of texting while driving was also measured and Americans reported significantly higher levels of personal acceptance toward drivers who type texts or emails while driving (8.1% for the United States, versus 4.1% for

Table 2. Proportions of Self-Declared Behaviors for All Three Regions

	Canada	United States	Europe
Hand-held mobile use while driving at least once during the past 12 months	25.0%[a]	49.7%[b]	37.8%[c]
Sending a text message or email while driving at least once during the past 12 months	24.0%[a]	35.3%[b]	27.2%[a]
Reading a text message or email while driving at least once during the past 12 months	29.4%[a]	41.6%[b]	36.4%[c]

Note: (1) Weighted means were applied for Canada, the United States, and Europe, which took into account individual population sizes from Austria, Belgium, Czech Republic, Denmark, Finland, France, Germany, Greece, Hungary, Ireland, Italy, Norway, Poland, Portugal, Slovenia, Spain, Sweden, Switzerland, the Netherlands and the United Kingdom; (2) Each row proportion with the same superscript letter denotes a non-significant difference. When the superscript letter is different, then the row proportions are significantly different from each other at the 0.05 level.

Table 3. Proportions of Personal Acceptability of Unsafe Traffic Behaviors for All Three Regions

	Canada	United States	Europe
I find it personally acceptable for a driver to talk on a hand-held mobile phone while driving	6.0%[a]	12.6%[b]	7.2%[a]
I find it personally acceptable for a driver to type a text message or email while driving	4.1%[a]	8.1%[b]	4.1%[a]

Note: (1) Weighted means were applied for Canada, the United States, and Europe, which took into account individual population sizes from Austria, Belgium, Czech Republic, Denmark, Finland, France, Germany, Greece, Hungary, Ireland, Italy, Norway, Poland, Portugal, Slovenia, Spain, Sweden, Switzerland, the Netherlands and the United Kingdom; (2) Each row proportion with the same superscript letter denotes a non-significant difference. When the superscript letter is different, then the row proportions are significantly different from each other at the 0.05 level.

Canada, and 4.1% for Europe) (Pearson's chi-squared = 37.24; $df = 2$; p-value = .000).

Attitudes Toward Unsafe Traffic Behaviors. Attitudes toward unsafe traffic behaviors were measured and participants were asked if they agreed with several statements concerning hand-held mobile use while driving (see Table 4). The agreement with a statement was measured on a five-point Likert scale ranging from disagree to agree. For the purpose of the analysis, values 1 to 3 were coded as disagree/neutral, and values 4 and 5 were coded as agree. Americans were significantly less likely to agree that talking on a hand-held device while driving decreases their attention to traffic (65.1% for the United States, versus 71.1% for Canada, and 74.5% for Europe) (Pearson's chi-squared = 40.33; $df = 2$; p-value = .000). All three regions felt similarly, and agreed that people talking on a hand-held mobile phone while driving were at a higher risk of getting involved in an accident (non-significant differences between 83.2% for Canada, 80.9% for the United States, and 82.1% for Europe) (Pearson's chi-squared = 1.80; $df = 2$; p-value = .407). Canadians were significantly less likely than Americans and Europeans to agree that almost all car drivers occasionally talk on a hand-held mobile phone while driving (35.1% for Canada, versus 52.5% for the United States, and 51.2% for Europe) (Pearson's chi-squared = 98.17; $df = 2$; p-value = .000). Participants were also asked to indicate their level of support for road safety policy measures that would implement zero tolerance for any mobile use while driving, and Canadians demonstrated the highest level of support, followed by the United States and Europe. The proportions in support of zero tolerance for mobile use were significantly different between all regions (60.3% for Canada, 50.5% for the United States, and 46.4% for Europe) (Pearson's chi-squared = 78.12; $df = 2$; p-value = .000).

Multivariate Analysis

Two dependent variables were selected for further analysis: self-declared talking on a hand-held mobile and self-declared sending a text message or email. Two binary logistic regression models are used to study the association between a set of explanatory variables and a dependent variable. For each model, the dependent variable is binary, and indicates the absence (0 = never) or presence (1 = at least once) of the self-declared behavior within the last 12 months.

Demographics (region, gender, age, and education level), frequency of driving, personal acceptability of unsafe traffic behaviors, attitudes toward unsafe traffic behaviors, and support for road safety measures were treated as the explanatory variables. Odds ratios (OR) and 95% confidence intervals (CI) were used to estimate the strength of the relationship between the variables. The reference categories are listed in brackets following each independent variable in the model. The goodness of fit was assessed with the Nagelkerke R square test.

Factors Predicting Self-Declared Talking on a Hand-Held Mobile Device While Driving. The logistic regression model for self-declared talking on a hand-held mobile device while driving

Table 4. Attitudes Towards Unsafe Traffic Behaviors for All Three Regions

	Canada	United States	Europe
I agree that my attention to traffic decreases when talking on a hand-held mobile phone while driving	71.1%[a]	65.1%[b]	74.5%[a]
I agree that people who talk on a hand-held mobile phone while driving are at higher risk of getting involved in an accident	83.2%[a]	80.9%[a]	82.1%[a]
I agree that almost all car drivers occasionally talk on a hand-held mobile phone while driving	35.1%[a]	52.6%[b]	51.2%[b]
I support zero tolerance for using any type of mobile device while driving	60.3%[a]	50.5%[b]	46.4%[c]

Note: (1) Weighted means were applied for Canada, the United States, and Europe, which took into account individual population sizes from Austria, Belgium, Czech Republic, Denmark, Finland, France, Germany, Greece, Hungary, Ireland, Italy, Norway, Poland, Portugal, Slovenia, Spain, Sweden, Switzerland, the Netherlands and the United Kingdom; (2) Each row proportion with the same superscript letter denotes a non-significant difference. When the superscript letter is different, then the row proportions are significantly different from each other at the 0.05 level.

is presented in Table 5. In comparison with Europe, the likelihood of drivers reporting that they have talked on a hand-held mobile device while driving in Canada decreased by 52.4% when controlling for all other factors (OR = 0.476; 95% CI: 0.313–0.725; $p < 0.01$). No significant difference in self-declared talking on a mobile was observed between the United States and Europe. This means Canadians are least likely to report talking on a mobile device while driving, whereas Europeans and Americans are most likely to report this behavior. These results are in line with the descriptive analysis showing that Canadian drivers report the lowest proportions of talking on a hand-held mobile device, whereas Europeans and Americans report higher proportions of this behavior.

Hand-held mobile use was associated with respondents' gender, as females were 23.1% less likely to report talking on a mobile while driving, in comparison with male drivers (OR= 0.769; 95% CI: 0.714–0.829; $p < 0.01$). Age was also a significant factor in mobile use while driving. When comparing drivers in the youngest age group (18 to 34 years) to those aged 35 to 54 and those over the age of 55, the likelihood of talking on the phone while driving decreased as age increased (35 to 54 years OR= 0.776; 95% CI: 0.709–0.829; $p < 0.01$; 55 years+ OR= 0.451; 95% CI: 0.409–0.497; $p <$ 0.01).

In comparison with those with a primary school education or lower, the likelihood of talking on a hand-held mobile device while driving increased with years of education. To illustrate, those with a secondary education were 43.6% more likely to report this behavior (OR = 1.436; 95% CI: 1.197–1.724; p < 0.01). Those with a bachelor's or master's degree were even more likely to report this behavior, with an increased likelihood of 66 % and 99.3% respectively (bachelor's degree OR = 1.660; 95% CI: 1.378–2.000; $p < 0.01$; master's degree OR = 1.993; 95% CI: 1.646–2.414; $p < 0.01$).

Driving frequency was also significantly associated with self-declared mobile use. In comparison with those who only drive a few days a year, individuals who report a driving frequency of a few days a month are 57.1% more likely to talk on a hand-held device while driving (OR = 1.571; 95% CI: 1.115–2.212; $p < 0.05$). Those who report a driving frequency of 1 to 3 days a week are 131.3% more likely to report talking on a phone while driving (OR = 2.313; 95% CI: 1.695–3.156; $p < 0.01$), and those who drive at least 4 days a week have a 248.2% increased likelihood of reporting this behavior (OR = 3.482; 95% CI: 2.576–4.708; $p < 0.01$).

Those who feel that it is acceptable to talk on a hand-held device while driving were 223.1% more likely to report having performed this behavior in comparison with others who find it unacceptable or hold a neutral opinion (OR = 3.231; 95% CI: 2.755–3.790; $p < 0.01$).

When comparing attitudes toward unsafe traffic behaviors, drivers who agree that attention to traffic decreases when talking on a hand-held mobile device were 43.7% less likely to report this behavior (OR = 0.563; 95% CI: 0.512–0.619; $p < 0.01$). Drivers who believe that almost all car drivers occasionally talk on a hand-held mobile device were 146% more likely to report talking on a mobile device themselves (OR = 2.460; 95% CI: 2.277–2.658; $p < 0.01$). The odds of talking on a hand-held mobile device while driving decreased by 63.3% when drivers agreed that people who talk on a hand-held mobile device have a higher risk of accident (OR = 0.367; 95% CI: 0.328–0.411; $p<0.01$). Drivers who support measures of zero tolerance for the use of any mobile device while driving were 48.9% less likely to report having used a mobile device while driving (OR = 0.511; 95% CI: 0.474–0.551; $p < 0.01$).

The main effect for personal acceptance of talking on a hand-held mobile phone while driving was significant across all regions; however, the effect of personal acceptability on self-declared talking on a hand-held mobile device while driving was larger in the United States than in all regions combined by a factor of 2.40. A significant interaction between region and personal acceptability indicated that U.S. drivers were 675.2% more likely to self-report talking on a hand-held mobile phone if they felt it to be personally acceptable to talk on a hand-held mobile phone while driving (OR = 2.402; $p <$ 0.05). The main effect for attitudes toward unsafe traffic behaviors (agreement with the attitude that drivers who talk on a hand-held mobile phone while driving are at higher risk of accident)

Table 5. Logistic Regression Models for Self-Declared Talking on a Hand-Held Mobile Phone While Driving

Factors (reference category)	Dependent variable: self-declared behavior (0 = never; 1 = at least once) All regions (Model 1)
	Odds Ratio (CI 95%)
Region (Europe)	
Canada	.476** (.313–.725)
USA	.958 (.625–1.469)
Gender (Male)	
Female	.769** (.714–.829)
Age (18 to 34 years)	
35 to 54 years	.776** (.709–.829)
≥55 years	.451** (.409–.497)
Educational level (primary education or none)	
Secondary education	1.436** (1.197–1.724)
Bachelor's degree or similar	1.660** (1.378–2.000)
Master's degree or higher	1.993** (1.646–2.414)
Frequency of driving (a few days a year)	
A few days a month	1.571* (1.115–2.212)
1 to 3 days a week	2.313** (1.695–3.156)
At least 4 days a week	3.482** (2.576–4.708)
Personal acceptability (unacceptable/neutral)	
Talk on a hand-held mobile phone (acceptable)	3.231** (2.755–3.790)
Attitudes toward unsafe traffic behaviors (disagree/neutral)	
Attention to traffic decreases when talking on a hand-held mobile phone while driving (agree)	.563** (.512–.619)
Almost all car drivers occasionally talk on a hand-held mobile phone while driving (agree)	2.460** (2.277–2.658)
People talking on a hand-held mobile phone while driving have a higher risk of accident (agree)	.367** (.328–.411)
Support for road safety measures (oppose/ no opinion)	
Zero tolerance for using any type of mobile phone while driving for all drivers (support)	.511** (.474–.551)
Region (Canada) × Personal acceptability (unacceptable/neutral)	
Talk on a hand-held mobile phone (acceptable)	1.152 (.612–3.735)
Region (United States) × personal acceptability (unacceptable/neutral)	
Talk on a hand-held mobile phone (acceptable)	2.402* (1.078–5.353)
Region (Canada) × Attitudes toward unsafe traffic behaviors (disagree/neutral)	
People talking on a hand-held mobile phone while driving have a higher risk of accident (agree)	1.403 (.874–2.254)
Region (United States) × Attitudes toward unsafe traffic behaviors (disagree/neutral)	
People talking on a hand-held mobile phone while driving have a higher risk of accident (agree)	1.705 * (1.069–2.719)
$R^2_{\text{Nagelkerke}}$ (pseudo R^2)	.267

Note: Weighted means were applied for Canada, the United States, and Europe, which took into account individual population sizes from Austria, Belgium, Czech Republic, Denmark, Finland, France, Germany, Greece, Hungary, Ireland, Italy, Norway, Poland, Portugal, Spain, Sweden, Switzerland, the Netherlands and the United Kingdom.
*$p < .05$; **$p < .01$.

was significant across all regions, however, the effect of drivers' attitudes on self-declared hand-held mobile use was larger in the United States than in all the regions combined by a factor of 1.705. A significant interaction between region and attitudes indicated that U.S. drivers were 37.4% less likely to self-report talking on a hand-held mobile phone while driving if they agreed with the attitude that drivers who talk on a hand-held mobile phone while driving have a higher risk of accident (OR = 1.705; $p < 0.05$).

Factors Predicting Self-Declared Sending of a Text Message or Email While Driving. The logistic regression model for

Table 6. Logistic Regression Models for Self-Declared Sending of a Text Message While Driving

Factors (reference category)	Dependent variable: self-declared behavior (0 = never; 1 = at least once)
	All regions (Model 2)
Region (Europe)	Odds Ratio (CI 95%)
Canada	.830 (.633–1.087)
USA	1.450** (1.154–1.824)
Gender (Male)	
Female	.853** (.791–.921)
Age (18 to 34 years)	
35 to 54 years	.477** (.438–.519)
≥55 years	.128** (.115–.143)
Educational level (Primary education or none)	
Secondary education	1.210 (.990–1.478)
Bachelor's degree or similar	1.374** (1.121–1.684)
Master's degree or higher	1.389** (1.128–1.711)
Frequency of driving (A few days a year)	
A few days a month	1.901** (1.331–2.714)
1 to 3 days a week	2.343** (1.688–3.252)
At least 4 days a week	3.756** (2.733–5.161)
Personal acceptability (Unacceptable/neutral)	
Type text messages or emails while driving (acceptable)	4.122** (3.270–5.197)
Check or update social media while driving (acceptable)	3.022** (2.364–3.862)
Support for road safety measures (Oppose/no opinion)	
Zero tolerance for using any type of mobile phone while driving for all drivers (support)	.489** (.450–.531)
Region (Canada) × Personal acceptability (Unacceptable/neutral)	
Check or update social media while driving (acceptable)	3.320 (.790–13.953)
Region (United States) × Personal acceptability (Unacceptable/neutral)	
Check or update social media while driving (acceptable)	3.213* (1.103–9.354)
Region (Canada) × Support for road safety measures (Oppose/no opinion)	
Zero tolerance for using any type of mobile phone while driving for all drivers (support)	.936 (.640–1.368)
Region (United States) × Support for road safety measures (Oppose/no opinion)	
Zero tolerance for using any type of mobile phone while driving for all drivers (support)	.659* (.463–.939)
$R^2_{Nagelkerke}$ (pseudo R^2)	.265

Note: Weighted means were applied for Canada, the United States, and Europe, which took into account individual population sizes from Austria, Belgium, Czech Republic, Denmark, Finland, France, Germany, Greece, Hungary, Ireland, Italy, Norway, Poland, Portugal, Spain, Sweden, Switzerland, the Netherlands and the United Kingdom.
*$p<.05$; **$p<.01$.

self-declared sending of a text message while driving is presented in Table 6. In comparison with Europe, Americans were 45% more likely to report sending a text message/email while driving (OR = 1.450; 95% CI: 1.154–1.824; $p < 0.01$). No significant difference in likelihood of self-declared sending a text message/email while driving was observed between Canada and Europe. These results agree with the descriptive analysis showing that Americans report significantly higher proportions of this behavior versus Canadians or Europeans.

When comparing self-declared texting/emailing while driving, women were 14.7% less likely to report this behavior than males (OR = 0.853; 95% CI: 0.791–0.921; $p < 0.01$). Age also had a significant effect on self-declared rates, as reports of this behavior decreased with increasing age. In comparison with the youngest age group (18 to 34 years old), those in the age group of 35 to 54 years old were 52.3% less likely to text or email while driving (OR = 0.477; 95% CI: 0.438–0.519; $p<0.01$). Those over the age of 55 were 87.2% less likely to report this behavior (OR = 0.128; 95% CI: 0.115–0.143; $p < 0.01$).

In comparison with those with a primary education or lower, rates of self-reported texting/emailing while driving were not significantly different for those with a secondary education. However, when comparing the rates of those with a primary education to those with a bachelor's or master's, the self-reported rates of texting/emailing while driving increased by 37.4% and 38.9% respectively (bachelor's degree OR = 1.374; 95% CI: 1.121–1.684; $p < 0.01$; master's degree OR = 1.389; 95% CI: 1.128–1.711; $p < 0.01$).

Frequency of driving was also associated with rates of self-declared texting/emailing while driving. Compared with those who only drive a few days a year, driving a few days a month increased the likelihood of reporting this behavior by 90.1% (OR = 1.901; 95% CI: 1.331–2.714; $p<0.01$). Those who drive 1 to 3 days a week are 134.3% more likely to report texting or emailing while driving (OR = 2.343; 95% CI: 1.688–3.252; $p < 0.01$). Driving at least 4 days a week translates into a 275.6% increase in the odds of reporting this behavior. Therefore, self-declared texting or emailing increases with the frequency of driving (OR = 3.756; 95% CI: 2.733–5.161; $p < 0.01$).

Those who felt it to be personally acceptable to type text messages/emails while driving were 312.2% more likely to report sending a text message/email while driving than those who indicated that this behavior is unacceptable or had no opinion (OR = 4.122; 95% CI: 3.270–5.197; $p < 0.01$). Similarly, those who felt it was acceptable to check or update social media while driving were 202.2% more likely to report sending a text message/email while driving (OR = 3.022; 95% CI: 2.364–3.862; $p < 0.01$).

In addition, drivers who support zero tolerance for using any type of mobile device while driving were 51.1% less likely to report sending a text message while driving than those who were against this or had no opinion (OR = 0.489; 95% CI: 0.450–0.531; $p < 0.01$).

The main effect for personal acceptance of checking or updating social media while driving was significant across all regions, however, the effect of personal acceptability of checking or updating social media while driving on self-declared sending a text message while driving was larger in the United States than in all regions combined by a factor of 3.213. This significant interaction between region and personal acceptability indicated that U.S. drivers were 871% more likely to self-declare sending of a text message while driving if they felt it to be personally acceptable to check or update social media while driving (OR = 3.213; $p < 0.05$). The main effect of support for zero tolerance for using any type of mobile phone while driving for all drivers was significant across all regions, however, the effect of support for this road safety measure on self-declared sending a text message while driving was smaller in the United States than in all regions combined by a factor of .659. This significant interaction between region and support for zero tolerance road safety measures indicated that U.S. drivers were 67.8% less likely to self-declare sending of a text message while driving if they support zero tolerance for using any type of mobile phone while driving for all drivers (OR = 0.659; $p < 0.05$).

Discussion

The objective of this study was to examine and compare the rates of self-reported mobile use while driving in Canada, the United States, and Europe, to determine whether any significant quantitative differences exist. The rates of self-declared mobile use by drivers in this study are similar to percentages found in other studies. In the current study, respondents from Canada, the United States, and Europe report concerning rates of talking on a hand-held mobile and sending text messages or emails while driving. Our descriptive analysis shows that 1 in 4 Canadians reported talking on a hand-held mobile device at least once during the past 12 months. Higher proportions were reported in Europe, as 37.8% of respondents declared this behavior while driving. The highest proportions were reported in the United States, where close to half (49.7%) of respondents reported that they had talked on a hand-held mobile phone while driving. When asked how often respondents had sent a text message or email while driving within the last 12 months, just under a quarter of Canadians and 27.2 % of Europeans had done so at least once. The highest proportions were reported in the United States, where 35.3% of respondents reported that they had sent a text message or email while driving. Based on these indicators of self-reported mobile use, these behaviors are quite prevalent in all three regions, and highest in the United States.

An in-depth look at the elevated rates of self-reported mobile use demonstrates that although similar proportions of drivers in Canada, the United States, and Europe agree that using a mobile device while driving puts you at higher risk of an accident (Canada: 82.3%, United States: 80.9%, Europe: 82.1%), the rates of self-declared mobile use do not reflect this same trend. As mentioned above, the U.S. respondents report significantly higher rates for all self-declared mobile use while driving (talking on a hand-held device and sending text messages/emails), in comparison with Canada or Europe. Furthermore, when asked to what extent they agree with the statement that talking on a hand-held device while driving decreases attention to traffic, the respondents in the United States had significantly lower rates of agreement than Canada or Europe (United States: 65.1% versus Canada: 71.1% and Europe: 74.5%). American respondents also had significantly higher rates of personal acceptance for drivers who talk on hand-held mobiles (United States: 12.6% versus Canada: 6.0% and Europe: 7.2%), and drivers who type text messages or emails (United States: 8.1% versus Canada: 4.1% and Europe: 4.1%). To summarize, drivers in the United States agree that using a phone while driving increases the risk of collision, but they report the highest rates of the self-declared behaviors and hold more permissive attitudes and acceptance of mobile use while driving.

When considering the results from the multivariate model, those in Canada were 52.4% less likely than Europeans to report talking on a hand-held mobile phone while driving, and no significant difference could be observed between the United States and Europe. Concerning texting, those in the United States were 45% more likely than Europeans to report sending a text message while driving, and no significant difference could be observed between Canada and Europe.

When examining personal acceptability and attitudes toward hand-held mobile use while driving across all three regions, those who agreed that it was acceptable to type a text message or check/update social media while driving were 4.1 and 3.02 times more likely to self-declare sending a text message while driving. Furthermore, those who agreed that almost all car drivers occasionally talk on a hand-held mobile while driving were 146% more likely to report doing so themselves. This behavior can be explained by the bandwagon effect, which suggests that individuals behave a certain way because they believe that other people are doing it as well (*19*). These results are in line with previous findings demonstrating a bandwagon effect to explain the dangerous behaviors that road users engage in (*20*).

The effects of personal acceptability and attitudes were examined by region. In the United States, personal acceptability of talking on a hand-held mobile phone while driving, and personal acceptability of checking or updating social media while driving had a larger effect on rates of self-declared hand-held mobile use in the United States in comparison with all three regions. However, attitudes of U.S. drivers had a lesser effect on rates of self-declared hand-held mobile use in the United States in comparison with all three regions.

Overall, the descriptive analysis and the multivariate models both suggest that self-declared hand-held mobile use while driving is least likely to be reported by Canadian drivers. Both analyses also indicate that U.S. drivers are more likely than European drivers to self-declare sending a text message while driving. To this point, the odds of U.S. drivers self-declaring hand-held mobile use while driving was significantly affected by personal acceptability, attitudes, and support for zero tolerance of these behaviors. It is hypothesized that the elevated rates and higher odds of U.S. drivers self-reporting sending a text message while driving could be the result of several factors, such as differences in public education and prevention strategies. It is possible that there is much greater variety in the educational messages directed at U.S. drivers who are distracted, and the lack of consistency could be affecting the personal acceptability and attitudes of drivers. Research has shown that repeated and consistent messaging about the risks of distracted driving may help to reduce this behavior (*21*). The current research highlights the quantitative differences between self-reported hand-held mobile use while driving in Canada, the United States, and Europe. The troubling rates of self-reported mobile use in all three countries will require a concerted effort to tackle this prevalent road safety issue (*1*).

To summarize, the current research demonstrates that there are large differences in the self-reported rates of mobile use by drivers in Canada, the United States, and Europe. However, as this current research relies on self-reported data, we acknowledge the limitations of these findings. As the rates of mobile use were self-reported, there can be no assumptions about their direct impact on road safety and crash risk (*22*). To expand upon the current findings, more research is needed to determine the factors (e.g., legislation, enforcement, public education, and campaigns) contributing to the significant disparities in the elevated rates and higher odds of mobile distraction.

Author Contributions

The authors confirm contribution to the paper as follows: study conception and design: HW-F, WV, RR, KT, WK, WVdB, UM; data collection: WV, RR, KT, WVdB, UM; analysis: Heather Woods-Fry, Ward Vanlaar, Katrien Torfs, Uta Meesmann; interpretation of results: HW-F, WV, KT, WK, UM; draft manuscript preparation: HW-F, WV, KT, WK, WVdB, UM. All authors listed here have reviewed the results and approved the final version of the manuscript.

References

1. Bowman, K., and R. D. Robertson. Preventable Injuries/fatalities Due to Distracted Driving: A Call for a Coordinated Action. *Journal of Orthopaedic & Sports Physical Therapy*, Vol. 46, No. 10, 2016, pp. 818–821.
2. Dingus, T. A., F. Guo, S. Lee, J. F. Antin, M. Perez, M. Buchanan-King, and J. Hankey. Driver Crash Risk Factors and Prevalence Evaluation Using Naturalistic Driving Data. *Proceedings of the National Academy of Sciences*, Vol. 113, No. 10, 2016, pp. 2636–2641.
3. Fitch, G. M., S. A. Soccolich, F. Guo, J. McClafferty, Y. Fang, R. L. Olson, M. A. Perez, R. J. Hanowski, J. M. Hankey, and T. A. Dingus. *The Impact of Hand-Held and Hands-Free Cell Phone Use on Driving Performance and Safety-Critical Event Risk*. Report No. DOT HS 811 757. National Highway Traffic Safety Administration, Washington, D.C., 2013.
4. Klauer, S. G., F. Guo, B. G. Simons-Morton, M. C. Ouimet, S. E. Lee, and T. A. Dingus. Distracted Driving and Risk of Road Crashes Among Novice and Experienced Drivers. *New England Journal of Medicine*, Vol. 370, No. 1, 2014, pp. 54–59.
5. Redelmeier, D. A. and R. J. Tibshirani. Association between Cellular-telephone Calls and Motor Vehicle Collisions. *New England Journal of Medicine*, Vol. 336, No. 7, 1997, pp. 453–458.
6. McEvoy, S. P., M. R. Stevenson, A. T. McCartt, M. Woodward, C. Haworth, P. Palamara, and R. Cercarelli. Role of Mobile Phones in Motor Vehicle Crashes Resulting in Hospital Attendance: A Case-Crossover Study. *British Medical Journal*, Vol. 331, No. 7514, 2005, p. 428.
7. Jonah, B., and V. Todd. Use of Electronic Communication Devices by Canadian Drivers in Urban Areas. Presented at 93rd Annual Meeting of the Transportation Research Board, Washington, D.C., 2014.
8. Pickrell, T. M., R. Li, and S. KC. *Driver Electronic Device Use in 2015*. Traffic Safety Facts Research Note. Report No. DOT HS 812 326. National Highway Traffic Safety Administration, Washington, D.C., 2016.
9. World Health Organization. *Mobile Phone Use: A Growing Problem of Driver Distraction*. World Health Organization, 2011. www.who.int/violence_injury_prevention/publications/road_traffic/en/index.html.
10. AAA Foundation for Traffic Safety. *2016 Traffic Safety Culture Index*. AAA Foundation for Traffic Safety, Washington, D.C., 2017.

11. Marcoux, K. D., W. G. M Vanlaar, and R. D. Robertson, *The Road Safety Monitor 2011: Distracted Driving*. Traffic Injury Research Foundation, Ottawa, Ontario, 2012.
12. Vinci. *European Responsible Driving Barometer*, http://fondation.vinci-autoroutes.com/fr/article/european-responsible-driving-barometer-survey. Accessed July 2017.
13Meesmann, U., S. Boets, and P. Silverans. *Appendix Methodology & Questionnaire. Results of the BIVV/IBSR Three-Yearly Road Safety Attitude Survey*. Belgian Road Safety Institute (BIVV/IBSR) - Road Safety Knowledge Centre, Brussels, 2014.
14. Belgian Road Safety Institute. *Nationale VerkeersONVeiligheidsenquête 2014 (NVOV)*. http://enquetebivv.be/files/folder_nl.pdf. Accessed July 28, 2017.
15. Cestac, J., and P. Delhomme. (eds.). *European Road Users' Risk Perception and Mobility*. The SARTRE 4 Survey. Public Imprim, Lyon, 2012.
16. AAA Foundation for Traffic Safety. *2015 Traffic Safety Culture Index*. AAA Foundation for Traffic Safety. Washington, D.C., 2016.
17. UNdata. *United Nations Statistics Division*, 2017. http://data.un.org/Explorer.aspx?d=POP&f=tableCode%3a22. Accessed July 2017.
18. IBM Corp. *IBM SPSS Statistics for Windows, Version 22.0*. IBM Corp. Armonk, New York, 2013.
19. McAllister, I., and D. T. Studlar. Bandwagon, Underdog, or Projection? Opinion Polls and Electoral Choice in Britain, 1979-1987. *The Journal of Politics*, Vol. 53, No. 3, 1991, pp. 720–741.
20. Vanlaar, W., H. Simpson, and R. Robertson. A Perceptual Map for Understanding Concern about Unsafe Driving Behaviors. *Accident Analysis & Prevention*, Vol. 40, No. 5, 2008, pp. 1667–1673.
21. Maheshwari, S. K., and K. A. D'Souza. *Impact of Education and Awareness Programs on the Usage and Attitudes towards Texting While Driving among Young Drivers*. Eastern Seaboard Intermodal Transportation Applications Center, Hampton, Va., 2014.
22. National Traffic Law Center. Investigation and Prosecution of Distracted Driving Cases. Report No. DOT HS 812 407. National Traffic Law Center, National Highway Traffic Safety Administration. Washington, D.C., 2017.

The Standing Committee on Vehicle User Characteristics (AND10) peer-reviewed this paper (18-01564).

Article

TRR
JOURNAL OF THE TRANSPORTATION RESEARCH BOARD

Transportation Research Record
2018, Vol. 2672(37) 84–93

DOI: 10.1177/0361198118797483
journals.sagepub.com/home/trr
SAGE

Voice Control Tasks on Cognitive Workload and Driving Performance: Implications of Modality, Difficulty, and Duration

Erika E. Miller[1], Linda Ng Boyle[2], James W. Jenness[3], and John D. Lee[4]

Abstract

The effects of an in-vehicle voice control system (VCS) on cognitive workload and driving performance were evaluated using a driving simulator study with 24 participants. Participants were asked to perform two types of in-vehicle tasks while driving: voice-command based radio and navigation tasks. The tasks were of two difficulty levels (easy, hard) with half of the tasks audio only and the other half audio with a visual display. Cognitive workload was measured using a tactile detection response task (TDRT) and a revised remote detection response task (RDRT). Driving performance was measured using standard deviation of lateral position (SDLP) and standard deviation of vehicle speed. An analysis of covariance for standard deviation of speed was used to examine the effects of VCS display with task duration as the covariate. Three separate mixed models were used to evaluate the variation in SDLP, response time, and misses given VCS tasks and detection response task (DRT) type. No significant differences were observed between TDRT and RDRT, suggesting both methods capture congruent measures of workload and that neither interfere with driving performance. Engagement with each hybrid VCS task was significantly associated with an increase in DRT misses above baseline driving. Additionally, hybrid display tasks had larger variations in vehicle speed and were completed more quickly than equivalent audio only tasks. Increasing values of task duration were associated with greater variation in lateral vehicle position. The results of this study suggest that design of VCSs should consider the modality, temporal components, and difficulty of tasks to reduce cognitive load.

Advancements in technology are expanding the capabilities of systems introduced into vehicles. Information, entertainment, and communication that has traditionally not been available to the driver can now be embedded within in-vehicle information systems (IVIS). Although many of these systems have the potential to increase safety, these technologies can distract the drivers' attention from activities critical for safe driving (*1*). IVIS often include an auditory interface to help drivers maintain visual attention on the road. These auditory interfaces coupled with voice interaction (i.e., voice control system, VCS) have the perceived advantage of alleviating distraction.

Studies have shown that VCSs degrade driving performance less when compared with similar systems with visual–manual interactions. In a test track study, Ranney et al. reported that drivers had lower standard deviation of lateral position (SDLP) and faster peripheral detection response times for tasks with a voice interface when compared with a visual–manual interface (*2*). Other studies have observed similar enhanced performance with a voice interface in relation to fewer long eyes-off-road glances (*3*) and less steering variability (*4*). However, Garay-Vega et al. did not find significant differences between voice and handheld in terms of hazard anticipation and lateral position (*3*). Depending on the task, some voice-based tasks can take longer than their manual-input counterpart. For example, Yager observed longer task times for speech-to-text tasks compared with manual texting (*5*). Ranney et al. reported longer completion times for voice dialing tasks compared with manual dialing (*2*). Garay-Vega et al. noted longer completion times

[1]Department of Mechanical Engineering, Systems Engineering Program, Colorado State University, Fort Collins, CO
[2]Department of Industrial & Systems Engineering, Department of Civil & Environmental Engineering, University of Washington, Seattle, WA
[3]Westat, Rockville, MD
[4]Department of Industrial & Systems Engineering, University of Wisconsin-Madison, Madison, WI

Corresponding Author:
Address correspondence to Linda Ng Boyle: linda@uw.edu

for music retrieval tasks using voice control over a handheld device (3) and Mehler et al. showed that increased task duration can negatively impact safety (6).

VCS designs can vary greatly across vehicles. Some systems recognize only specific phrases, some can chain multiple commands, others can accommodate natural speech, and some rely on the center console to convey visual information (6–8). These previous studies have explored the presence of visual information provided by VCSs, however few studies have actually quantified the effects of increasing visual information.

There are some discrepancies related to the safety benefits that VCSs afford relative to visual–manual tasks. The dual demand needed to drive while also engaged in a secondary task, regardless of the task modality, can cause cognitive interference of the working memory (9–11). Although the premise of VCSs is to alleviate visual–manual demand, the advantages of these systems can be compromised when commands require too much memory, tasks include visual information, durations are excessive, and information/decisions are complex. The primary goal of this study is to understand whether the cognitive demand induced by various voice control tasks will affect the attentional resources needed for driving. This was examined using a detection response task (DRT), which is a reliable and sensitive measure of cognitive workload while driving (12). In more recent literature, tactile detection response tasks (TDRT) have been favored as they are minimally intrusive and do not require visual attention. However, it is not clear how sensitive they are for tasks that require higher cognitive workload. In this study, two types of DRT (tactile and visual) were considered, with the goal of examining whether there are differences in detecting task workload under various demands.

Method

This driving simulator study was conducted at the University of Washington in Seattle, WA. Cognitive workload was induced using various VCS tasks, and the analysis compared measures of workload between DRT methods.

Participants

The study included 24 participants, balanced equally between males and females. Participants were between 25 and 54 years old, had a valid US driver's license, drove at least 3,000 mi annually, and were native English speakers. None of the participants reported having any hearing problems. Participants were compensated US$50 for their time.

Apparatus

Driving Simulator. The study was conducted using a fixed-base driving simulator (National Advanced Driving Simulator miniSim). The simulator was equipped with a 7-in. monitor to the right of the steering console, which provided a visual display for the hybrid (audio + visual) VCS tasks.

Voice Control System. The VCS application was developed using Visual Studio, and was modified from a previous version used by Jenness et al. (13). The application followed a Wizard-of-Oz protocol, where an experimenter cued the responses of the system while participants interacted with it. Participants did not know the system was controlled by the experimenter. Wizard-of-Oz methodology has been used in previous studies evaluating the effects of engagement with in-vehicle VCSs (14). Participants interacted with the system using only voice (e.g., no manual input).

This VCS was designed to be consistent with a system that could be observed in an actual vehicle that includes radio and navigation interfaces. The VCS application had two display modes, audio only (i.e., no visual display) and audio + visual (hybrid). There were two levels of task difficulty (easy and hard). Easy tasks were designed to require less information processing, whereas hard tasks conveyed more information. The tasks used in this study were similar to the tasks developed in Jenness et al. (13). However, they were revised based on task performance and workload. The final set of tasks used in this current study were pilot tested on eight undergraduate students to verify differentiation between task difficulties.

Detection Response Task. Cognitive workload was measured using two types of DRT. For both types, participants responded to a stimulus by pressing a button with their finger. The stimulus randomly vibrated (TDRT) or illuminated (RDRT) once every 3 to 5 s.

The TDRT used a small vibrating tactor (stimulus) taped to the participants neck and followed ISO 17488:2016 standard (15). The RDRT used a small red light (stimulus) mounted such that it was very difficult to see in the periphery when looking at the 7-in. monitor. The location of the tactor, red LED, and interstimulus interval were similar to approaches used by Engstrom and Mardh (16) and Bruyas and Dumont (17).

The modified RDRT protocol designed for this study used a small white filter (1.5-in. diameter) placed in front of the red LED. This allowed for a purer measure of attention to the forward roadway as it sufficiently reduced the detectability of the light to reduce obvious peripheral detection. The white filter also allowed for a constant background surrounding the light. Previous studies have mounted the LED in front of the driving scene, causing the area immediately behind the LED to constantly change. These RDRT revisions were evaluated in this study to examine possible differences in workload that TDRT versus RDRT could detect, as RDRT is particularly useful in identifying drivers' ability to detect forward roadway events.

Protocol

After participants signed an informed consent, they completed one practice drive and four study drives. Each drive was approximately 15 min with a 10-min break in between. All of the drives were on a four-lane (two lanes in each direction), undivided highway with no horizontal or vertical curves. The posted speed limit was 50 mph with a lead vehicle traveling at 50 mph plus sinusoidal variations. Drivers were instructed to maintain a safe following distance and stay in the right lane (same lane as lead vehicle). Cognitive workload was measured using TDRT for two drives and RDRT for the other two drives.

Within each drive, participants completed three radio easy, three radio hard, three navigation easy, and three navigation hard tasks. The order of these tasks were randomly generated by the VCS application within each drive. There were 24 unique tasks (12 audio only and 12 hybrid), such that each participant completed the same 24 tasks with the TDRT and RDRT. Half of the participants were randomly assigned to complete the two TDRT drives first, the other half were randomly assigned to complete the two RDRT drives first, with the order of VCS display mode also randomly assigned to control for order effects.

For the radio tasks, participants were instructed to select a radio station that most closely matched the genre described by the system. For example, the system would say "Please tune to a sports radio station" followed by a list of 12 radio station names, for which one would satisfy the genre condition (e.g., "Football News"). For the hybrid tasks, the VCS also displayed the list of stations on the monitor. The radio display had two pages of stations, with six stations on each page. For the radio easy tasks, the correct station was always within the first six stations and the station name contained at least one key word from the genre. For the radio hard tasks, the correct station was always within the last six stations and the station name contained a synonym rather than the exact wording of the genre. Participants selected their response to the task by speaking the station name or number.

For the navigation tasks, participants were instructed to navigate to an American restaurant with the highest star rating, cheapest price, and closest distance. For these tasks, the system would say "What would you like to do?" and participants were trained to say "Find American restaurants." The system would then provide a list of five restaurants, giving the name, cuisine, number of reviews, star rating, price, and distance. For the hybrid mode, the list was also displayed on the monitor. The options for the navigation easy tasks included two American and three non-American restaurants; of the two American restaurants, one would be better on all three criteria (e.g., star rating, price, and distance). The options for the navigation hard tasks included four American and only one non-American restaurant, for which one restaurant was better on two of the three criteria. Participants selected their response to the task by speaking the restaurant name or line number.

In drives 1 and 3, participants also completed a delayed digit recall task (1-back), once for each TDRT and RDRT. The 1-back task was performed after finishing the 12 VCS tasks. During the 1-back task, the system would say a list of 14 single digits with a short pause between each, and the participant repeated the number that was one step previous to the number they had just heard. Previous studies have also used this working memory task to evaluate cognitive load while driving (*18*).

Dependent Variables

Standard Deviation of Lateral Position. Lateral position of the vehicle was measured as the deviation from lane center (in inches), which was collected at 60 Hz. SDLP was computed for each task across all participants using the equation defined in SAE J2944 (*19*).

Standard Deviation of Vehicle Speed. Vehicle speed in the driving simulator was measured in mph at 60 Hz. The standard deviation of vehicle speed was aggregated by VCS task on the participant level. Previous research suggests that larger values of standard deviation of vehicle speed are indicative of driving performance degradations, specifically while engaging in voice control tasks (*6, 9*).

DRT Response Time. Reaction time, in milliseconds (ms), was measured from the onset of the DRT stimulus until the participant pressed the button. Only responses considered a correct hit (see below) were included in the analysis of response time, to isolate changes in reaction time versus missing a stimulus. Larger values of response time are associated with higher cognitive workload (*12*).

DRT Miss Count. A miss was classified as a response time shorter than 100 ms, longer than 2,500 ms, or more than one button press for a stimulus (*15*). The number of misses was then aggregated for each task by participant.

The total number of stimulus events for the task was also computed to account for exposure and was incorporated into the model. Increases in misses are associated with higher cognitive workload (*12*). Miss rate was used as opposed to percent correct to keep cognitive measures consistent with each other, in which increasing values suggest increasing cognitive workload.

Independent Variables

All independent variables were within-subject variables. *DRT Type* had two levels (tactile and visual) and *VCS Display Mode* had two levels (audio only, audio + visual). For audio only, the 7-in. monitor displayed a black screen. For the hybrid mode, the same audio was provided, but the information was repeated visually on the monitor. All of the radio and navigation tasks were completed with both audio only and hybrid modes, however the 1-back tasks were always audio only. *VCS Task Type* included five levels 1) radio easy, 2) radio hard, 3) navigation easy, 4) navigation hard, and 5) 1-back. *VCS Task Duration* was used as a covariate in the driving performance models. This was measured as the time (in seconds) to complete each task, in which the task started when the VCS chimed to cue the participant to begin. The end of the task was signaled with a verbal confirmation (e.g., "Now starting navigation").

Data Analysis

Four models were used to evaluate the effects of DRT type and VCS tasks (mode × type) on cognitive workload and driving performance. The models evaluated DRT type, VCS tasks, and participant gender as fixed effects, and a random intercept was fit for each participant. The two models of cognitive workload included baseline driving as a task, whereas the driving performance tasks did not include baseline driving. This is because the cognitive models accounted for individual differences in the cognitive demands of driving. The models for driving performance focused specifically on differences within the manipulated tasks as opposed to differences from baseline driving. This was considered appropriate because all drivers were able to maintain lateral and longitudinal control (e.g., minimal SDLP and maintain the speed limit) during baseline driving tasks and variations between participants were minimal.

Model 1. A mixed effects analysis of covariance (ANCOVA) on SD vehicle speed was first performed to understand the effects of VCS display mode. Included in the analysis were task duration, VCS mode (three levels: audio only, hybrid, 1-back), DRT type (two levels: tactile, visual), gender (two levels: male, female), and random intercept on participant. As this was a balanced design, Type I sum of squares were used for the mixed ANCOVA (*20*). Tukey's honest significant difference (HSD) test was then used to compare means between VCS display modes.

Model 2. A linear mixed model was used to examine SDLP. The general form of the model is provided in Equation 1, with the random intercept defined in Equation 2, which is adapted from the two-stage random-effects model given by Laird and Ware (*21*). This model provided an estimate for task duration because it did not include driving as a task. The nine tasks considered were: radio easy audio, radio easy hybrid, radio hard audio, radio hard hybrid, navigation easy audio, navigation easy hybrid, navigation hard audio, navigation hard hybrid, and 1-back, for which radio easy audio was the baseline task:

$$SDLP_{ij} = \beta_{0i} + \beta_1 DRT + \beta_{2k} Task_k + \beta_3 Gender + \beta_4 TaskTime + e_{ij}, \quad (1)$$

$$\beta_{0i} = \gamma_{00} + \delta_{0i}, \quad (2)$$

where

DRT = detection response type (two levels): tactile or visual;

$Task$ = VCS task type by VCS display (eight levels), with radio easy audio as reference;

$TaskTime$ = time to complete task, in seconds;

i = participant i where i is 1, ..., 24;

j = trial, repeated observations of each participant; and

k = unique predictor for each task where k is 1, ..., 8.

Model 3. A linear mixed model was fit for DRT response time. The general form for the model is provided in Equation 3 and the intercept in Equation 2 still holds. The response time was log transformed to account for the normality assumption. The model accounted for 10 tasks: driving, 1-back, radio easy audio, radio easy hybrid, radio hard audio, radio hard hybrid, navigation easy audio, navigation easy hybrid, navigation hard audio, and navigation hard hybrid. Driving only (with no secondary task) was the baseline task:

$$y_{ij} = \beta_{0i} + \beta_1 DRT + \beta_{2k} Task_k + \beta_3 Gender + e_{ij}, \quad (3)$$

where

y = log (response time);

DRT = DRT type (two levels): tactile or visual;

$Task$ = VCS task type by VCS display (nine levels), with driving as reference;

i = participant i where i is 1, ..., 24;

j = trial, repeated observations of each participant;

k = unique predictor for each task where k is 1, ..., 9.

Model 4. A negative binomial mixed model was fit on the outcome variable of DRT miss count, using the

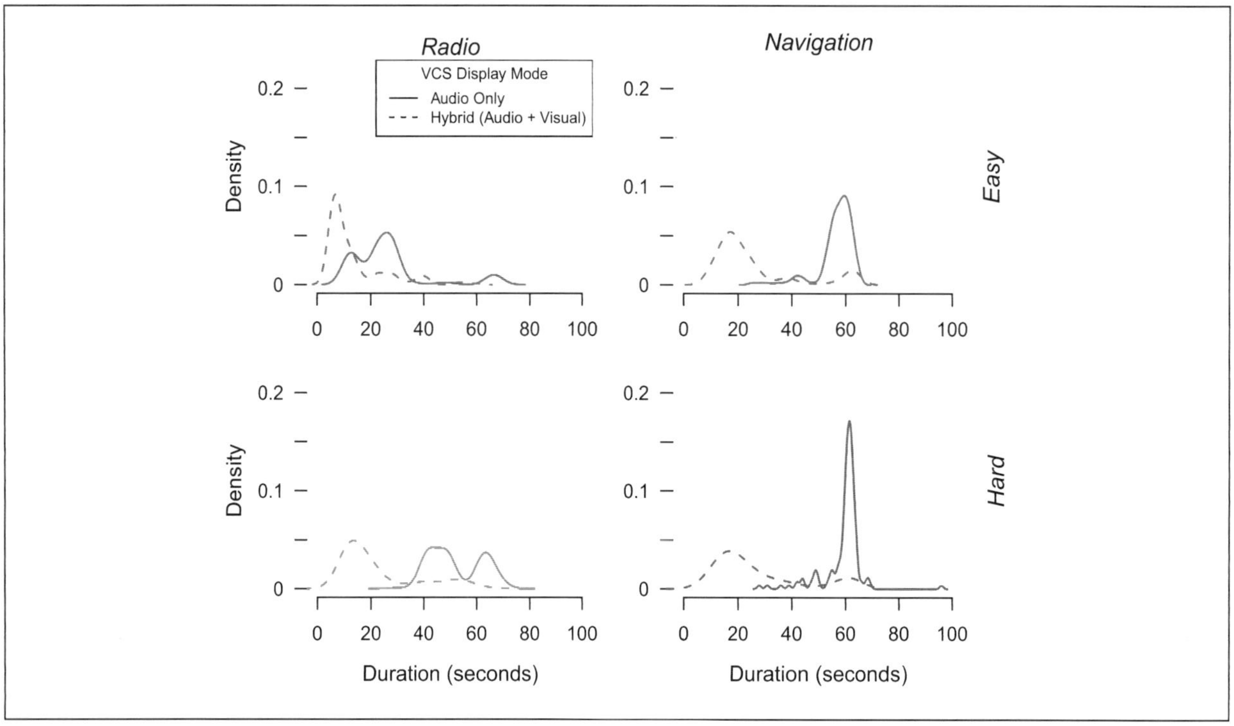

Figure 1. Probability density functions (relative count) of task durations.

generalized linear model with the log-link function. An extension of the model presented in Equation 3 was used with the outcome miss count, the exposure offset of the total number of stimulus events within trial *j* for participant *i*, and the error term was gamma distributed.

Data analysis was conducted using the statistical software program R (version 3.4.0), with packages nlme, lme4, and multcomp. Statistical significance was assessed at $\alpha = 0.05$.

Results

This study included 12 males, mean age 33.9 years old (SD = 7.7) and 12 females, mean age 34.6 years old (SD = 9.9). Participants self-reported that they had driven between 0 and 1,000 mi in the previous week, with a mean of 234 mi (SD = 248).

Task Performance

The VCS tasks were a priori designed, and confirmed through pilot testing, to induce different levels of cognitive workload, with radio easy intended to be the simplest and navigation hard intended to be the most difficult. It was also hypothesized that tasks with the hybrid display could be completed more quickly and more accurately than their audio only counterpart.

The time it took to complete each task for all participants was computed and aggregated by task type (Figure 1). Within each task type, the majority of the hybrid display tasks were completed more quickly than the audio only tasks. When sequenced by mean task completion time within each respective display mode, the order of task types was the same: radio easy were completed quickest, followed by radio hard, then navigation easy, and lastly navigation hard. The mean completion times for this sequence under audio only was 25.9 s, 51.9 s, 56.5 s, and 59.3 s; and for hybrid mode it was 13.7 s, 22.8 s, 25.9 s, and 27.3 s.

Table 1 shows the aggregation of task completion accuracy by task type, display mode, and DRT type. Participants were able to perform the hybrid tasks with greater accuracy than the audio only tasks, $t(313) = 1.81$, $p = 0.04$. They also performed the easy tasks with greater accuracy than the hard tasks, $t(313) = 4.04$, $p < 0.01$. For the radio easy task, TDRT generated higher accuracy than RDRT, $t(70) = 2.17$, $p = 0.02$.

Driving Performance

Standard Deviation of Vehicle Speed. After accounting for task duration and within-subject differences, the results of the ANCOVA suggest that there were significant differences in mean SD of vehicle speed between VCS

Table 1. Percent of Tasks Completed Accurately

Task	VCS mode	TDRT (%)	RDRT (%)
Radio easy	Audio	95.9	93.1
	Hybrid	100.0	91.7
Radio hard	Audio	89.0	90.3
	Hybrid	90.3	87.5
Navigation easy	Audio	93.1	93.0
	Hybrid	95.8	95.8
Navigation hard	Audio	80.6	75.1
	Hybrid	87.5	91.7

Note: VCS = voice control system; TDRT = tactile detection response task; RDRT = remote detection response task.

modes, $F(2, 403) = 10.59$, $p < 0.01$. Tukey's HSD test showed that each pairwise display mode was significantly different, where the mean SD of speed was larger for hybrid compared with audio ($p = 0.05$), larger for hybrid compared with 1-back ($p < 0.01$), and larger for audio compared with 1-back ($p = 0.05$). There were no significant differences in mean SD of speed between DRT types, $F(1, 403) = 0.04$, $p = 0.85$, or between gender, $F(1, 22) = 2.10$, $p = 0.16$.

Standard Deviation of Lateral Position. The linear mixed model for SDLP (see Table 2) showed no significant differences for DRT types and gender. However, task duration was significant, in which each additional second of task duration was associated with a 0.05-in. increase in SDLP for the task. Additionally, radio hard hybrid and both navigation hybrid tasks were associated with an increase in SDLP compared with the radio easy audio task.

The driving performance data were aggregated to the participant level and the SDLP was plotted by task for each DRT (Figure 2). TDRT and RDRT values are comparable across all tasks, with a slightly noticeable deviation observed for the radio hard tasks.

Cognitive Workload

DRT Response Time. The linear mixed model (see Table 3) suggests that there were no significant differences in response time between genders. The method for measuring DRT was not significant, thus suggesting that the two DRT methods provided similar measures of cognitive workload. Within the VCS tasks, navigation hard audio, 1-back, and both radio hybrid tasks induced higher workload above baseline driving.

DRT Miss Count. The negative binomial mixed model for miss count is provided in Table 4. Recall, the total number of DRT events was included to account for exposure, such that the model yields insight on the rate for number of misses per total number of events. There was no significant difference by gender or DRT type. However, all of the hybrid tasks, the navigation hard audio, and 1-back tasks showed a significant increase in miss count above baseline driving ($p < 0.001$).

Comparison of DRT Measures on Cognitive Workload. A side-by-side comparison of response time and miss rates was conducted (see Figure 3). These two plots correspond to the data in tables 3 and 4. The figure shows similar measures of cognitive workload within the respective metric (e.g., response time and miss rate) regardless of DRT type across the driving, VCS audio, and VCS hybrid tasks. However, for response time and miss rate, the TDRT method appears more sensitive to the demands of the 1-back.

Table 2. Linear Mixed Model of Standard Deviation of Lane Position (SDLP) (inches)

Variable	Estimate	SE	*t*-statistic	*p*-value
(Intercept)	4.98	0.715	6.96	<0.001
DRT type (reference tactile)				
Visual DRT	−0.21	0.175	−1.19	0.236 (ns)
VCS task (reference radio easy audio)				
Navigation easy audio	0.07	0.434	0.16	0.876 (ns)
Navigation hard audio	0.18	0.447	0.39	0.695 (ns)
Radio hard audio	0.52	0.414	1.25	0.211 (ns)
1-back	−0.83	0.509	−1.63	0.104 (ns)
Radio easy hybrid	−0.63	0.372	−1.68	0.092 (ns)
Radio hard hybrid	1.07	0.358	3.00	0.003
Navigation easy hybrid	1.04	0.357	2.91	0.004
Navigation hard hybrid	1.68	0.356	4.73	<0.001
Male	0.86	0.897	0.96	0.958 (ns)
Task duration	0.05	0.008	6.05	<0.001

Note: Observations = 1198; clusters = 24; random intercept by participant. DRT = detection response task; VCS =voice control system; SE = standard error; ns = not significant.

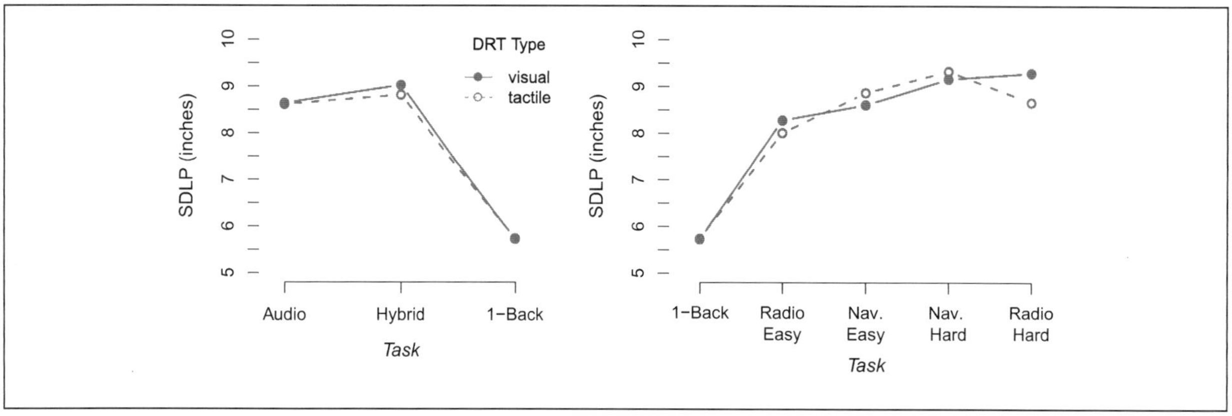

Figure 2. Interaction of task and DRT on SDLP.

Table 3. Linear Mixed Model of Log (Response Time [ms])

Variable	Estimate	SE	*t*-statistic	*p*-value
(Intercept)	6.14	0.087	70.43	<0.001
DRT type (reference tactile)				
Visual DRT	0.01	0.018	0.69	0.493 (ns)
VCS task (reference driving)				
Radio hard audio	0.04	0.053	0.73	0.465 (ns)
Navigation easy audio	0.08	0.053	1.53	0.127 (ns)
Navigation easy hybrid	0.09	0.054	1.62	0.106 (ns)
Radio easy audio	0.09	0.053	1.64	0.101 (ns)
Navigation hard hybrid	0.09	0.054	1.62	0.104 (ns)
Navigation hard audio	0.12	0.053	2.29	0.022
Radio hard hybrid	0.12	0.054	2.15	0.031
Radio easy hybrid	0.18	0.054	3.31	<0.001
1-back	0.39	0.066	5.95	<0.001
Male	−0.01	0.104	−0.07	0.941 (ns)

Note: Observations = 1227; clusters = 24; random intercept by participant. ms=milliseconds; DRT = detection response task; VCS = voice control system; SE = standard error; ns = not significant.

Table 4. Negative Binomial Mixed Model of Miss Count per Total Events

Variable	Estimate	SE	*t*-statistic	*p*-value
(Intercept)	−2.99	0.227	−13.20	<0.001
DRT type (reference tactile)				
Visual DRT	0.11	0.070	1.53	0.126 (ns)
VCS task (reference driving)				
Radio hard audio	0.05	0.141	0.36	0.716 (ns)
Navigation easy audio	0.23	0.133	1.75	0.079 (ns)
Radio easy audio	0.25	0.160	1.54	0.124 (ns)
Navigation hard audio	0.50	0.127	3.94	<0.001
Radio hard hybrid	0.69	0.151	4.56	<0.001
Radio easy hybrid	0.76	0.174	4.39	<0.001
Navigation hard hybrid	0.83	0.141	5.88	<0.001
Navigation easy hybrid	0.90	0.140	6.41	<0.001
1-back	1.33	0.157	8.47	<0.001
Male	0.13	0.297	0.45	0.654 (ns)

Note: Observations = 479; clusters = 24; random intercept by participant. DRT = detection response task; VCS = voice control system; SE = standard error; ns = not significant.

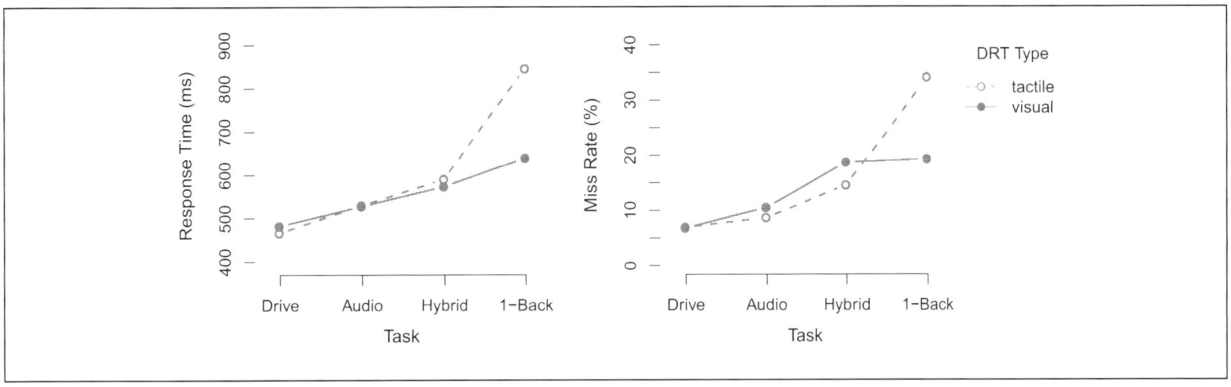

Figure 3. Interaction of task and DRT on response time *(left)* and miss rate *(right)*.

Discussion

VCSs are often perceived as a safer alternative to visual–manual secondary tasks because they can reduce eyes-off-road glances. However, these systems are often implemented such that tasks take longer to complete, impose a greater memory load, and still include a visual component (*6*, *8*). Therefore it is important to evaluate the impact of duration, difficulty, and modality of voice control tasks on mental workload and driving performance.

In this study, tasks with a hybrid display were completed more quickly than the audio only equivalent. Radio easy tasks were designed to be the least difficult and were the quickest to complete. The navigation hard tasks took the longest to complete and were designed to be the most difficult. These results suggest that redundant information across modality (e.g., auditory and visual) and lower task difficulty (e.g., less memory load and fewer decisions) can decrease the time required to complete a task. This is an important design consideration, as previous research has suggested that task duration can affect safety and system usability (*18*).

Easy tasks had higher accuracy as compared with hard tasks within each radio and navigation task type. This supports the classification of tasks as either easy or hard. Hybrid tasks were completed accurately more consistently than audio tasks for all TDRT radio, TDRT navigation, and RDRT navigation tasks. Participants had higher accuracy for audio only radio tasks with RDRT, as compared with the hybrid radio tasks. This suggests that the participant's ability to complete the tasks was not compromised by the DRT method used; however for the simpler and shorter tasks (e.g., radio), the dual demand on the visual sensors (e.g., RDRT with visual display) may have confounded their ability to complete the task accurately.

There were no significant differences in DRT type with respect to SDLP or standard deviation of speed. Nor were there differences in DRT type with respect to response time or miss count, suggesting that both types of DRT captured similar measures of workload. However, the TDRT did show a longer response time and greater miss rates for the 1-back task when compared with the RDRT. Ranney et al. also showed that TDRT was marginally more sensitive than RDRT while driving, but noted that both were comparable measures when participants were not driving (*22*).

This current study used a revised RDRT method to minimize peripheral visibility of the LED and provided a consistent backdrop around the LED. This revised protocol was evaluated to better capture attention to the forward roadway (e.g., ability to detect a roadway hazard). The white filter around the LED reduced the conspicuity of the LED when viewed in the periphery. Reducing conspicuity could eliminate the abrupt onset cue that automatically captures attention. Although there were no significant differences observed across DRT, additional research should evaluate differences in DRT with respect to eye glances. This could be important, as RDRT likely provides further insight for eyes-on-road glances, whereas TDRT captures workload regardless of eye glance position.

There were no significant differences observed between males and females for workload or driving performance. However, this study focused on drivers 25–54 years old, and larger deviations between gender have been associated with drivers who were younger and older than the tested age group. Guo et al. quantified crash risk for driver distraction within age groups and only found a meaningful difference between genders for drivers 16–20 years old (*23*). Similarly, Reed and Green reported no gender effects on lane keeping or speed control after controlling for age, but that older women had higher steering variance (*24*). Further research should consider broader age ranges to evaluate the differences across age and gender.

The 1-back, navigation hard audio (e.g., task with most memory demand), and all four hybrid tasks significantly increased miss count, above that of baseline driving. The navigation hard audio, both hybrid radio, and

1-back tasks each had a significant effect above baseline driving for response time. The differences between effects on miss rate and response time could be confounded by the variations in time required to complete these tasks, which could not be controlled for with the inclusion of driving as baseline. This is consistent with previous literature that suggests task time can influence workload (*6*). Further research should consider tasks of similar temporal demand to parse out differences of information difficulty versus information modality.

The analysis on driving performance focused on observations within secondary task involvement, thus capturing the effect of task duration. The results of this analysis indicated that increasing task completion times were significantly associated with increased SDLP. Both navigation hybrid tasks and the hard radio hybrid tasks were associated with higher SDLP as compared with radio easy audio. After accounting for task duration on SD of vehicle speed, hybrid tasks had higher variations in speed.

This study evaluated radio tuning tasks and navigation tasks, which were developed to be similar to tasks found in various in-vehicle systems currently on the road. However, further research should consider a broader range of VCS task types (e.g., other phone tasks, media controls, or point of interest tasks). Validating these findings across different VCS devices could also be useful. This study was conducted in a highly controlled laboratory environment with limited roadway distractions, and future research should consider potential confounding effects of real-world driving with traffic and various geometric roadway configurations.

More generally, this study focused on cognitive load rather than distraction. This distinction is important because drivers may adapt the load they are willing to accept from a non-driving task based on the demands of the driving situation. As a consequence, increased cognitive load might not correspond to increased crash risk. This limitation should be carefully considered in weighing measures of cognitive load in guiding design. High cognitive load indicates a greater risk of distraction, and should prompt additional testing to assess whether drivers will be able to shed the load if roadway demands require it.

The increasing market penetration and broadening capabilities of in-vehicle VCSs emphasizes the importance of design considerations. The results of this study suggest that engagement with voice tasks is associated with increased cognitive workload and decreased driving performance. Information difficulty, sensory modality, and temporal duration of VCS tasks influence the cognitive demands and compensatory effects on driving performance, which should be considered to reduce cognitive load.

Acknowledgments

This study is based on work that was partially supported by the U.S. Department of Transportation (U.S. DOT), National Highway Traffic Safety Administration (NTHSA).

Author Contributions

The authors confirm contribution to the paper as follows: study conception and design: EM, LNB, JJ, JDL; data collection: EM; analysis: EM; interpretation of results: EM, LNB, JJ, JDL; draft manuscript preparation: EM, LNB, JJ, JDL. All authors reviewed the results and approved the final version of the manuscript.

References

1. Lee, J. D., M. A. Regan, and K. L. Young. Defining Driver Distraction. In *Driver Distraction: Theory, Effects, and Mitigation* (Regan, M. A., J. D. Lee, and K. Young, eds.), CRC Press, Boca Raton, FL, 2008, pp. 31–40.
2. Ranney, T. A., J. L. Harbluk, and Y. I. Noy. Effects of Voice Technology on Test Track Driving Performance: Implications for Driver Distraction. *Human Factors*, Vol. 47, No. 2, 2005, pp. 439–454.
3. Garay-Vega, L., A. K. Pradhan, G. Weinberg, B. Schmidt-Nielsen, B. Harsham, Y. Shen, G. Divekar, M. Romoser, M. Knodler, and D. L. Fisher. Evaluation of Different Speech and Touch Interfaces to In-Vehicle Music Retrieval Systems. *Accident Analysis & Prevention*, Vol. 42, No. 3, 2010, pp. 913–920.
4. Owens, J., S. McLaughlin, and J. Sudweeks. On-Road Comparison of Driving Performance Measures When Using Handheld and Voice-Control Interfaces for Mobile Phones and Portable Music Players. *SAE International Journal of Passenger Cars - Mechanical Systems*, Vol. 3, No. 1, 2010, pp. 734–743.
5. Yager, C. *An Evaluation of the Effectiveness of Voice-to-Text Programs at Reducing Incidences of Distracted Driving*. Report 600451-00011-1. Texas A&M Transportation Institute, College Station, TX, 2013.
6. Mehler, B., D. Kidd, B. Reimer, I. Reagan, J. Dobres, and A. McCartt. Multi-Modal Assessment of On-Road Demand of Voice and Manual Phone Calling and Voice Navigation Entry across Two Embedded Vehicle Systems. *Ergonomics*, Vol. 59, No. 3, 2016, pp. 344–367.
7. Cooper, J. M., H. Ingebretsen, and D. L. Strayer. *Mental Workload of Common Voice-Based Vehicle Interactions across Six Different Vehicle Systems*. AAA Foundation for Traffic Safety, Washington, D.C., 2014.
8. Reagan, I. J., and D. G. Kidd. Using Hierarchical Task Analysis to Compare Four Vehicle Manufacturers' Infotainment Systems. *Proc., Human Factors and Ergonomics Society 57th Annual Meeting*, San Diego, CA, SAGE, Thousand Oaks, CA, 2013, pp. 1495–1499.
9. He, J., A. Chaparro, B. Nguyen, R. Burge, J. Crandall, B. Chaparro, R. Ni, and S. Cao. Texting While Driving: Is Speech-Based Text Entry Less Risky Than Handheld Text Entry? *Accident Analysis & Prevention*, Vol. 72, 2014, pp. 287–295.

10. Pashler, H. Dual-Task Interference in Simple Tasks: Data and Theory. *Psychological Bulletin*, Vol. 116, No. 2, 1994, pp. 220–244.
11. Wickens, C. D. Multiple Resources and Performance Prediction. *Theoretical Issues in Ergonomics Science*, Vol. 3, No. 2, 2002, pp. 159–177.
12. Victor, T. W., J. Engstrom, and J. L. Harbluk. Distraction Assessment Methods Based on Visual Behavior and Event Detection. In *Driver Distraction: Theory, Effects, and Mitigation* (Regan, M. A., J. D. Lee, and K. Young, eds.), CRC Press, Boca Raton, FL, 2009, pp. 135–165.
13. Jenness, J. W., L. N. Boyle, J. D. Lee, C. C. Chang, V. Venkatraman, M. Gibson, K. E. Riegler, and D. Kellman. *In-Vehicle Voice Control Interface Performance Evaluation*. Publication DOT HS 812 314. NHTSA, Washington, D.C., 2016.
14. Simmons, S. M., J. K. Caird, and P. Steel. A Meta-Analysis of In-Vehicle and Nomadic Voice-Recognition System Interaction and Driving Performance. *Accident Analysis & Prevention*, Vol. 106, 2017, pp. 31–43.
15. ISO 17488:2016. *Road Vehicles – Transport Information and Control Systems – Detection-Response Task (DRT) for Assessing Attentional Effects of Cognitive Load in Driving*. International Organization for Standardization, Geneva, Switzerland, 2016.
16. Engstrom, J., and S. Mardh. *SafeTE Final Report*. Report No. 2007:36. Swedish Road Agency, Gothenburg, Sweden, 2007.
17. Bruyas, M.-P., and L. Dumont. Sensitivity of Detection Response Task (DRT) to the Driving Demand and Task Difficulty. *Proc., 7th International Driving Symposium on Human Factors in Driver Assessment, Training, and Vehicle Design*, Bolton Landing, NY, University of Iowa, Iowa City, IA, 2013, pp. 64–70.
18. Mehler, B., B. Reimer, and J. F. Coughlin. Sensitivity of Physiological Measures for Detecting Systematic Variations in Cognitive Demand from a Working Memory Task: An On-Road Study across Three Age Groups. *Human Factors*, Vol. 54, No. 3, 2012, pp. 396–412.
19. SAE J2944 201506. *Operational Definitions of Driving Performance Measures and Statistics*. Society of Automotive Engineers, Detroit, MI, 2015.
20. Edwards, L. K. (ed.). Applied Analysis of Variance in Behavioral Science. In *Statistics: Textbooks and Monographs*, Vol. 137, Marcel Dekker, NY, 1993.
21. Laird, N. M., and J. H. Ware. Random-Effects Models for Longitudinal Data. *Biometrics*, Vol. 38, No. 4, 1982, pp. 963–974.
22. Ranney, T. A., G. H. Baldwin, L. A. Smith, E. N. Mazzae, and R. S. Pierce. *Detection Response Task (DRT) Evaluation for Driver Distraction Measurement Application*. Publication DOT HS 812 077. NHTSA, Washington, D.C., 2015.
23. Guo, F., S. G. Klauer, Y. Fang, J. M. Hankey, J. F. Antin, M. A. Perez, S. E. Lee, and T. A. Dingus. The Effects of Age on Crash Risk Associated with Driver Distraction. *International Journal of Epidemiology*, Vol. 46, No. 1, 2017, pp. 258–265.
24. Reed, M. P., and P. A. Green. Comparison of Driving Performance On-Road and in a Low-Cost Simulator using a Concurrent Telephone Dialing Task. *Ergonomics*, Vol. 42, No. 8, 1999, pp. 1015–1037.

The Standing Committee on Vehicle User Characteristics (AND10) peer-reviewed this paper (18-05086).

Any opinions, findings, conclusions, or recommendations expressed in this publication are those of the authors and do not necessarily reflect those of the U.S. DOT or NHTSA.

Article

TRR
JOURNAL OF THE TRANSPORTATION RESEARCH BOARD

Transportation Research Record
2018, Vol. 2672(37) 94–100

DOI: 10.1177/0361198118777382
journals.sagepub.com/home/trr
SAGE

The Effects of Voice System Design Components on Driver Workload

Douglas Getty[1], Francesco Biondi[1], Shae D. Morgan[2], Joel M. Cooper[1], and David L. Strayer[1]

Abstract

In-vehicle voice control systems are standard in most new vehicles. However, despite auditory-vocal interaction allowing drivers to keep their hands on the steering wheel and eyes on the forward roadway, recent findings indicate the potential for these systems to increase levels of workload and lead to lengthy interaction times. Although many studies have examined the distraction potential of interacting with in-vehicle voice control systems, more research is needed to understand the relationship between different system design components and workload. In this study, we investigate the role of system delay, system accuracy, and menu depth in determining the overall level of demand and interaction times on eight different 2017 model-year vehicles. Voice system accuracy was measured via playback of a pre-recorded sample of voice commands through a studio monitor mounted near the headrest. Menu depth and system delay were calculated by measuring, respectively, the number of interaction steps and total system processing time required to access common infotainment functions. These measures were validated through linear and multiple regression analyses with workload and task time collected in an on-road study. We found system delay and system accuracy to be significant predictors of task time and subjective measures of workload from the NASA Task Load Index and the Driving Activity Load Index. A In addition to providing valuable information on the role of separate voice control system design components on resulting levels of workload, these results extend past research by generalizing findings to multiple current auditory-vocal systems.

Voice interfaces were originally implemented in vehicles with the intention of reducing driver distraction (*1*). By keeping drivers' eyes on the road and hands on the wheel, it was believed that drivers could remain attentive to the primary driving task while completing secondary tasks with the voice system. However, recent findings on in-vehicle voice control systems (VCS) raise serious concerns about the safety of such technologies (*2–4*). Studies of VCS interactions have found increased levels of workload (*2*), increases in number of glances and eyes-off road time (*2*), and lengthy task times, sometimes exceeding 90 s (*2–4*). At the same time, voice interfaces in vehicles have become increasingly prevalent, with the majority of vehicles sold since 2012 having some form of on-board VCS (*4*). Given the demonstrated potential for high workload and lengthy interactions during the use of VCS, identifying the features of VCS interfaces that affect workload is an important line of research to facilitate safe driving conditions. Specifically, a better understanding of how to design a low-demand VCS interface will allow manufacturers to design systems that are safer for use while driving.

There is evidence that interface-level features of in-vehicle VCS can affect driver workload. In an on-road study by Cooper, Ingebretsen, and Strayer (*5*), researchers measured participants' heart rate and subjective workload while driving on a residential street. This study also took direct measures of workload using a head-mounted LED detection response task, which is a task designed by the International Standards Organization to measure cognitive workload in driving contexts (*5, 6*). Participants carried out common phone and music-related tasks via the VCS, such as calling a specific contact (e.g., "Call Aaron Jones") or changing the radio station (e.g., "Tune to 99.5 FM"). Even though the tasks were constant between all six vehicles, the results showed that workload varied significantly between vehicles. Furthermore, this study found that, among these vehicles, voice system accuracy was significantly correlated with mental workload (*5*). This suggests that differences in the design components in different vehicle VCS may be responsible for the driving workload associated with their use.

A set of three National Highway Traffic Safety Administration-funded studies on VCS evaluation by Jenness et al. identified design components that may be responsible for driving workload in VCS systems (*4*). Study 2 utilized a "Wizard

[1]Department of Psychology, University of Utah, Salt Lake City, UT
[2]Communication Sciences and Disorders, University of Utah, Salt Lake City, UT

Corresponding Author:
Address correspondence to Douglas Getty: doug.getty@utah.edu

Table 1. Voice Command Types

Radio	Calling
<frequency> <band>	Call <phone number>
<band> <frequency>	Dial <phone number>
Tune to <frequency> <band>	Call <contact name>
Radio tune to <frequency> <band>	Call contact <contact name>
<frequency>	Call <contact name> mobile
Tune the radio to <frequency> <band>	Call <contact name> cell
	Call <contact name> home
	Call <contact name> work

of Oz" methodology—a method by which the VCS system is a mock-up controlled entirely by a hidden research assistant—in which the researchers manipulated the presence of a recognition error, the length of the voice system delay, and type of functions accessed. They found that when their mock VCS had a long delay (8 s) or was less accurate, participants had more off-road glances and longer task durations. Interestingly, this study also found that a long (8 s) system delay resulted in lower levels of cognitive load as measured by the tactile detection response task (TDRT) than interactions with no delay. These findings together suggest a longer system delay may result in a trade-off between visual demand and cognitive demand. Cognitive demand may go down when the driver has adequate time to task-switch to the primary driving task; at the same time visual demand will go up, as indicated by more off-road glances.

Another study using a Wizard of Oz methodology manipulated the accuracy of the voice system, however they did not manipulate system delay (*7*). Using physiological, behavioral, and subjective measures, they found that voice systems with lower fidelity are associated with higher levels of cognitive workload.

Similar results were found in a usability evaluation of a Volkswagen engineered in-vehicle VCS (*8*). Analyzing video recordings of user interactions with the VCS, they determined that 28 out of 30 study participants had difficulty identifying the appropriate system-specific voice command, resulting in system recognition errors (*1, 8*). The authors concluded that voice interfaces ought to be designed to accept voice commands that are consistent with users' dialects. All these findings taken together, it follows that VCS that respond accurately to a wider range of typical voice commands, with fewer steps, and with shorter system delay will result in shorter time on task and lower levels of visual demand than systems without these characteristics. While the literature has identified that system delay, accuracy, and menu depth appear to be important drivers of workload, these factors have not all been addressed directly by a study on multiple production-level vehicles. Wizard of Oz methodologies have been consistently utilized for their ease of implementation, however they may lack ecological validity as they are fully constrained by the controlling researcher. By extending these lines of research to multiple production-level vehicles, we can get a better sense of the design components of VCS that contribute to driver distraction in vehicles.

The present study attempts to identify the design components of VCS that are important factors in predicting driver workload associated with use of VCS. Specifically, we collected data on the number of steps (menu depth) associated with common VCS tasks, the system delay between steps, and the accuracy of the voice system under ideal conditions. To gather accuracy without human participants, a pre-recorded voice command methodology was used. To the authors' knowledge, this is the first time such a method has been used. As such, we were interested in whether there were significant voice system accuracy differences between vehicles. The importance of VCS accuracy, menu depth, and system delay was examined by testing whether they were predictive of time on task and subjective workload data gathered as part of a separate validation study.

Methods

Materials

Voice Recordings. For this project, recordings of a set of 28 unique vehicle voice commands were created. The set of commands were generated from common commands accepted by VCS seen in past studies by Strayer et al. (*2, 9*). There are six different command structures for radio controls and eight different command structures for calling controls (Table 1). These different command structures are populated with two sets of frequency/band pairs ("1020 AM" and "99.5 FM") and two contact names ("Jessica Day" and "Aaron Jones"), for a total of 12 radio commands and 16 phone calling commands. Voice recordings were gathered from a 23-year-old female talker. The talker was a native English speaker determined to have a typical American-English accent and normal hearing by self-report.

The talker was recorded in a sound-treated booth using a Shure SM100 head-mounted microphone located at a 45° angle to the mouth at a distance of approximately 1 in. The microphone was connected to a Marantz digital audio recorder and sampled at a rate of 44.1 kHz. The talker was given a brief paragraph to read aloud to get accustomed to speaking in the recording booth before recording the commands. Two experimenters situated outside the booth monitored all productions of the commands. Any disfluent, erroneous, or otherwise unwanted productions observed by the listeners were noted. Recording level was also monitored during production to assure that no peak-clipping of the audio file occurred. After the talker finished the recording task, they were instructed to repeat commands that were marked by the listeners. These repetitions were similarly monitored for errors, and repetitions continued until no errors in the production were observed by either listener.

The recorded files were segmented into individual files using ELAN (*10*) and MATLAB (*11*). Within ELAN, boundaries were marked at the onset and offset of each sentence. ELAN output a file with the onset and offset locations (in seconds) of each target sentence within a given recording, which was used by a custom MATLAB script to excise the individual commands from the original long file recording into separate short files for playback. Each short file (i.e., each command) was scaled to the same average RMS amplitude, to assure that variation in recorded intensity between commands was normalized. A 50 ms buffer of silence was added to the beginning and end of each command after amplitude normalization to aid playback.

Vehicles. The accuracy and interface-level data were gathered from eight model-year 2017 vehicles equipped with VCS. Vehicles were acquired from multiple sources, including a car rental company and vehicle purchases. The vehicles assessed were: a Nissan Rogue equipped with Nissan Connect, a Tesla Model S equipped with software version 8.1, a Chrysler 300 equipped with UConnect, a Buick Enclave equipped with IntelliLink, a GMC Yukon equipped with IntelliLink, a Honda Ridgeline equipped with HondaLink version 4.2.2, a Kia Sportage equipped with UVO, and a Dodge Durango equipped with UConnect.

These vehicles were selected because they are representative of some of the most popular vehicle manufacturers in the United States. The models we evaluated account for 510,837 of the vehicles sold in the first three fiscal quarters of 2017. Further, the VCS in these vehicles are also used by many other models (e.g., many Hondas are equipped with similar versions of HondaLink). The current study was carried out to identify design components of VCS that lead to varying levels of workload, and not as a benchmark of the performance of these vehicles' VCS. As such, the vehicles are hereafter anonymized (in random order) as System-A, System-B, System-C, System-D, System-E, System-F, System-G, and System-H.

Audio Playback. Voice commands were played back via an M-Audio BX5 studio monitor with a frequency response of 56–22,000 Hz. The speaker was positioned 50 cm away from the center of the steering wheel, such that the monitor occupied the same area in space next to the headrest as the head of a driver of roughly average height. Recordings were played through the monitor at an average volume of 60 dB as measured 3 ft from the source.

Video Recording Device. A Garmin VIRBXE camera was used to make recordings for the system delay calculations. Video was filmed at 30 frames/s from a flexible mount attached to the vehicle headrest, aimed to capture the steering wheel and the touchscreen on the center stack.

Independent Variables and Design. The independent variables were system and task type. The dependent variables gathered were recognition accuracy, number of steps, and lag time. There were two task types (calling, audio) and eight vehicles (listed above). There were 28 unique VCS commands, 12 for radio tasks and for calling tasks. Each command was repeated three times, for a total of 84 commands per vehicle.

Dependent Measures

Voice System Accuracy. Hits were defined as interactions where the system completed the desired command or where the system repeated the user's command correctly. For example, if the input was "Call Aaron Jones" and the system said, "Calling Aaron Jones. Is this correct?" the trial was recorded as a hit even though the system had not yet called Aaron Jones. Misses were defined as interactions where the system either did not understand the voice input and thus did not carry out the command (i.e., "I did not understand. Please repeat."), or when the system carried out a command incorrectly (e.g., changing the radio station to 98.5 FM instead of 99.5 FM).

The data collector played commands in counterbalanced blocks containing both task types. The order of commands within each block was randomized via a random number generator. The data collector played one test voice command to the VR system at the start of each new task to ensure the data collector understood the cadence necessary for playback. The procedure for each trial was as follows: (1) begin on the home screen of the in-vehicle infotainment system, (2) press VCS activation button, (3) play voice command immediately after system begins listening, (4) end voice system interaction when action has been completed or VCS system commits an error. Vehicle engines and air conditioning systems were turned off during voice recordings to eliminate ambient noise from affecting VCS system accuracy.

Menu Depth. The menu depth variable is defined as the number of steps required to complete a task if the system understands the commands perfectly. This includes all button presses and voice commands. For example, if the default setting of the vehicle was to begin calling a contact immediately after the user says "Call Aaron Jones," that vehicle would be considered to have two steps—one button press to activate the voice system and one voice command.

Total System Delay. The system delay variable is defined as the sum of the windows of time between each step. The window of time between each step was quantified by taking the difference between the last frame of video containing the user voice command and the next frame containing the system's auditory response. In cases where a step in the voice command process was not demarcated by an auditory response from the system, visual feedback

Table 2. Menu Depth and System Delay

Voice system	Menu depth		Total system delay (ms)	
	Radio	Calling	Radio	Calling
System-A	2	2	2,508	2,966
System-B	2	2	2,733	2,666
System-C	2	2	2,858	3,258
System-D	3	3	3,841	3,891
System-E	2	2	6,725	7,441
System-F	2	2	4,491	4,491
System-G	NA	3	NA	3,925
System-H	NA	2	NA	8,875

Note: NA = not applicable, function was not accessible using the voice system.

on the touchscreen was used instead. The frames were then converted to milliseconds and the milliseconds per step was added together.

On-Road Validation Study

In the validation study, participants carried out common calling and radio VCS tasks while driving vehicles down a residential street using a similar protocol as that in a study by Strayer et al. (*9*). A total of 60 unique participants (mean age 25.3 years, 30 female, 30 male) drove the vehicles used in this study and all filled out Institutional Review Board-approved informed consent documents. Of these participants, 19 drove only one vehicle, four participants drove all eight vehicles, and the average participant drove 3.5 vehicles. No participants indicated that the tested vehicles were the same as their personal vehicle.

Participants were trained on how to use the VCS immediately before the study, and researchers gave them practice trials to ensure their sound understanding of the system. Participants then completed common voice tasks that were given to them as scenarios (e.g., "You want to listen to your favorite FM radio station") in order to prohibit participants from "parroting" the voice commands exactly as researchers told them. The voice commands that drivers utilized in the validation study overlapped with many of the voice commands tested in the present study. Researchers gathered time on task data and subjective workload data using the NASA Tax Load Index (NASA-TLX) and the Driving Activity Load Index (DALI) measurement scales.

The version of the NASA-TLX is a revision of the original, and consists of eight items on 21-point Likert scales designed to probe mental demand, physical demand, perceived pace of task, perceived success of task, effort, frustration, ease of use, and complexity of the system. The DALI used in the on-road study consists of seven items on a 6-point Likert scale designed to assess overall demand, visual demand, auditory demand, tactile demand, temporal demand, and how much the secondary voice tasks interfered with the driving task. Time on task was defined as the time between when the researcher tells the participant to begin the trial and when the participant completes the trial.

Data Analysis

Voice recordings were processed using ELAN (*10*) and MATLAB software (*11*). All other data were analyzed using R open source software (*12*).

Results

Menu depth and system delay data for the eight voice systems evaluated are presented below (Table 2). Most systems required only two steps for both tasks, however System-D notably required three steps per both tasks. The total amount of system delay for radio tasks varied between a low of 2.5 s to a high of 6.7 s, with an average of 3.9 s. For calling tasks, the shortest total delay was 2.7 s whereas the high was 8.9 s, with a mean of 4.7 s. A univariate ANOVA on system delay per vehicle revealed no significant differences between vehicles.

Accuracy data were analyzed to test whether the prerecorded voice command method is sensitive to differences between vehicles. Two systems were not equipped with functionality to change radio stations using the VCS, so only six vehicles are included in the analysis for radio tasks. Shapiro–Wilk normality tests indicated that the accuracy data did not violate the assumption of normality. An omnibus univariate ANOVA test with vehicle as the independent variable and accuracy as the dependent variable revealed significant differences for systems within the radio task, $F(5, 66) = 4.911, p < .001$. A post hoc Tukey test revealed significant differences between System-A and System-B ($p < .05$), System-E and System-B ($p < .001$), and System-E and System-C ($p < .001$). No significant differences were found for the same comparison of systems within the calling task, $F(7, 120) = 1.436, p = .197$. Accuracy data are represented below (Figure 1). Because the menu depth variable only yielded two different values, it was excluded from the rest of the analysis.

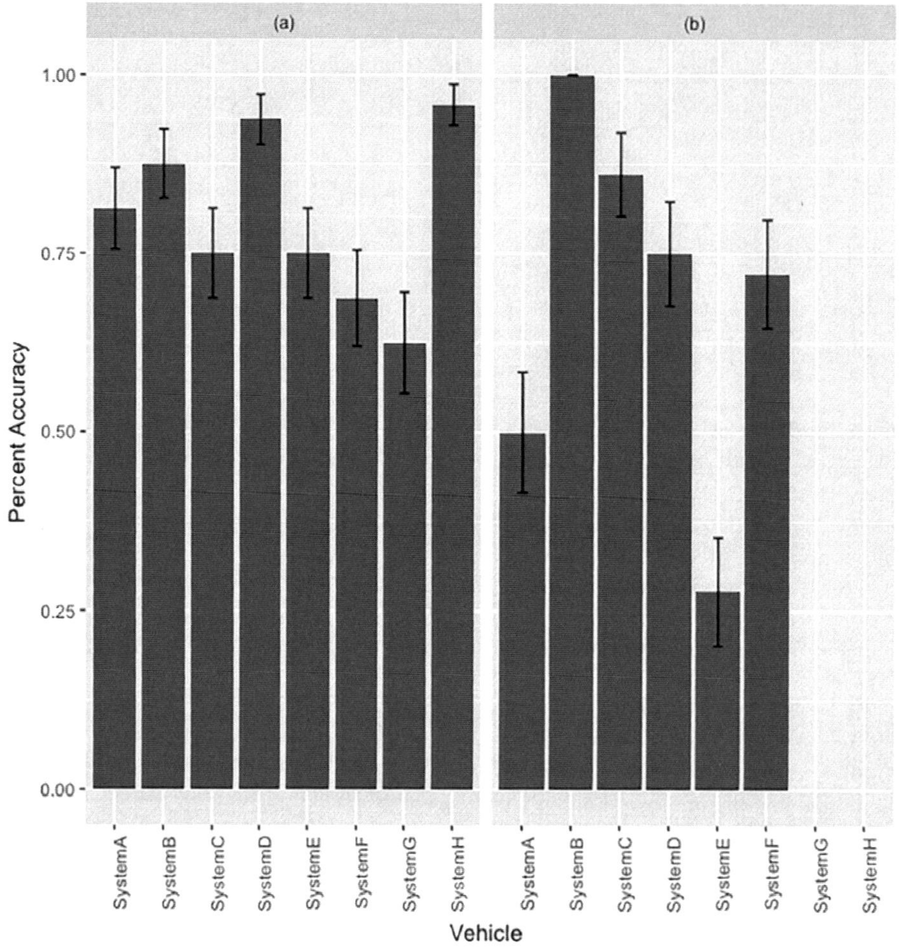

Figure 1. System accuracy between vehicles: (*a*) calling; (*b*) radio.

Differences in accuracy between these systems can be explained by the ability of some VCS to accept a wider range of commands than others. All command structures were repeated six times, three repetitions for each frequency/band pair or contact name. If a system correctly understood a command once, it was almost certain to understand it for all repetitions. For all voice command repetitions across all vehicles, 98% had a hit rate of either 0/6 or 6/6. In the radio condition, for example, System-E had such a low accuracy because it only responded correctly to two voice command structures tested ("Radio tune to <Frequency><Band>" and "Tune the radio to <Frequency><Band>").

The mean score per vehicle of each item of the DALI and the NASA-TLX was calculated in order to match the dimensions of the system delay and accuracy datasets for correlation analysis. Pearson correlations were conducted between design components measured in the current study and the subjective data collected in the on-road validation study. We examined items from the DALI questionnaire and found that system delay is significantly correlated with visual demand, $r(6) = .79$, $p < .05$, system delay is significantly correlated with stress, $r(6) = .82$, $p < .05$, and accuracy is significantly correlated with auditory demand, $r(6) = -.81$, $p < .05$ for calling voice commands. For radio voice commands, we found that system delay and accuracy are significantly correlated with auditory demand, $r(4) = .85$, $p < .05$ and $r(4) = -.9$, $p < .05$, respectively. Pearson correlations were also conducted for items from the NASA-TLX scale. We also found significant correlations between system delay and effort for calling voice commands, $r(6) = .81$, $p < .05$.

Next, linear regression was used to examine whether system delay and voice system accuracy together are more predictive of the subjective items identified above as significantly correlated. For the items from the DALI that are predictive of calling commands, both system delay, $b = 0.002$, $t(5) = 6.1$, $p < .01$ and accuracy $b = -2.19$, $t(5) = -3.02$, $p < .05$ were found to be predictive of stress. System delay and accuracy also explained a significant proportion of the variance in stress for the calling commands, $F(2, 5) = 19.59$, $R^2 = .84$, $p < .01$. System delay and accuracy together were not significant predictors of visual demand or auditory demand for calling commands. For the effort item from the NASA-TLX scale, system delay was a significant predictor, $b = .0006$, $t(5) = 4.47$, $p < .01$, however accuracy was not significant, $b = -5.54$, $t(5) = -1.94$, $p < .11$, for calling commands. Together, they explained a significant amount of variance in effort, $F(2, 5) = 10.36$, $R^2 = .73$, $p < .05$. The above results demonstrate that system delay and accuracy are important factors in various measures of subjective measures of driver workload for calling commands.

For radio commands, system delay and accuracy were not significant predictors of mental demand for radio conditions, nor were they significant predictors for auditory demand. Potential explanations for this will be discussed below.

A hierarchical linear regression method was used to examine whether accuracy or system delay were significant predictors of time on task for either task type. For calling commands, system delay, $b = 3.097$, $t(5) = 4.31$, $p < .01$ and voice system accuracy, $b = -0.0005$, $t(5) = -4.287$, $p < .01$ were found to be significant predictors of time on task. Further, they explained a significant amount of variance in time on task, $F(2, 5) = 14.77$, $R^2 = .8$, $p < .001$. For radio commands, system delay and voice system accuracy were not significant predictors of time on task.

Discussion

Past work has suggested that system delay, accuracy, and menu depth play an important role in workload associated with using VCS (*2–4*). Voice systems with lower accuracy have been shown to be problematic for even for experienced users (*8*) and to increase workload associated with their use (*5, 6*). A study by Jenness et al. manipulated system delay, and found that a longer system delay is associated with longer eyes-off road time, and, interestingly, reduced cognitive load (*4*). However, it is currently unclear if these findings generalize, as none of these studies have attempted to examine these variables in multiple production-level vehicles.

The results above demonstrate that voice system accuracy and the total amount of system delay in a voice interaction are associated with both subjective measures of workload as well as time on task. In particular, longer delay times were found to be associated with increases in visual demand, stress, and auditory demand. Increases in the accuracy of voice systems is associated with a decrease in auditory demand in both calling and radio voice commands. These relationships are consistent with research that found that decreases in accuracy are associated with increases in behavioral and subjective measures of workload, respectively (*1, 3*).

We also found that increases in system delay are associated with an increase in self-reported visual demand, auditory demand, and stress. These findings push back on research by Jenness et al. (*3*) by showing evidence that system delay can raise workload. However, the study by Jenness et al. dictated an 8-s delay between voice commands whereas most of the vehicles in the present study were under 8 s. In fact, the total system delay (sum of the delay between all steps) for all but one of the vehicles tested was under 8 s. The delay between steps in all vehicles tested was well below 8 s. The increases in workload found in this study may not therefore be comparable to the decreases found by Jenness et al. because the delay varies. Furthermore, differences in workload could be accounted for by the different measures used: the present study utilized subjective measures of workload and Jenness et al. measured workload using a TDRT.

The above findings show that system delay is a predictor of time on task. Shutko and Tijerina (*13*) argue that time on task represents the length of time during which a driver may be exposed to an unexpected hazard. Regardless of the demand of the primary task, attentional resources are shared between the primary task and the secondary task and the risk of a hazard occurring during the secondary task compounds over time. Designing systems to have minimal or no delay can limit time on task thereby reducing the risk of adverse events occurring while attention is divided. Burns et al. (*14*) suggest using time on task as a tool to assess where more careful design is needed. In the present study, although menu depth was not analyzed for association with time on task, it follows that voice dialogues with more steps will have more delay associated with them over the course of the interaction. Reducing the number of interaction steps then, also has a reasonable connection to reducing overall time on task.

Interestingly, the set of radio voice commands used to measure accuracy was demonstrated to pick up significant variability between systems. However, system delay and accuracy together were not predictive of subjective workload measures or time on task associated with radio interactions. One potential reason for this is that, given the small sample of systems with radio capability, there was not enough power to detect significant predictors. The lack of significant results could also indicate that delay does not play a strong role in time on task and for radio conditions in the vehicles tested, as their delay tended to be less than in the calling condition.

Results verified that the set of radio voice commands tested is sensitive to differences between voice systems of different vehicles. One reason the calling voice commands may not have varied significantly is the relative lack of different voice command structures utilized. Most of the calling commands were of the form "Call <contact name>" which most vehicle VCS are likely to be able to detect. For the future, more varied and conversational commands, such as "Contact <contact name>", "Ring <contact name>", or "Let's call <contact name>" may be considered for testing, particularly as VCS begin to adopt a more personal assistant-like logic.

The findings of the present study extend past research in two ways. First, results demonstrate that voice system accuracy and system delay are associated with driver workload in multiple production-level vehicle VCS. Past research, has utilized Wizard of Oz methodologies (*1, 3*), but the present study shows that their findings generalize to production-level VCS.

Second, the methodology used has many practical benefits as well. While it is important to require human participants to measure workload associated with IVIS interactions, such studies are costly and time intensive. They require a large number of paid human participants, the means to acquire vehicles for an extended period of time, and a formidable research staff to maintain the vehicles and collect data. This works for well-funded assessment groups or for smaller scale studies, but it does not scale up easily to the en masse testing and comparison of the full market of production-level vehicles. Given the renewed interest in driver distraction following a number of years of traffic fatality increases (*15*), it is likely the demand for large-scale testing of potentially distracting vehicle systems will only increase.

One of the largest limitations of the proposed methodology is the fact that the current voice commands cannot be used to test VCS that require commands with different command dialogues. For example, systems that require the individuals to first say "Call" and "Aaron Jones" as separate commands are not testable using the current set of commands. Future work may be done to add to the current body of recorded commands. Follow-up studies should also include commands from multiple talkers to better capture variability in drivers' voices that arise from gender, accents, or other factors that can influence vocal inputs to VCS systems.

Future work should be undertaken to compare the accuracy measured by the proposed set of commands with the accuracy of drivers interacting with VCS in a naturalistic setting in order to further validate the methodology. Additionally, more vehicles should be tested to ensure the widespread generalizability of the method to other vehicle systems.

Acknowledgments

This research was generously supported by a grant from AAA Foundation for Traffic Safety.

Author Contributions

The authors confirm contribution to the paper as follows: study conception and design: D.G., F.B., D.L.S., J.M.C.; data collection: D.G., S.D.M.; analysis and interpretation of results: D.G., F.B.; draft manuscript preparation: D.G., F.B., S.D.M. All authors reviewed the results and approved the final version of the manuscript.

References

1. Burnett, G. E., and A. M. Parkes. The Benefits of "Pre-Information" in Route Guidance Systems Design for Vehicles. In *Contemporary Ergonomics* (E. J. Lovesey, ed.), Taylor & Francis, London, 1993, pp. 397–402.
2. Strayer, D. L., J. M. Cooper, J. Turrill, J. R. Coleman, and R. J. Hopman. *Measuring Cognitive Distraction in the Automobile III: A Comparison of Ten 2015 In-Vehicle Information Systems*. AAA Foundation for Traffic Safety, Washington, D.C., 2015.
3. Reimer, B., and B. Mehler. *The Effects of a Production Level "Voice-Command" Interface on Driver Behavior: Summary Findings on Reported Workload, Physiology, Visual Attention, and Driving Performance*. White Paper 2013-18A. MIT AgeLab, Cambridge, MA, 2013.
4. Jenness, J. W., L. N. Boyle, J. D. Lee, C-C. Chang, V. Venkatraman, M. Gibson, K. E. Riegler, and D. Kellman. *In-Vehicle Voice Control Interface Performance Evaluation*. Final Report No. DOT HS 812 314. National Highway Traffic Safety Administration, Washington, D.C., 2016.
5. Cooper, J. M., H. Ingebretsen, and D. L. Strayer. *Mental Workload of Common Voice-Based Vehicle Interactions across Six Different Vehicle Systems*. AAA Foundation for Traffic Safety, Washington, D.C., 2014.
6. ISO. *Road Vehicles – Transport Information and Control Systems – Detection Response Task (DRT) for Assessing Selective Attention in Driving*. ISO TC 22 SC 13 N17488 (Working Draft). Under development by Working Group 8 of ISO TC22, SC 13, 2012.
7. Strayer, D. L., J. Turrill, J. R. Coleman, E. V. Ortiz, and J. M. Cooper. *Measuring Cognitive Distraction in the Automobile II: Assessing In-Vehicle Voice-Based Interactive Technologies*. AAA Foundation for Traffic Safety, Washington, D.C., 2014.
8. Chang, J. C., A. Lien, B. Lathrop, and H. Hees. Usability Evaluation of a Volkswagen Group In-Vehicle Speech System. *Proc., 1st International Conference on Automotive User Interfaces and Interactive Vehicular Applications – Automotive UI 09*, September 21–22, ACM, Essen, Germany, 2009, pp. 137–144. doi:10.1145/1620509.1620535
9. Strayer, D. L., J. M. Cooper, R. M. Goethe, M. M. McCarty, D. Getty, and F. Biondi. *Visual and Cognitive Demands of Using In-Vehicle Infotainment Systems*. AAA Foundation for Traffic Safety, Washington, D.C., 2017.
10. Sloetjes, H., and P. Wittenburg. Annotation by Category – ELAN and ISO DCR. *Proc., 6th International Conference on Language Resources and Evaluation*. Max Planck Institute for Psycholinguistics, Nijmegen, The Netherlands, The Language Archive, 2008.
11. The Mathworks, Inc. *MATLAB and Simulink R2017a*. Natick, Massachussets, The Mathworks Inc., 2017.
12. R Core Team. *R: A Language and Environment for Statistical Computing*. R Foundation for Statistical Computing, Vienna, Austria, 2016.
13. Shutko, J., and L. Tijerina (2006). Eye Glance Behavior and Lane Exceedences during Driver Distraction. *Presented at 3rd International Driver Metrics Workshop*, Ottawa, October, 2006. http://ppc.uiowa.edu/drivermetricsworkshop/.

14.Burns, P., J. Harbluk, J. P. Foley, and L. Angell. The Importance of Task Duration and Related Measures in Assessing the Distraction Potential of In-Vehicle Tasks. *Proc., 2nd International Conference on Automotive User Interfaces and Interactive Vehicular Applications*, ACM, 2010, pp. 12–19.

15National Safety Council. *NSC Motor Vehicle Fatality Estimates*. National Safety Council Statistics Department, Itasca, IL, 2017.

The Standing Committee on User Information Systems (AND20) peer-reviewed this paper (18-03749).

Article

TRR
JOURNAL OF THE TRANSPORTATION RESEARCH BOARD

Transportation Research Record
2018, Vol. 2672(37) 101–112

DOI: 10.1177/0361198118795004
journals.sagepub.com/home/trr
SAGE

In-Vehicle Passing Collision Warning System for Two-Lane Highways Considering Driver Characteristics

Udai Hassein[1], Maksym Diachuk[1], and Said M. Easa[1]

Abstract

Passing collisions are one of the most dangerous traffic safety problems. These head-on collisions occur when the driver of the passing vehicle is distracted or does not assess the situation appropriately. The purpose of this study is to develop a passing collision warning system (PCWS) for drivers on two-lane highways to prevent passing collisions and improve road safety. This paper presents a framework and algorithm design for a PCWS that ensures that drivers have an adequate sight distance for a safe passing maneuver. The system uses an available radar sensor to detect the closest opposing vehicle traveling in the left lane and calculates its position, speed, and acceleration rate to estimate the time to collision and compare it with the time required for the passing vehicle to clear the lane. Realistic initial time and passing time models were established using actual experimental field data collected using a global positioning system (GPS) data logger device that was installed in the passing, impeding, and opposing vehicles and used to record the position and speed of different passing vehicles at 1-s intervals. The MATLAB simulation was developed and used to replicate real-life passing maneuvers and was also used to create the algorithm for the proposed warning system. The passing maneuver parameters were selected from probability distribution curves based on field data. The simulation model determines the relative distance and speed of the opposing vehicle at four different time intervals. The different factors that impact system accuracy were also examined.

Adequate passing sight distance (PSD) is necessary to ensure the safety of passing (overtaking) drivers on two-lane highways. Passing collisions occur on two-lane highways when a fast-moving vehicle attempts to pass a slow-moving vehicle by traveling in the left lane. In Canada, passing (head-on) collisions accounted for approximately 30% of all collisions reported in 2010 (*1*). In the U.S.A. passing collisions accounted for approximately 20% of all collisions in 2004 (*2*). Safe driving requires drivers to be alert, constantly scanning their environment and properly responding to the movements of opposing vehicles. Driving errors contribute to collisions that cause the loss of human lives as well as costly property damage. Numerous driver performance factors can contribute to passing collisions. These factors include incorrect driver reaction time, following too closely, and driver inattention (*3*). These factors could be eliminated with the aid of a passing collision warning system (PCWS), improving roadway safety.

Vehicle warning systems can be categorized as kinematics-based or perceptual-based. Kinematics-based warning systems make use of the principal laws of motion and then assume driver reaction times to activate the warnings. A kinematics-based PCWS generates warnings when the passing vehicle is within the specified minimum passing distance. Perceptual-based warning systems activate the warnings using estimated driver reaction times along with time-to-collision (TTC) tolerance (*4*). The warning algorithm is an important component of these systems. A poorly designed algorithm may initiate the warning too early or too late, affecting the reliability of the system (*5*, *6*). A PCWS can help drivers avoid passing collisions by reducing the chance of human error.

The purpose of this study was to develop a prototype of a PCWS for drivers on two-lane straight highways to prevent passing collisions and improve road safety. Experimental field data were collected using a global positioning system (GPS) data logger device that was installed in the passing, impeding, and opposing vehicles

[1]Department of Civil Engineering, Ryerson University, Toronto, ON, Canada

Corresponding Author:
Address correspondence to Udai Hassein: uhassein@gmail.com

and used to record the positions and speeds of different passing vehicles at 1-s intervals. Regression models were developed to determine the necessary initial time and passing time for passing vehicles using data collected from the field. The proposed system was tested using the MATLAB (MATrix LABoratory) Simulink model (*7*) which was developed to replicate real-life passing maneuvers. The passing maneuver parameters were generated from frequency distributions established from the field data. The next sections present the literature review, the proposed PCWS, data collection and parameter estimation, Simulink model development, and application examples, followed by concluding remarks.

Literature Review

Advanced driver assistance systems (ADAS) have been developed for various purposes, including collision warning systems (*8–11*) and control systems for lane departure, lane change, and merge controls (*12–14*). An example of a collision warning system for intersections is the one developed by Calspan Corporation (*15*). This system makes use of an in-vehicle positioning system along with two 77 GHz (later replaced with 24 GHz) radar sensors with rotating antennae to identify the speed of approaching vehicles (using the Doppler effect) and warn the driver of any potential conflict.

Several collision warning systems have been developed for traffic on two-lane highways. Isermann et al. (*16*) proposed a warning system for passing ("overtaking") maneuvers to inform drivers when an overtaking maneuver is too dangerous due to opposing traffic. In that project, the passing vehicle could detect the distances and velocities of the opposing and impeding vehicles using RADAR and camera sensors. The data obtained from the two sensors underwent a sensor fusion with Kalman filters, and the overtaking maneuver was predicted using the data from all three vehicles. If the overtaking maneuver became too dangerous, the system would give a warning to the driver. In case the driver failed to respond, automatic braking would be initiated to allow the passing vehicle to return to its former position behind the impeding vehicle. A BMW 540i was used in the experiment, equipped with CSF200 camera system and ARS300 long-range RADAR (operating at 77 GHz). However, in that study the safety margins and overtaking model were not calibrated.

In another study, a mathematical model along with an adaptive controller for overtaking was presented by Petrov and Nashashibi (*17*). The authors applied standard robotic nomenclature to establish rotational and translational velocities and displacements. A general kinematic model for the vehicles throughout the overtaking maneuver was also proposed. The approach used in that study consisted of recognizing virtual reference points that represented the desired theoretical known distances from the impeding vehicle as well as one point attached to the passing vehicle.

The present study considers an autonomous overtaking maneuver independent of any information from the road infrastructure or the impeding vehicle. The only information used by the passing vehicle is from sensors that give its orientation and current position relative to the opposing and impeding vehicles. Based on this information, the unknown velocity of the opposing vehicle is estimated and the required and available passing times are compared, providing the basis for the warning system.

Proposed Collision Warning System

System Technology

The in-vehicle PCWS uses two radar sensors with rotating antennae, which are utilized to identify the speed of opposing vehicles (using the Doppler effect) and warn the driver of any potential conflicts. The radar sensors use a range of no less than 1000 m to detect all the vehicles traveling in the opposite direction which could potentially collide with the passing vehicle. The proposed system uses a radar sensor, such as the Symeo LPR-1D24 (*18*). A pair of the radar sensors installed on the right and left sides of the front bumper of the passing vehicle within a distance of 0.618 m from the center. The radar sensors can provide information such as the speed, location, and distance of the impeding and opposing vehicles. The radar sensors send a signal at every time interval, Δt, to scan the left lane for any opposing traffic. In this study, the time interval is 0.01 s based on the 100 Hz radar sensors used. The closest object detected is recorded at a distance d and azimuth angle θ with polar coordinates used with the origin point to correspond with the detector's location.

PCWS Algorithm

The system algorithm provides an estimation of the distance and time needed for the passing vehicle to accelerate and complete the passing maneuver. The algorithm also considers the time needed for the driver of the passing vehicle to detect the message displayed by the PCWS and react to it. The algorithm involves the following procedures (see Figure 1):

1. The driver's characteristics are entered in the system (e.g., age, gender, experience).
2. The system takes four successive measurements of both the location and speed of the nearest vehicle detected in the left lane and the time interval

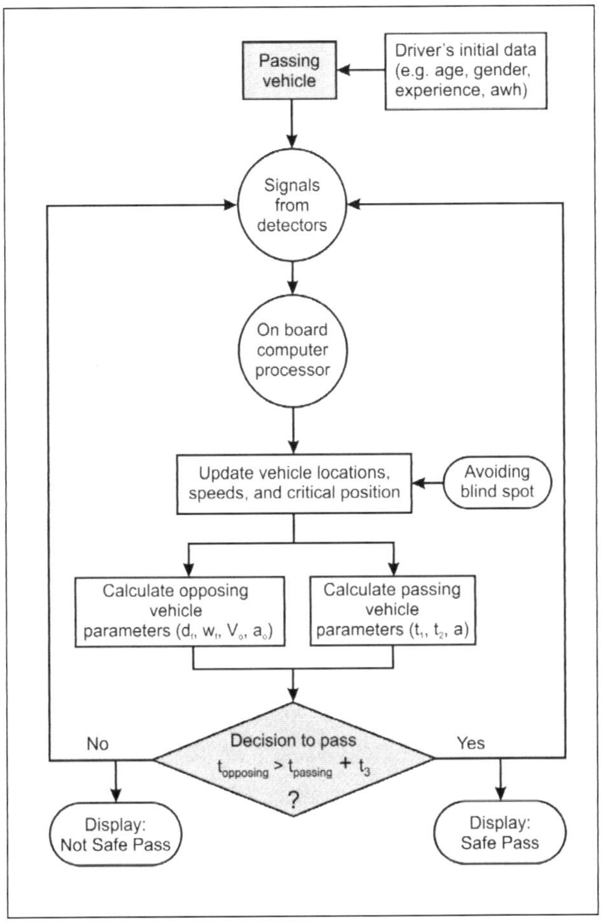

Figure 1. Logic of the PCWS.

between two measurements equal to the inverse of the detector's frequency.

3. Using these measurements, the system estimates the relative distance and speed of the vehicle that has been detected.
4. The system then determines the time needed for the opposing vehicle to reach the safe point, $t_{opposing}$.
5. This information is then used to determine the time ($t_{passing}$) needed for the passing vehicle to complete the maneuver, which is equal to the sum of t_1 (the perception–reaction time and initial acceleration of the passing driver) and t_2 (passing time) as shown in Figure 2*a*.
6. The system compares times $t_{opposing}$ (the sum of $t_{passing}$ and t_3) where t_3 represents a safety margin of 2 s, and decides based on one of the following criteria:

   (a) If $t_{opposing}$ is greater than $t_{passing}$, the "Safe to Pass" message is turned on and the driver may begin the passing maneuver.

   (b) If $t_{opposing}$ is smaller than $t_{passing}$, the "Not Safe to Pass" message stays on and the system replicates the algorithm to find a safe time gap.

The PCWS uses trigonometric functions based on detection sensor readings to calculate the available PSD. Based on the initial part of the model, the range and azimuth angles for the opposing vehicles are calculated at four 0.01-s intervals (T; $T + t$; $T + 2t$; $T + 3t$) where T equals the start time and t equals the particular time interval of the detection sensors, as shown in Figure 2*b*. The range and azimuth angle are included in the PCWS system and are calculated from the initial part of the model.

Required Passing Vehicle Time

The total time needed for the driver of a passing vehicle to complete the maneuver, $t_{passing}$, is the sum of the following:

(a) The driver's perception–reaction time and initial acceleration (t_1), which is the time needed by the driver to notice the "Safe to Pass" signal and then take the necessary action (to activate throttle).

(b) The vehicle's travel time (t_2), which is the time needed to accelerate the vehicle and clear the path for the oncoming opposing vehicle. The time needed to cross the offset distance involving the passing vehicle and the opposing vehicle is considered, along with the length of the passing vehicle itself.

The total time required for the passing vehicle to complete the pass, $t_{passing}$, is calculated using the following formula (see Figure 2*a*):

$$t_{passing} = t_1 + t_2 \tag{1}$$

Available Opposing Vehicle Time

The calculation of the distance and angle at each time interval is illustrated in Figure 2*b*. The radar sensor detectors produce the detection beams at time T to scan the left lane of the two-lane highway. If nothing is detected, a "Safe to Pass" message is displayed. Otherwise, the closest detected object is measured at distance d_1 and azimuth angle θ_1 where the polar coordinates used with the origin point coexist with the detector's location. At time $T + \Delta t$, the detector produces another detection beam. The Δt represents the detector's time interval (at 100 Hz frequency, the inverse of the frequency = 0.01 s). The new location of vehicle

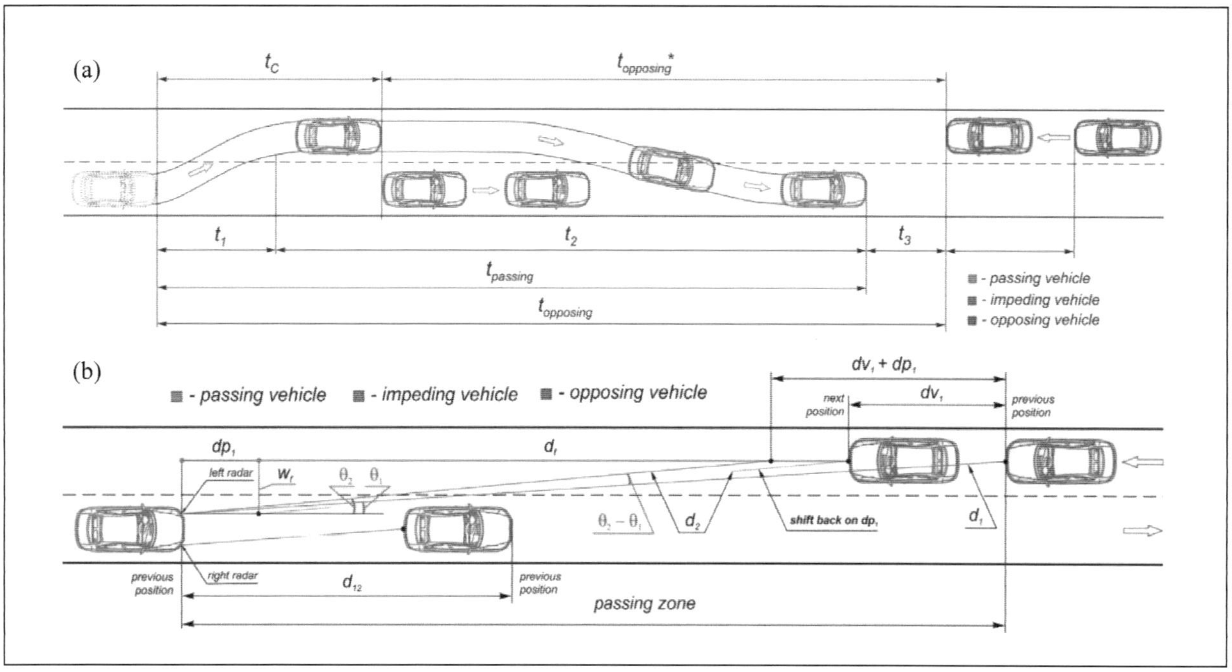

Figure 2. Geometry of passing maneuver: (*a*) definition of passing maneuver times; (*b*) calculation of distance and angle at each time interval.

A is recorded at distance d_2 and azimuth angle θ_2. Similarly obtained are: d_3, θ_3, and d_4, θ_4. Next step requires a shift on one sample time forward, keeping in memory the four last measured values. The distance that is crossed by the vehicle after four time intervals can be calculated using the following equation:

$$dv_i = \sqrt{d_i^2 + d_{i+1}^2 - 2 \cdot d_i \cdot d_{i+1} \cos(\theta_i - \theta_{i+1})} - dp_i, \quad i = 1, 2 \ldots n \tag{2}$$

The speed of the opposing vehicle, $V_{T+3\Delta t}$, is numerically calculated using the four-point formula of first derivatives based on Berezin et al. (*19*):

$$V_3 = \frac{-2 \cdot Dv_0 + 9 \cdot Dv_1 - 18 \cdot Dv_2 + 11 \cdot Dv_3}{6 \cdot \Delta t} + \frac{\Delta t^3}{4} \cdot f^{(IV)}(\xi) \tag{3}$$

where Dv_0, Dv_1, Dv_2, Dv_3 = function values in specified nodes *0*, *1*, *2*, *3*, respectively, Δt = constant step, $f^{(IV)}(\xi)$ = fifth derivative and ξ = time. The function values are given by:

$$Dv_0 = dv_T \tag{4}$$

$$Dv_1 = dv_T + dv_1 \tag{5}$$

$$Dv_2 = dv_T + dv_1 + dv_2 \tag{6}$$

$$Dv_3 = dv_T + dv_1 + dv_2 + dv_3 \tag{7}$$

where dv_T = initial value in first contact with radar beam. For small Δt the last team of Equation 4 is negligible. Subsuming Dv_0, Dv_1, Dv_2, Dv_3 for Equations 5 to 8 into Equation 4, then:

$$V_3 = V_{opp} = (1/3 \cdot dv_1 - 7/6 \cdot dv_2 + 11/6 \cdot dv_3)/\Delta t \tag{8}$$

The side offset between the opposing vehicle and the passing vehicle, w_f, is obtained using:

$$w_f = s_l + d_{i+1} \cdot \sin(\theta_{i+1}) \tag{9}$$

The offset may be used as a gauge to ascertain whether the opposing vehicle is traveling in the left lane. This is accomplished by comparing the calculated value with the sum of the setback distance (between the detector and the edge of the left lane) and the width of the left lane, where s_l represents the location of the left radar sensor on the passing vehicle, which is equal to 0.618 (m). The distance between the opposing vehicle and the detector, d_f, is obtained using:

$$d_f = d_{i+1} \cdot \cos(\theta_{i+1}) \tag{10}$$

The time needed for the opposing vehicle to reach the end of its part of the distance to the critical point is calculated using the following formula:

$$t_{opposing} = d_f / (v_{opp} + v_p) \quad (11)$$

where d_f can be calculated from Equation 10; v_p represents the passing vehicle speed; v_{opp} represents the instantaneous velocity of an opposing vehicle, which is determined for each value of the time interval Δt, and v_{opp} can be calculated from the speed of an opposing vehicle (Equation 8). The blind spot is detected using the following procedure. First, the coordinates of the intersection of the point of two lines are determined. One line is defined by the segment joining a passing vehicle's left sensor point and an opposing vehicle's front bumper midpoint and the other line is defined as perpendicular to the longitudinal axis of an impeding vehicle and passing through the outermost point of its rear bumper. If the point is within the boundaries of the impeding vehicle's width (Figure 2*b*), it follows that the signal does not pass and two sensors measure the distance to a slow vehicle. Moreover, the rapprochement of passing and slow cars, when moving strictly in track, enhances the blind spot effect. The primary maneuver toward the separating dash line is thus provided by the simulation control algorithm. After that, the intersection point of ray and bumper goes out of impeding vehicle width and the passing car's left sensor can see an opposing vehicle. The algorithm then turns on the warning system.

Decision-Making Process

The last step in the algorithm (see Figure 1) compares the time needed for the opposing vehicle to reach a conflict point, $t_{opposing}$, with the time needed for the passing vehicle to complete the passing maneuver, $t_{passing}$ (see Figure 2*a*). The following condition is used for the initial decision:

$$t_{opposing} > t_{passing} + t_3 \quad (12)$$

where $t_{opposing}$ can be calculated from Equation 11, $t_{passing}$ can be calculated from Equation 1, and t_3 is the safety margin (s). If the above condition is met for the opposing vehicle detected, a "Safe to Pass" message is displayed for the driver. If not, a "Not Safe to Pass" message will be displayed until the above condition is met. The subsequent timing of the passing and opposing vehicles will be updated at each time interval during the passing maneuver (see Figure 2*a*). The algorithm tracks these times for subsequent decision making. The time needed by the passing vehicle to complete the passing maneuver can be calculated using the following formula:

$$t_{passing}^{*} = t_1 + t_2 - t_C \quad (13)$$

where $t_{passing}$* = the actual passing time during the passing maneuver (s), t_1 = initial time (s), t_2 is the passing time (s), and t_C is the current sensor's signal processing time (s). The t_C value is zero before the initial decision. During $t_{passing}$*, the data processing continues using the radar sensor and the measurement of the distance between the passing and opposing vehicles is provided. The system continuously monitors overtaking safety until the passing maneuver is complete (see Figure 2*a*). The following condition is used for subsequent decisions:

$$t_{opposing}^{*} > t_{passing}^{*} + t_3 \quad (14)$$

where $t_{opposing}$* = the opposing vehicle time during the passing maneuver that can be calculated from Equation 11, $t_{passing}$* can be calculated from Equation 13, and t_3 is the safety margin (s). If the above condition is met for the opposing vehicle detected, a "Safe to Pass" message is displayed for the driver. If not, a "Not Safe to Pass" message will be displayed until the above condition is met.

Data Collection and Parameter Estimation

Data Collection

The field data were collected using a total of 25 drivers who repeatedly drove on the two-lane highways. However, the situation involving the combination of passing, impeding, and opposing equipped vehicles occurred only 105 times. There were many instances when the passing driver did not follow an impeding, sensor-equipped vehicle. The information about driver characteristics was collected using a questionnaire completed by each participant before the study commenced (*20*).

A GPS data logger device installed on-board the passing, impeding, and opposing vehicles was used to record the position (latitude, longitude, and altitude) and speed of the vehicles at 1-s intervals. The GPS data logger used in this study (Holux RCV-3000) measures speed with a precision level of 0.1 m/s (0.36 km/h) (*21*). Since this system is normally applied to passing zones on two-lane highways with speed limits of 80 km/h, the relative error is expected to be insignificant. The analysis included a total of 105 passing maneuvers (acceleration profiles) obtained from 25 different drivers and vehicles. This data was then loaded onto a computer. The sample was randomly selected from each group of passing drivers and included 17 male and 8 female drivers of ages ranging between 20 to 63 years.

Table 1 shows the means and standard deviations of passing maneuver parameters based on the field study. The table shows the passing maneuver parameters for the field study (design speed of 80 km/h), where *Gender* = passing driver gender (0 for males and 1 for females), *Age* = passing driver age (yrs), *Exp* = passing driver experience (yrs), *Awh* = passing driver weekly driving hours (h), *m* = speed differential (km/h), Vp, Vi, and Vo

Table 1. Passing Maneuver Parameters of the Field Study (Design Speed 80 km/h)

Sample variable	Field data		
	Distribution	Mean	SD
Age (year)	Weibull	34	13
Exp (year)	Weibull	12	10
Awh (h)	Weibull	21.4	10.2
Vp (km/ h)	Normal	72.4	8.6
Vi (km/h)	Weibull	58.9	7
Vp–Vi, m (km/h)	Lognormal	13.4	5.9
Vp–Vo (km/h)	Normal	10.3	10.8
D (m)	Weibull	451.3	109.5
d_{12} (m)	GEV[a]	52.5	17.1
GT (s)	Lognormal	8.1	2.3
GD (m)	Lognormal	161	45.8
t_1 (s)	GEV	3.6	0.6
t_2 (s)	GEV	9.6	2.5
TTC (s)	Nakagami	6.1	4.7

Note: GEV = generalized extreme value.

= speeds of passing, impeding, and opposing vehicles (km/h), respectively, D = initial distance between passing and opposing vehicles (m); d_{12} = distance headway between passing and impeding vehicles (m), GT = passing gap time (s), GD = passing gap distance (m), t_1 = initial passing time (s), t_2 = passing time (s), and TTC = time to collision (s).

Passing Vehicle Parameter Estimation

Linear regression models for the initiation of passing time (t_1) and passing time (t_2) which are used to calculate the required passing time, $t_{passing}$, were developed using MATLAB program (*7*) based on the collected field data. The practical nature of the PSD problem was well suited to regression analysis which provided important tools such as analysis of variance, statistical significance, correlation analysis, and various ways of predictor selection (e.g., stepwise selection). Twenty-five drivers participated in this study and a total of 105 observations were collected. The estimated models for t_1 and t_2 were as follows:

$$t_n = \sum_i k_i \cdot f_i; n = 1, 2 \quad (15)$$

where Gender = gender of passing driver (0 for males and 1 for females), Age = age of passing driver (yrs), Exp = experience of passing driver (yrs), Awh = weekly driving hours recorded for the passing driver (h), and m = speed differential (m/s). The statistical characteristics of the model variables are presented in Table 2. The model considered the initial passing time parameter (t_1) and the passing time (t_2) to ensure acceptable performance of the passing maneuver.

Simulink Model Description

The Simulink model was implemented to replicate a real-life passing situation and to assess the system functions and performance. The present level of transport process precision modeling involves the integration of the vehicle's dynamics and controlled numerical simulation. The proposed model can be customized for any vehicle and takes into account the design, physical properties, and geometric properties of the vehicle. The simplest objective is shown in Figure 3*a*, which allows for the combination of several car models in one simulation scenario to reflect overtaking processes on two-lane highways.

The basis of a single vehicle model shown in Figure 3*a* is block 1, in which all of the necessary 2D-steerability components are implemented via the Simulink library. Blocks 2 and 3 are vectors that transmit initial positions (plane location and yaw rotation) and velocities (longitudinal, lateral, and yaw rate) according to degrees of freedom. Block 4 allows for the adjustment of the traction dynamics by following the desired course velocity, and the current steering wheel angle is provided through input port 5. The outputs of block 1 are as follows: in port 6, accelerations (longitudinal and lateral) in the vehicle's local coordinate system; in port 7, velocities (longitudinal and lateral); in port 8, all absolute displacements relative to the global coordinate system; in port 9, steered wheels turning angles. Subsystem 10 sets the initial conditions of motion and control laws, and subsystem 11 obtains graphical representations of solutions. The model makes it possible to start an animation script that visually reflects the realism of the simulation.

The system dynamics is represented in the vehicle's local coordinate system via longitudinal, lateral, and rotational around axis OZ dynamics. m represents the vehicle's mass, I represents the moment of inertia relative to the OZ axis, V_{Cx} and V_{Cy} represent the velocities along the axes of the local coordinate system associated with mass center C, δ_r represents the coefficient that takes into account the inertia of the transmission's rotating masses, ϕ represents the angle of the vehicle's rotation around the OZ axis, and $F^{(e)}{}_x$, $F^{(e)}{}_y$, M_z signify the generalized external force factors along the x- and y-axes and around z (respectively). The vehicle control is carried out through the change of power factors by magnitude and direction, that is, traction and the lateral reactions of the tires. The system dynamics is mathematically represented by:

Table 2. Estimated Regression Models for Passing Parameters

Variable	Coeff. name	Coeff. value	Std. error	t-value	p-value
Initial passing time (t_1): regression model					
Constant	k_1	9.54364	2.48294	3.84369	0.00234
Gender	k_2	0.06821	0.19221	0.35485	0.72886
Age (year)	k_3	−0.13411	0.11004	−1.21872	0.24637
EXP (year)	k_4	−0.14564	0.10693	−1.36196	0.19823
Awh (h)	k_5	−0.25548	0.12289	−2.07896	0.05973
m (m/s)	k_6	−1.57177	0.66245	−2.37267	0.03523
EXP·AWH	k_7	0.01511	0.00533	2.83713	0.01498
Age·m	k_8	0.04043	0.03162	1.27873	0.22517
EXP·m	k_9	0.03062	0.03061	1.00021	0.33695
AWH·m	k_{10}	0.05952	0.03593	1.65660	0.12349
EXP·AWH·m	k_{11}	−0.00382	0.00140	−2.72329	0.01849
Estimated Dispersion		0.077			
F-statistic versus constant model		3.2			
p-value		0.0303			
R^2		0.7271			
Model LogLikelihood		4.33636			
RMSE		0.2049			
Passing time (t_2): regression model					
Constant	k_1	22.56487	4.91494	4.59107	0.00051
Gender	k_2	−1.22921	0.83955	−1.46414	0.16692
Age (year)	k_3	−0.22458	0.28334	−0.79261	0.44223
Age^2	k_4	0.00459	0.00406	1.13077	0.27858
EXP (year)	k_5	−0.19853	0.08951	−2.21799	0.04498
Awh (h)	k_6	0.25114	0.26677	0.94141	0.36366
Awh^2	k_7	−0.00601	0.00602	−0.99922	0.33593
m (m/s)	k_8	−7.66867	4.08507	−1.87724	0.08310
m^2	k_9	1.64940	1.00521	1.64085	0.12479
m^3	k_{10}	−0.10899	0.07503	−1.45252	0.17006
Estimated Dispersion		0.847			
F-statistic versus constant model		3.81			
p-value		0.0146			
R^2		0.7252			
Model LogLikelihood		24.16990			
RMSE		0.7076			

$$\begin{cases} m\cdot\delta_r\cdot\left(\dfrac{dV_{Cx}}{dt}-\dfrac{d\phi}{dt}\cdot V_{Cy}\right)=\sum F_x^{(e)} \\ m\cdot\left(\dfrac{dV_{Cy}}{dt}+\dfrac{d\phi}{dt}\cdot V_{Cx}\right)=\sum F_y^{(e)} \\ I\cdot\dfrac{d^2\phi}{dt^2}=\sum M_Z \end{cases},\quad \begin{bmatrix}\dfrac{dV_{Cx}}{dt}\\ \dfrac{dV_{Cy}}{dt}\\ \dfrac{d\omega}{dt}\end{bmatrix}=\begin{bmatrix}\dfrac{1}{m\cdot\delta_r} & 0 & 0\\ 0 & \dfrac{1}{m} & 0\\ 0 & 0 & \frac{1}{I}\end{bmatrix}\cdot\begin{bmatrix}\sum F_x^{(e)}\\ \sum F_y^{(e)}\\ \sum M_C\end{bmatrix}+\underbrace{\frac{d\phi}{dt}}_{\omega}\cdot\begin{bmatrix}V_{Cy}\\ -V_{Cx}\\ 0\end{bmatrix} \tag{16}$$

The left side of Equation 16 represents the system of dynamics equations while the right side represents the derivative equations using vectors. This equation is easy to implement using Simulink tools, making it simple and obtainable without the necessity of any special commercial software. Computational procedures are performed almost instantly, making this a promising model for the simulation of traffic situations involving several vehicles. The main feature of the proposed model is its four-point (by all wheels) road contact instead of the bicycle models that are still widely used. The distribution of the vertical forces on the wheels, which is the most influential factor on a tire slip process, is provided. This distribution takes into account the effect of the inertia's pseudo-forces at the vehicle's mass center. The model takes into account non-linear tire-road contact and degree of adhesion with road surface. The distribution of steering angles is represented functionally as a result of the steering trapezoid's optimized design.

Figure 3*b* shows the Simulink structure of the vehicle dynamics solver, which coherently integrates the system of six ordinary differential equations that were reduced to Cauchy's form. This structure represents a stepwise solver for the differential equation system presented in

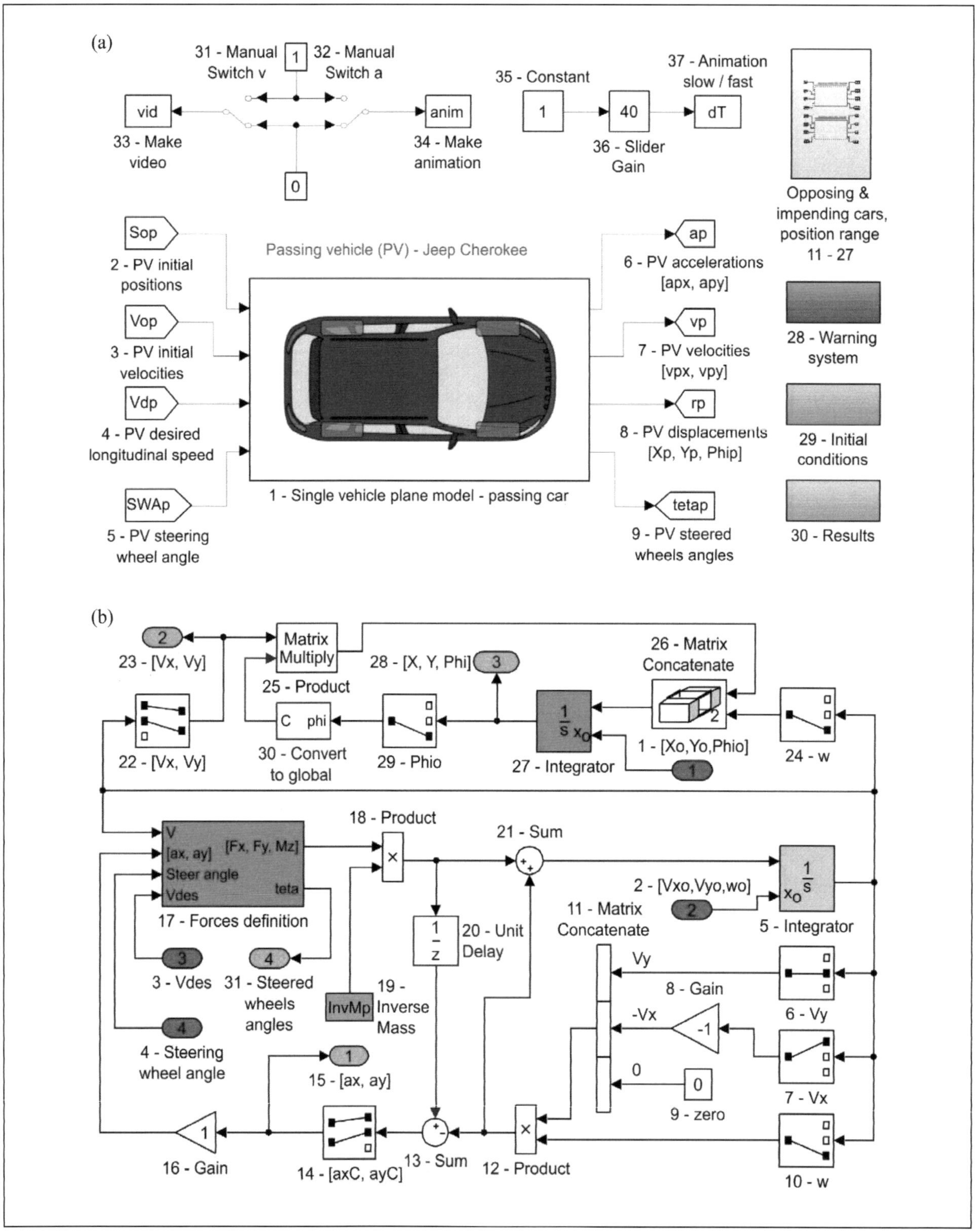

Figure 3. Simulink model for development of collision warning system: (*a*) general interface of three-vehicle model; (*b*) Simulink structure of vehicle dynamics solver.

Equation 16. The vehicle model calculates force factors (aligning moments and forces in the tires' contact spots) and kinematic parameter (acceleration, velocity, and displacement). Computational procedures are performed almost instantly, making this a promising model for the simulation of traffic situations involving several vehicles. Many virtual tests were conducted to demonstrate the quality and accuracy of the model's representation of the maneuvering process of vehicles.

The basic functional idea behind this warning system is the preventive forecasting of mutual locations of moving toward one another vehicles. Two radar sensors are used via detectors that are mounted symmetrically on the vehicle's front bumper. These detectors operate using the Doppler effect. This allows the warning system to record simultaneously the positions of two vehicles. This paper does not cover device simulation and signal processing due to the complicated nature of the general model. The calculation of the distance from the sensor to the opposing vehicle is obtained geometrically from the solutions of differential equations. The StateFlow (*7*) chart was used for computational procedures and the arrangement of the warning system. A poll of the sensors is provided at the time interval of Δt = 0.01 s and new vehicle route distances are defined after each assessment. Once the first three measurements data are stored, finite difference formulas are used to determine the kinematic characteristics of the opposing vehicles. The kinematics of a passing vehicle are determined by reading devices that allow for the use of these parameters directly from motion equation solutions.

Application Example

An application example of one loop is provided to illustrate the methodology used in this study. To show the uniqueness of the overtaking conditions for each new simulation before calculations, random variables of the model parameters were generated according to the distributions established from field data. The input data for the Simulink model were based on Table 1.

The passing maneuver parameters based on the field data were consistent with those obtained in previous research conducted by AASHTO (*22*). When analyzing the passing maneuver parameters using AT RISK software (*23*), the distribution was generated based on chi-square and Kolmogorov-Smirnov goodness-of-fit statistics. The application example assumes that the advisory system is using a detector with a frequency of 100 Hz and precision levels of 0.1 m and 0.1 for the reading distance and azimuth angle, respectively. The first detector was installed on the right side of the front bumper of the passing vehicle. The setback distance (d_{12}) was approximately 58.13 m from the front bumper of the passing vehicle to the front bumper of the impeding vehicle. The second detector was installed on the left side of the front bumper of the passing vehicle and the initial distance (D) was approximately 451.11 m from the front bumper of the passing vehicle to the front bumper of the opposing vehicle.

To illustrate the application of Simulink, the results of a single loop are presented. The initial detection measurements for the opposing vehicle were obtained at a distance (d_1) of 446.4 m, a sensor ray angle (θ_1) of 0.3991°, and a speed of approximately 16.52 m/s. The second detection measurements for the same vehicle were obtained at a distance (d_2) of 446.0 m, a sensor ray angle (θ_2) of 0.3973°, and a speed of 16.43 m/s. The third detection measurements for the same vehicle were obtained at a distance (d_3) of 445.7 m, a sensor ray angle (θ_3) of 0.3977°, and a speed of 16.41 m/s. The final distance between the opposing vehicle and the detector ($d_4 = d_f$), calculated using Equation 10, was 445.3 m, a sensor ray angle (θ_4) of 0.398°, a speed of 16.40 m/s, and the distance between the opposing vehicle and the conflict point was 150.40 m. Since the impeding vehicle was traveling at a near-constant speed, the time required for the passing vehicle to overtake the impeding vehicle ($t_{passing} = t_1 + t_2$), computed using Equation 1, was 8.29 s.

The time required for the opposing vehicle to reach the conflict point ($t_{opposing}$), computed using Equation 11, was 11.02 s. The difference between $t_{oppossing}$ and $t_{passing}$, the 2.73 s safety margin, was larger than the critical 2.0 s (95% confidence interval). This safety margin was significantly larger than the measurement error associated with the precision level (0.02 s). The warning message displayed to the passing driver can therefore be deactivated. The scenario is organized in conformity with the conditions corresponding to the statistics of the experimental data. In Figure 4*a*, the curves of vehicles' longitudinal velocities are shown. If there are sufficient initial distances between the vehicles and the positive signal from the warning system, the passing vehicle starts accelerating with a smooth approach to the dashed line. When the time exceeds t_1 = 3.3 s the passing vehicle moves onto the oncoming traffic lane and becomes visible to both the opposing and impeding vehicles through the rear-view mirror.

To facilitate mutual safety, the opposing and impeding vehicles slightly reduce their speeds, giving the passing vehicle more time to complete the maneuver. After reaching the peak speed of Vp = 25.12 m/s at 6.17 s, the passing vehicle arrives to the critical point, from which the maneuver must be completed. The reserve of the vehicles' speed difference is sufficient and the passing vehicle decreases speed with the simultaneous return to its lane. By the time $t_{passing}$ = 8.5 s (Figure 4*b*), the passing vehicle almost returns to the left lane, freeing up space for the

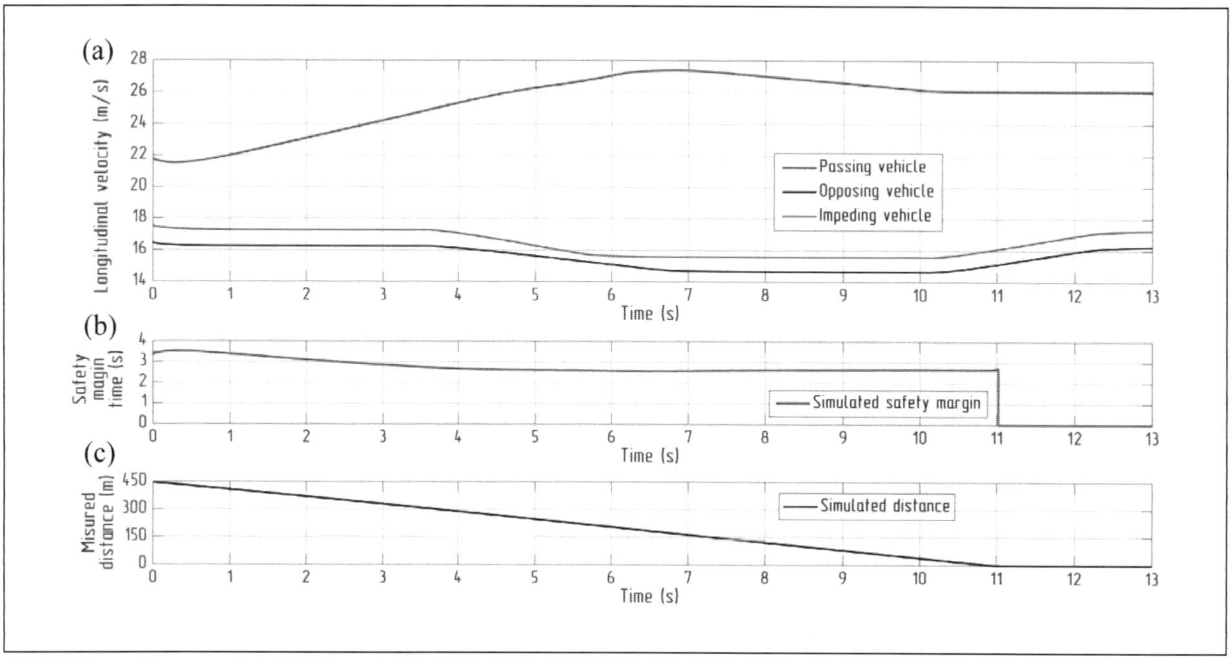

Figure 4. Example of numerical data processing.

opposing vehicle; from this moment the opposing and impeding vehicles gain speeds again. The distance between the approaching vehicles at this moment is about 100 m, but the danger disappears, therefore the safety time margin continues to remain at the level of 2.6 s. At the hypothetical collision point of approaching vehicles (df = 0, Figure 4*c*) the maneuver is safely completed and the time margin calculation is stopped.

Conclusions

This paper has presented a framework and algorithm for a collision warning system that is intended to help drivers perform passing maneuvers safely on two-lane highways. The system makes use of a detector (radar sensor) to measure the relative distance and speed of the closest opposing vehicle on a two-lane highway at four successive time intervals to determine its distance to the conflict point and the speed. A mathematical model was developed to estimate the driver's perception–reaction time. The initial time and passing time of the passing driver were obtained using data collected from the field in which a GPS data logger device recorded the position and speed of various vehicles at 1-s intervals. This study investigated driver behavior and its effect on the passing maneuver.

A Simulink scenario was developed to evaluate the performance of the proposed system for different precisions of the distance and the azimuth angle to ensure acceptable performance under different conditions. There are several advantages to traffic modeling using such integrated vehicle dynamics simulation. Various factors can be considered, including the technical parameters of the vehicle and a combination of various road situations, roadway curvature, type of friction between the tires and the road, control algorithm programming, and decision making, and instant animation of scenarios for visual verification. The results of the Simulink model showed that 95% of the drivers had a safety margin of 2 s or more. The inclusion of the safety margin in the system's algorithm increased the reliability of the system by decreasing the amount of mistimed warnings.

Similar to other collision warning systems, the driver's response to the interface with the PCWS is a very important consideration. Some preliminary guidelines relevant to the driver–vehicle interface have been presented in Campbell et al. (*24*). However, there is a need to investigate more fully the human factors associated with the effective implementation of the collision warning system interface. Another important aspect is related to integrating the presented system with other systems. The PCWS is one of the ADAS options that will use a common on-board embedded system. Ideally, the proposed system should be designed to have multiple functional features. Beyond detection and warning for passing maneuvers, the system could be designed to detect pedestrians and large animals in the vehicle's path and to provide assistance to drivers at intersections for safe left- and right-turn maneuvers. However, more research is needed to explore the challenges of this integration.

This stage of the study has only involved passenger cars and straight highway alignments. Future research should include trucks and other vehicle types, overtaking multiple vehicles, and horizontal and vertical alignments, allowing for the development of a universal PCWS algorithm. In addition, more research is required to identify various hardware components that meet the minimum requirements of the proposed system. Future studies may also include a variable in the regression models regarding the physical conditions of the driver.

Acknowledgments

The authors are grateful to six anonymous reviewers for their thorough and most helpful comments. The financial support of the Natural Sciences and Engineering Research Council of Canada (NSERC) is acknowledged. Since this study involved humans, it has been reviewed and approved by the university Research Ethics Board.

Author Contributions

The authors confirm contribution to the paper as follows: study conception and design: Udai Hassein, Maksym Diachuk, Said M. Easa; data collection: Udai Hassein; analysis and interpretation of results: Udai Hassein, Maksym Diachuk, Said M. Easa; draft manuscript preparation: Udai Hassein, Maksym Diachuk, Said M. Easa. All authors reviewed the results and approved the final version of the manuscript.

References

1. Road Safety in Canada. *Transport Canada*, 2010. www.tc.gc.ca/eng/roadsafety/tp-tp15145-1201.htm. Accessed July 28, 2016.
2. Persaud, B. N., R. A. Retting, and C. A. Lyon. Crash Reduction Following Installation of Centerline Rumble Strips on Rural Two-Lane Roads. *Accident Analysis and Prevention*, Vol. 36, No. 6, 2004, pp. 1073–1079.
3. Brown, T. L., J. D. Lee, and D. V. McGehee. Human Performance Models and Rear-End Collision Avoidance Algorithms. *Human Factors*, Vol. 43, No. 3, 2001, pp. 462–482.
4. Nilsson, L., H. Alm, and W. H. Janssen. *Collision Avoidance Systems: Effects of Different Levels of Task Allocation on Driver Behaviour*. Driver Project V1041 Haren, The Netherlands: Generic Intelligent Driver Support System, Deliverable GIDS / Man3, Traffic Research Center, 1991.
5. Dabbour, E. *Development of In-Vehicle Collision Warning System for Intersections*. PhD thesis. Ryerson University, Toronto, Canada, 2009.
6. Mehmood, A. *Development of Framework for In-Vehicle Rear-End Collision Warning System Considering Driver Characteristics*. PhD thesis. Ryerson University, Toronto, Canada, 2010.
7. Mathworks. *Stateflow Documentation - MathWorks*, 2016. www.mathworks.com/help/stateflow/. Accessed July 28, 2016.
8. Hamid, U. Z. A., H. Zamzuri, M. A. A. Rahman, and W. J. Yahya. A Safe-Distance Based Threat Assessment with Geometrical Based Steering Control for Vehicle Collision Avoidance. *Journal of Telecommunication, Electronic and Computer Engineering*, Vol. 8, No. 2, 2016, pp. 53–58.
9. Bevly, D., X. Cao, M. Gordon, G. Ozbilgin, D. Kari, B. Nelson, J. Woodruff, M. Barth, C. Murray, A. Kurt, and K. Redmill. Lane Change and Merge Maneuvers for Connected and Automated Vehicles: A Survey. *IEEE Transactions on Intelligent Vehicles*, Vol. 1, No. 1, 2016, pp. 105–120.
10. Raksincharoensak, P., T. Hasegawa, and M. Nagai. Motion Planning and Control of Autonomous Driving Intelligence System Based on Risk Potential Optimization Framework. *International Journal of Automotive Engineering*, Vol. 7, 2016, pp. 53–60.
11. Dabbour, E., and S. M. Easa. Proposed Collision Warning System for Right-Turning Vehicles at Two-Way Stop-Controlled Rural Intersections. *Transportation Research Part C*, Vol. 42, 2014, pp. 121–131.
12. Son, J., H. Yoo, S. Kim, and K. Sohn. Real-time Illumination Invariant Lane Detection for Lane Departure Warning System. *Expert Systems with Applications*, Vol. 42, No. 4, 2015, pp. 1816–1824.
13. Miman, M., O. O. Akırmak, and H. C. Korkmaz. Lane Departure System Design Using IR Camera for Night-Time Road Conditions. *Technology Education Management Journal*, Vol. 4, 2015, pp. 54–59.
14. Bevly, D., X. Cao, M. Gordon, G. Ozbilgin, D. Kari, B. Nelson, and K. Redmill. Lane Change and Merge Maneuvers for Connected and Automated Vehicles: A Survey. *IEEE Transactions on Intelligent Vehicles*, Vol. 1, No. 1, 2016, pp. 105–120.
15. Pierowicz, J., E. Jocoy, M. Lloyd, A. Bittner, and B. Pirson. *Intersection Collision Avoidance Using ITS Countermeasures. Task 9: Final Report*. Report No. DOT HS 809 171. National Highway Traffic Safety Administration, Washington, D.C., 2000.
16. Isermann, R., R. Mannale, and K. Schmitt. Collision-avoidance Systems PRORETA: Situation Analysis and Intervention Control. *Control Engineering Practice*, Vol. 20, 2012, pp. 1236–1246.
17. Petrov, P., and F. Nashashibi. Modeling and Nonlinear Adaptive Control for Autonomous Vehicle Overtaking. *IEEE Transactions on Intelligent Transportation Systems*, 2014, pp. 1643–1656.
18. SYMEO GmbH. http://www.symeo.com/en/applications/distance-measurement/index.html. Accessed December 30, 2016.
19. Berezin, I. S., and N. P. Zhidkov. *Computing Methods*. Translated from Russian, 2nd ed. Addison-Wesley, Wesley, Boston, MA, 1962.
20. Hassein, U., M. Diachuk, and S. M. Easa. Evaluation of Dynamic Passing Gap Acceptance on Two-Lane Highways Using Field Data. *Canadian Journal of Civil Engineering*,

Vol. 44, 2017, pp. 871–880. https://dx.doi.org/10.1139/cjce-2016-0572.
21. Holux Technology Inc. *Holux RCV-3000 GPS Logger User Manual*, 2014.
22. American Association of State Highway and Transportation Officials. *A Policy on Geometric Design for Highways and Streets*. American Association of State Highway and Transportation Officials, Washington, D.C., 2004.
23. Palisade Corporation. *@RISK User Guide*, 2016. https://www.palisade.com/downloads/manuals/6/EN/PTree6_EN.pdf.
24. Campbell, J. L., C. M. Richard, J. L. Brown, and M. McCallum. *Crash Warning System Interfaces: Human Factors Insights and Lessons Learned*. National Highway Traffic Safety Administration, U.S. Department of Transportation, Washington, D.C., 2007.

The Standing Committee on User Information Systems (AND20) peer-reviewed this paper (18-03903).

Article

TRR

Transportation Research Record
2018, Vol. 2672(37) 113–122

DOI: 10.1177/0361198118773569
journals.sagepub.com/home/trr
SAGE

The Challenge of Advanced Driver Assistance Systems Assessment: A Scale for the Assessment of the Human–Machine Interface of Advanced Driver Assistance Technology

Francesco N. Biondi[1], Douglas Getty[1], Madeleine M. McCarty[1], Rachel M. Goethe[2], Joel M. Cooper[1], and David L. Strayer[1]

Abstract

Despite driver assistance systems being engineered to enhance safety, recent studies show the potential for some of these systems and deficient human–machine interfaces to cause unintended consequences on safety. The NHTSA, the Alliance of Automotive Manufacturers, and the European Commission have all released best practices and human factors guidelines for the development and assessment of function-aspecific interfaces. However, given their broad scope, none of these documents provides a rating and benchmarking tool for assessing design aspects pertinent to a wide spectrum of assistance systems, ranging from rearview cameras to lane keeping assist systems. In this article we detail the development of a scale for assessing the human–machine interfaces of 10 different assistance systems. The scale contains 59 items, developed through multiple iterations, in which a total of 94 distinct assistance systems available on vehicles of different makes and models underwent evaluation. For each system included, we provide a description of its characteristics, a list of items for assessment, and relevant references. Widely accepted industry (ISO, SAE) standards, design guidelines, and assessment methodologies were considered for the development of this scale. The adoption of this scale required at least two evaluators to rate each system against specific assessment items using the following 4-point scale: No Concern (3 points), Minor Concerns (2 points), Serious Concerns (1 point), Not Applicable (0 points). Final ratings resulting from this evaluation will aid evaluators in the benchmarking process, and in determining what specific design aspects of the systems assessed merit further attention.

Advanced driver assistance systems (ADAS) are being introduced into the market at a fast-moving pace. The adoption of assistance systems and vehicle automation technology comes with the promise of a reduction in road accidents by up to 90%, according to some estimates (*1*). Nonetheless, recent studies suggest that, if poorly designed or difficult-to-use, human–machine interfaces (HMI) of systems engineered to enhance safety [e.g., lane keeping assist system (LKAS) or adaptive cruise control (ACC)] may instead counteract their potential benefits, and lead to potential unintended consequences on safety (*2*,3).

The study by Stanton and Young (*2*) investigated the effect of ACC on driver behavior. Drivers drove a simulated vehicle either with or without ACC engaged, and driving demand and levels of system information provided to drivers were manipulated. Driving demand was manipulated by means of increasing traffic density, and three different feedback conditions were created with drivers receiving increasing levels of information from the vehicle HMI. Feedback conditions included: auditory signal warnings, visual information presented on the vehicle's instrument cluster display, visual information presented on a head-up display, or a combination of the three. Results showed that the level of feedback presented to the driver had a direct impact on driver behavior and awareness of the road environment. In particular, with the level of feedback being at its highest (i.e., auditory warnings + visual information on instrument cluster display + visual information on head-up display), this was found to increase levels of driver visual workload and diminish their awareness of the road environment (see Vollrath et al. (*3*) for similar findings). Findings from the investigation by Biondi et al. (*4*), in which participants drove an on-road SAE level-2 vehicle, also found interaction with LKAS to cause an increase in driver stress, with some participants feeling like they were "zoning-out" toward the end of the drive (*5*).

Driver interaction with warning signals represents another primary safety issue for ADAS. Recent studies

[1]University of Utah, Center for the Prevention of Distracted Driving, Salt Lake City, UT
[2]AAA National, Washington, D.C.

Corresponding Author:
Address correspondence to Francesco N. Biondi:
francesco.biondi@utah.edu

suggest that poorly designed warning signals presented by collision avoidance systems have the potential to endanger safety by startling drivers and, thus, counteracting the potential benefits of ADAS. In the studies by Biondi and colleagues (*6*, *7*), drivers drove a simulated vehicle equipped with a speed mitigation system. When approaching a dangerous road intersection, auditory warning signals were presented if the maintained speed was over a safety threshold. Although the presentation of auditory signals was observed to apparently benefit safety – a significant decline in speed was indeed found as a result of warning signals being presented to drivers – a more fine-grained analysis on the data unveiled a different phenomenon. The abrupt, unexpected onset of auditory warning signals startled drivers who, as a consequence, reduced their voluntary control over the vehicle [for similar results, see Adell et al. (*8*)]. Effects of concurrent warning signals presented by ADAS on driver behavior were investigated in the study by Biondi et al. (*9*). Compared with unimodal (vibrotactile or auditory) warnings, results showed bimodal warnings (vibrotactile + auditory) to aid in the execution of evasive maneuvers in emergency situations [see Scott and Gray (*10*)]. Nonetheless, additional findings (*11*) revealed that the benefits of bimodal warnings do not hold in high workload conditions (*12*–14), as a result of causing slower response times and high levels of annoyance.

Taken together, these findings suggest that the design and characteristics of the HMI of assistance systems has a direct impact on their use, the way drivers interact with them, and, in turn, their efficacy. In particular, poorly designed warnings and system HMIs have the potential to counteract the potential benefits of assistance systems on safety, and cause unintended consequences on driving.

Guidelines for Human–Machine Interface Assessment

Over the last 10 years, governmental institutions and transportation agencies have put their efforts into developing guidelines for the development and assessment of the HMI for in-vehicle systems, particularly in-vehicle infotainment systems or IVIS. IVIS are native vehicle systems via which the drive can, for example, place phone calls, listen to the radio, and set navigation destinations. Modern IVIS interfaces include touchscreens, instrument cluster displays, steering wheel commands, voice recognition systems, head-up displays, and so forth. In 2006, the European Commission released its recommendations on safe and efficient in-vehicle information and communication systems (*15*). Design principles and recommendations included in the document were then revised, expanded upon, and adapted to create the UK Transport Research Laboratory (TRL) Checklist for the assessment of in-vehicle information systems (*16*). More recently, the NHTSA has released its human factors design guidance (*17*) with the goal of providing guidelines for the design and development of driver–vehicle interfaces. The document contains a list of principles covering the design of, for example, visual messages, auditory messages, haptic messages, and driver inputs. For each principle, it includes a specific design guidance, appropriate reference, and possible design issues for assessing general, function-aspecific interfaces [for a similar document issued by the Alliance of Automobile Manufacturers, see (*18*)]. Although the adoption of the NHTSA guidance is of importance for the design and development of HMIs, its structure, per se, does not provide a measurement tool for rating the usability and demand of various in-vehicle systems.

The goal of this study is to provide a tool for the assessment of the HMI of a variety of advanced assistance systems. Our research team at the University of Utah was tasked with developing robust assessment tools for conducting on-road evaluations of in-vehicle interfaces. Although standardized metrics are available for measuring driver workload of IVIS (*18*), this is not the case for assessing the usability and demand of ADAS. The interaction with assistance systems has indeed different characteristics to that of accessing IVIS functions. First, some ADAS interactions are too short to afford a meaningful measurement (i.e., parking sensors providing a single beep when the distance to objects is too close). Second, the interaction does not always require an overt response by the driver (e.g., blinking light presented by the blind spot monitor). Moreover, the process of interacting with assistance systems controlling certain vehicle operations (i.e., ACC) may be extended in time, and the use of meaningful, objective human factors metrics for assessing driver demand (e.g., detection response task) has not yet been validated within this context. Finally, given the number of assistance systems available on the market, their interfaces may be widely different in characteristics and a one-fits-all approach may fall short in exhaustively assessing their usability and demand.

This study documents the development and structure of the Utah Vehicle Technology Assessment Scale (UTA-S). The development of this scale arises from the need to conduct human factors evaluations of a multitude of advanced assistance systems, and provide meaningful ratings for the benchmarking of a large number of vehicles. Studies in literature have investigated driver interaction with a limited number of ADAS (*2*, *3*, *9*), nonetheless, to the authors' best knowledge, the literature does not offer one tool for assessing the demand and usability of the HMI of a wide range of ADAS, and conducting meaningful comparisons among vehicles of different makes and models. Industry standards and design codes offering guidance on the development and assessment of function-aspecific systems are available, but a single checklist tailored to the assessment of assistance systems is not currently present. For such reasons, a scale for human factors evaluations of a broad range of ADAS HMI was developed. Assistance systems considered in this scale include ACC, rearview camera, rear-cross traffic alert (RCTA), lane departure warnings, lane keeping assist system (LKAS)/lane centering, rear parking sensors (RPS), forward collision warning system, blind spot monitor +

lane change assist (LCA), automatic lane change (ALC), and automatic parallel parking (APP).

Method

ADAS Evaluated

To date, 40 vehicles of different makes and models have undergone evaluation. All vehicles were 2017 model-year. Via an iterative process, the evaluation of the ADAS available on these vehicles has contributed to the development of the latest version of UTA-S. Thirty-eight vehicles (95% of the total) had a rear-view camera available on them, whose image was displayed on a screen with sizes ranging from 4.2 to 17 in. Seven vehicles had a screen smaller than 6 in. in size; 22 vehicles had a screen with sizes ranging from 6 to 9 in.; one vehicle had a screen larger than 9 in. in size. Twelve vehicles (30% of the total) were equipped with RPS and seven vehicles (17.5% of the total) had front parking sensors. Seven vehicles (17.5% of the total) were equipped with RCTA, ACC, LKAS (or lane centering), blind spot monitor, and LDWS. Only one vehicle was equipped with ALC and an APP system. The evaluation of a total of 94 distinct systems contributed to the development of UTA-S.

UTA-S Development Process

The evaluation of ADAS was conducted by six human factors experts. All evaluators hold a bachelor's degree in ergonomics or a related field, and three evaluators hold a doctorate in human factors and have extensive experience developing methodologies for the assessment of in-vehicle systems. Evaluators drove each vehicle extensively in order to become familiar with its assistance systems and relative HMIs. Vehicles were driven in different weather and lighting conditions to identify specific issues that could cause the interaction with the system to be particularly distracting or troublesome (e.g., display glare with low environmental light). Evaluators conducted formal evaluations of the HMI of assistance systems and, via an iterative process, the latest version of the scale was developed. The process of evaluating each system had a duration of 1 to 4 h, during which evaluators closely examined separate aspects of the HMI system. This phase aimed at identifying the most relevant aspects of driver interaction with assistance systems and their interfaces. Information and recommendations contained in design guidelines (*15*–18) and industry standards [SAE, ISO, e.g., (*19*–22)] were adopted in the development of this scale, and adapted to the evaluation of 10 different ADAS and their HMIs. The latest version of the scale is presented below.

Results

UTA-S Rating Scale

The rating scale used for the assessment of ADAS HMI expands upon the scale used in the TRL Checklist (*16*). Each item contained in UTA-S was rated by two human factors experts using a scale with four levels: No Concern, Minor Concerns, Serious Concerns, Not Applicable.

- "No Concern" should be selected when interaction with the HMI does not compromise the driver's ability to control the vehicle and operate it safely, for most drivers under most conditions; or the HMI is informative about the status and functioning of the system, and when the process of understanding its contents does not cause an increase in levels of distraction; or the system operates the vehicle in a manner that is considered pleasant, not stressful.
- "Minor Concerns" should be selected when interaction with the HMI could, sometimes, compromise vehicle control and operation for some drivers under some conditions; or some of the information provided by the HMI is not clear and could require the driver to divert attentional and visual resources away from the driving task for a limited amount of time; or the system sometimes operates the vehicle in a manner that is judged potentially stressful, unpleasant.
- "Serious Concerns" should be selected when interaction with the HMI is likely to prevent a significant number of drivers, under normal driving conditions, from maintaining full control and safe operation of the vehicle; or the HMI for ADAS is deemed highly distracting, or generates high levels of stress or frustration; or the system operates the vehicle in an abrupt, jolty manner that prevents the driver from using the system pleasantly and without stress.
- "Not Applicable" should be selected when the HMI does not have the particular attribute the question is addressing.

Numerical values were assigned to levels of the scale to allow for quantification of evaluations and comparisons across systems: 3 points are assigned when No Concern was selected; 2 points were assigned when Minor Concerns was selected; 1 point was assigned when Serious Concerns was selected; 0 points were assigned when Not Applicable was selected.

UTA-S

UTA-S has 10 sections, one for each of the systems under evaluation. Each section contains a number of items giving a total of 59 different items. Some items contain sub-items to guide in the evaluation, and aid evaluators in covering all aspects included in each of the items. Due to word limit restrictions, lists of sub-items have not been presented here. Each item is accompanied by a description that facilitates evaluators in understanding its specific contents. For each item, examples of good and bad design are presented as practical cases. A list of selected references is also presented for each item. Selected references are of design guidelines and industry standards (ISO, SAE), and relevant studies in the literature covering driver interaction with ADAS.

A list of evaluation items for each system, accompanied by a description of the technology and rationale for their selection, and relevant references are presented below.

A. Rearview Camera

Rearview cameras afford the driver a view of the area behind the vehicle, which facilitates detection of small objects that are not ordinarily visible with mirrors or looking over the shoulder (*23*, *24*). Given the number of injuries and deaths caused by backover collision – an estimated 221 fatalities and 14,000 injuries every year (*25*) – NHTSA will require automanufacturers to install rearview cameras on all vehicles from 2018 (*26*).

Scale Items

- A1 Is the camera providing useful information that is representative of the real world?
- A2 Is the screen displaying the video feed sufficiently large in size and with sufficiently high resolution to afford drivers a clear view of the vehicle rear zone?
- A3 Are points of reference or guidelines overlaid to the video feed?
- A4 Is the video feed presented on the screen effective and useful in a variety of weather and lighting conditions?
- A5 Are the colors adopted for the guidelines effective in all weather and lighting conditions?

Rationale. The image provided by the rearview camera should be clear and representative of the real world (*26*–28). The size of the screen should also be sufficiently large and have sufficiently high definition to afford the driver a clear view of the vehicle rear zone (*27*, *28*). If present, points of reference and overlaid guidelines will assist drivers in gauging the presence of obstacles and their distance to the vehicle (*17*, 29, 30). Furthermore, the view of the vehicle's rear zone offered by the rearview camera should be unaffected by adverse lighting and weather (e.g., rain, or glare) (*17*, *27*, *28*).

B. Rear-Cross Traffic Alert

RCTA is designed to detect vehicles that might be crossing the vehicle's rear path. The system may use a combination of visual or auditory warning signals to alert the driver about the presence of a rear-crossing vehicle.

Scale Items

- B1 Is the location of visual messages for RCTA effective in informing the driver about the location and direction of rear-cross traffic?
- B2 Is the design of the visual messages used for RCTA informative and easy to comprehend?
- B3 Do false alarms or misses occur?
- B4 Given their characteristics, are auditory warnings effective in providing information about rear-cross traffic?

Rationale. The characteristics of visual signals for RCTA should facilitate the detection of potential hazards, and be different from those used by other systems (*29*, *30*). Their location on the HMI of the vehicle should be consistent with that of the rearview camera (*31*). Visual messages for RCTA should not be positioned in a location that is distant from the screen where the rearview camera is displayed, as this may increase the potential for visual demand when reversing. Furthermore, because of the nature of the system, the occurrence of false alarms (i.e., when warning signals are presented in absence of a hazard) might reduce driver acceptance and trust in the technology (*32*, *33*).

C. Rear Parking Sensors

RPS alert the driver about the presence of obstacles located in the vehicle's reversing path. Given the high number of fatalities and injuries caused by backover collisions (*23*, *25*), RPS may contribute to a reduction in accidents of this nature. Nonetheless, despite the potential benefits of this system, findings suggest that the presence of RPS may make drivers pay less attention to the rearview camera image, as a result of them becoming over reliant on this assistance technology (*34*).

Scale Items

- C1 Is the location of visual messages for RPS effective for informing the driver about the location and direction of rear-cross traffic?
- C2 Is the design of the visual warnings used for RPS informative and easy to comprehend?
- C3 Are the characteristics of auditory warnings for RPS adequate and informative?
- C4 Are haptic warnings for RPS effective in informing the driver about the presence of obstacles?
- C5 Are warnings clearly detectable by drivers?
- C6 Are the RPS covering a sufficiently wide area of the vehicle's rear zone?
- C7 Effectiveness of multimodal warnings presented jointly.
- C8 Occurrence of false alarms and misses.

Rationale. Visual messages should be well integrated with the rearview camera image (*31*). The presence of auditory and haptic warning signals should assist the driver in detecting potential obstacles, without causing a startling reaction or going unnoticed (*29*, *35*–*37*). The presence of multimodal warnings should enhance drivers' awareness of obstacles.

Thus, drivers should perceive multimodal warnings to be necessary, and not redundant or confusing (*37–39*). Furthermore, the occurrence of false alarms or misses should be reduced to a minimum (*33*, *34*).

D. *Forward Collision Avoidance System*

In 2012, 1.7 million rear-end crashes resulted in more than 1,700 fatalities and 500,000 injuries (*40*). According to the National Transportation Safety Board, many of these crashes could have been mitigated or prevented as a result of a broader adoption of FCAS (*41*). Whenever an impending collision with the forward vehicle is detected, the system warns the driver by means of auditory, visual, or haptic signals.

Scale Items

D1. Is the location of the visual warnings for FCAS effective in informing the driver about the location of and distance to objects in the front field?
D2. Is the design of the visual warnings used for FCAS informative and easy to comprehend?
D3. Are the characteristics of auditory warnings for FCAS adequate and informative?
D4. Are haptic warnings for FCAS effective in informing the driver about the presence of obstacles?
D5. Are warnings clearly detectable by drivers?
D6. Do the forward collision sensors cover a sufficiently wide area of the vehicle's forward zone?
D7. Effectiveness of multimodal warnings presented jointly.
D8. Do false alarms or misses frequently occur?

Rationale. The characteristics and location of visual messages for forward collision warnings should aid drivers to detect obstacles in the forward path (*29*, *33*). In particular, the location of visual warning signals should be close to the forward roadway to increase detectability, and reduce the time of the motor response (*29*, *42*). Given the nature of the system, warning signals should be clearly detectable and not go unnoticed.

E. *Blind Spot Monitor + Lane Change Assist*

Blind spot monitor (BSM) warns the driver about the presence of vehicles in the vehicle's side blind spot. The characteristics of the warning vary (i.e., signals become more salient) whenever the driver is about to change lane and the turn signals are activated.

Scale Items

E1. Is the location of visual warnings for BSM+LCA informative about the location of the hazard?
E2. Are the characteristics of visual warnings adequate for BSM+LCA?
E3. Are auditory warnings effective in informing the driver about objects in the vehicle's side blind zone?

Rationale. The location chosen for visual warnings should enable drivers to detect obstacles in the vehicle's blind spot. Nonetheless, if visual warnings are too salient, this may increase levels of distraction and frustration, due to drivers' visual attention being captured by the frequent onset of the warning (*29*, *43*, *44*). Auditory warning signals for BSM should not be startling or generate an adverse impact on driving (*7*, *29*). Moreover, the characteristics of warning signals used by BSM should make them clearly distinguishable from signals used by other systems – in one of the vehicles tested in this assessment, BSM and RCTA utilized visual signals with the same characteristics, making their interpretation confusing for drivers.

F. *Adaptive Cruise Control*

ACC adjusts the vehicle speed to maintain a safe distance from the forward vehicle. The use of ACC is estimated to have the potential to reduce accidents by 17% in the following two scenarios: when an ACC equipped vehicle approaches a slower vehicle traveling at a constant speed; and when a forward vehicle decelerates in front of an ACC equipped vehicle (*45*). Nonetheless, despite estimated benefits, studies show that motorists may lack a full understanding of system limitations, and accurate information regarding the intended use of this system (*33*).

Scale Items

F1. Are controls used for ACC easy to reach?
F2. Are labeling and icons on controls informative, easy to understand?
F3. Is ACC easy to engage?
F4. Are visual messages used to inform about the system status easy to comprehend?
F5. Are decelerations and accelerations executed in a timely and natural manner?

Rationale. Controls should not be hidden nor their view obstructed, and labels and icons used on controls should be intuitive (*19*–21, *46*). By looking at the HMI, drivers should be able to easily understand whether the system is properly tracking the distance to the forward vehicle and following it, what the set distance to the forward vehicle is, what the set speed is. Visual messages and icons should be easy to comprehend and the process to understand them should not cause an increase in levels of driver visual or cognitive demand (*29*). Furthermore, when the system is operating, it should control the vehicle's longitudinal movements in a way that is pleasant and expected by the driver. The execution of abrupt maneuvers may cause an increase in driver stress and a reduction in acceptance (*22, 47, 48*).

G. Lane Departure Warning System

The LDWS monitors the position of the vehicle within the lane and informs the driver whenever the vehicle drifts out of the lane. According to an Insurance Institute of Highway Safety analysis conducted by Jermakian (*49*), the use of systems assisting the driver in maintaining the vehicle within the lane could generate an annual reduction in collisions equal to 179,000. Data from a field operational test conducted at NHTSA by Wilson, Stearns, Koopmann, and Yang (*50*), suggests LDWS could potentially reduce crashes by 1 to 8%, depending on the traffic scenario.

Scale Items

G1. Are the controls to engage/disengage the system easy to identify and reach?
G2. Are the visual messages providing information about whether or not the system is engaged and functioning easy to understand?
G3. Are all warnings presented in a timely manner?
G4. Are auditory warnings effective in informing the driver about the position of the vehicle within the lane?
G5. Are haptic warnings effective in informing the driver about the position of the vehicle within the lane?
G6. Effectiveness of multimodal warnings presented jointly.

Rationale. Visual messages provided by the system should inform the driver about whether or not the system is engaged, and whether or not the system tracks lane markings and functions properly (*51*). This process should not require extended periods of time, or increase levels of distraction (*29*). The presentation of multimodal signals should be well integrated so that one clear message is provided to the driver. The driver should not perceive signals as redundant or unnecessary (*7, 36, 37*). Moreover, all warning signals should be presented in a timely manner so that drivers are given sufficient time to operate evasive maneuvers if necessary.

H. Automatic Lane Change

ALC takes control of steering wheel movements and a vehicle's longitudinal control to execute a lane change when the driver activates the turn signal on a multi-lane road. The system detects the presence of vehicles in the adjacent lane and, if traffic conditions are deemed suitable, the vehicle autonomously moves over to the desired lane.

Scale Items

H1. Controls to engage ALC are easy to reach.
H2. Instructions about how to engage ALC are readily and safely accessible.
H3. Information about the system status is provided at all times.
H4. Does the system operate the vehicle in a smooth and pleasant manner?
H5. Overriding the system is quick and intuitive.
H6. Information regarding the system's limitations and driver's requirements are provided and readily accessible.

Rationale. Instructions about how to engage and operate ALC should be made readily available to the driver via the HMI of the system or a manual. If instructions are provided via the vehicle HMI, access to long strings of text should be restricted while the vehicle is in motion, as this may represent a source of distraction to the driver (*29*). Information about the system status should be provided at all times and be comprehensible to the driver (*9, 18*). Overriding the system should require a process that is intuitive and quick to execute (*52, 53*).

I. Lane Keeping Assist System

LKAS operates the steering wheel to maintain the vehicle within the lane. IIHS estimates that single-vehicle crashes, in which vehicles moved out of the roadway, accounted for 40% of fatal collisions and 21% of non-fatal-injury crashes in 2014. Drowsiness, fatigue, or, more generally, driver incapacitation are estimated to be major causes of lane-change crashes (*54*).

Scale Items

I1. Are controls used for LKAS easy to reach?
I2. Are labeling and icons on controls informative, easy to understand?
I3. Are visual messages about the system status easy to comprehend?
I4. Is information for LKAS presented in a coherent manner with that for LDWS?
I5. Is the implementation of LKAS perceived as safe and natural?
I6. Is information provided about the system's limitations and driver's requirements?
I7. The driver should be able to override LKAS with no difficulty.

Rationale. Operating LKAS should not cause an adverse impact on driving and safety (*19*–21, *46*). Visual messages used for LKAS should be easy to comprehend (*55–57*). With the system being operational, steering wheel movements should be perceived as pleasant and timely, and not increase levels of driver stress (*41*, *48*). The system should provide clear and accessible information about its limitations and driver requirements. Drivers should be reminded that they are the ones in charge of monitoring the driving environment and that they should stay alert at all times (*4*, *41*). Furthermore, the process of overriding the automated system should be intuitive and quick to execute (*56*).

J. Automatic Parallel Parking

APP is an autonomous system that moves a vehicle from a traffic lane into a parking spot to perform a parallel parking maneuver. According to an estimate by NHTSA, backover crashes result in at least 183 fatalities, and between 6,700 and 7,419 injuries every year (*58*).

Scale Items

J1. Are controls used to engage APP easy to reach?
J2. Are instructions provided to the driver about how to engage APP?
J3. The driver should be able to override APP with no difficulty.
J4. Does the system always function as expected?
J5. Is the behavior of the system smooth and pleasant?
J6. Is information provided about the system's limitations and driver's requirements?

Rationale. The system should provide the driver with accurate information about how to activate the system (*16*, *17*). If visual information is provided while the vehicle is in motion, this should not distract the driver. Operating APP should represent a sufficiently easy task, regardless of the characteristics of the driver (e.g., age). Information should be easy to comprehend, and the system should guide the driver through the different steps required by the system to complete the parking maneuver (*16*, *17*, *29*). The driver should also be able to override the system without difficulty. When in control of the steering and acceleration, the system should operate the vehicle in a pleasant manner and, importantly, in a way that matches the driver's expectations. Information about the system's limitations and driver requirements should be provided, and should be easy to access.

UTA-S Implementation

UTA-S was applied to the assessment of five 2017 vehicles equipped with a total of 27 distinct assistance systems. The 2017 model-year vehicles we considered were: Mercedes-Benz C300, Tesla Model S75, Honda Ridgeline RTL.E, Chrysler 300 C, and GMC Yukon SLT. In describing the process of assessment of assistance systems, vehicles were anonymized and refered to as vehicles A to E. Ratings for each system are presented in Table 1. An inter-rater reliability analysis was conducted to examine the degree of agreement among evaluators. A weighted Cohen's kappa index of 0.74 was obtained, suggesting a substantial level of agreement among coders. Statistical analysis was conducted using R software.

All rearview camera systems evaluated on these vehicles received a Minor Concerns rating overall. Despite all systems using large displays (item A2 of UTA-S), providing dynamic guidelines (A3) and producing images that are effective in daylight and at night (A4), some of them suffered from slight image distortions (A1) and used monocolor, non-color-coded guidelines (A5). Information provided by RCTA systems was well integrated with that from rearview cameras (B1), was easy to comprehend (B2), and, during our testing, no false alarms occurred (B3). Warnings presented by RPS were easy to detect and their message easy to comprehend (C2, C5). However, in some cases, the characteristics of auditory warnings (frequency, intensity) made them uninformative about the distance to obstacles (C3). Similar concerns were raised for some of the FCAS tested, with auditory warnings that might be more difficult to detect by certain segments of the population in high background noise conditions (D5). Overall Serious Concerns were observed for BSM on Vehicle E, in which visual warnings were presented in an unconventional location and used unnatural mapping (E1, E3). Although evaluators found standard labeling for ACC in Vehicle B, two main issues observed on Vehicle E related to controls not being directly accessible to the driver (F1), and the absence of information regarding current following distance on the instrument cluster display (F4). Two of the LDWSs evaluated caused Serious Concerns, mainly because of the controls being difficult to reach (G1), poor visual information about the status of the system (G2), and haptic warnings being easy to miss (G5). For ALC, the process of reading instructions on how to access the system was complicated and could distract drivers (H2), information about system limitations and driver responsibilities were not clear (H6), and movements could be jolty, abrupt at times (H4). Similar concerns were raised for LKAS on vehicles tested, with movements executed by the system generating discomfort (I5), controls being difficult to reach (I1) and non-standard labeling (I2). For APP, instructions on Vehicle E were difficult to access (J2) and might be hard to comprehend by certain segments of the population (J3).

Discussion

This article describes the development of the UTA-S. The scale provides a tool for the assessment of the usability and demand of the HMI of advanced assistance systems. The scale consists of 59 items developed for the assessment of 10 active and passive assistance systems: rearview camera, RCTA, RPS, FCAS, BSM + LCA, ACC, LDWS, ALC, LKAS, and APP. Each item has a list of sub-items to guide the evaluators in the assessment process. The scale builds

Table 1. UTA-S Ratings Expressed for 27 Systems on Five 2017 Vehicles

Systems	Veh. A	Veh. B	Veh. C	Veh. D	Veh. E
Rearview Camera	MiC	MiC	MiC	MiC	MiC
Rear-Cross Traffic Alert (RCTA)	na	NoC	NoC	na	na
Rear Parking Sensor (RPS)	na	MiC	NoC	na	MiC
Forward Collision Avoidance System (FCAS)	MiC	NoC	MiC	na	MiC
Blind Spot Monitor (with Lane Change Assist)	na	MiC	MiC	NoC	SeC
Adaptive Cruise Control (ACC)	na	na	MiC	na	SeC
Lane Departure Warning System (LDWS)	na	SeC	NoC	na	SeC
Autonomous Lane Change (ALC)	na	na	na	na	SeC
Lane Keep Assist System (LKAS)	na	SeC	MiC	na	SeC
Automatic Parallel Parking (APP)	na	na	na	na	SeC
Number of ADAS	*1*	*7*	*6*	*2*	*9*

Note: Veh. = vehicle; na = Not Applicable; MiC = Minor Concerns; SeC = Serious Concerns; NoC = No Concerns.

upon practices and standards developed for assessing usability and demand of in-vehicle HMI. The rating scale adopted in UTA-S is a revised version of that used in the TRL Checklist (*16*). NHTSA's *Human Factors Design Guidance for Driver–Vehicle Interfaces* (*17*) and Alliance for Automobile Manufacturers' design guidelines (*18*), as well as a number of ISO and SAE standards were considered in the development of this scale.

The scale was applied to the assessment of 27 systems available on five 2017 vehicles. For passive assistance systems, the main recurrent issues related to the characteristics of visual, auditory, and haptic warnings. In some instances, auditory warnings used by RPS were not indicative of the distance of the vehicle to obstacles present in the surrounding area (*17*, *42*), or visual warnings adopted by BSMs were located in unconventional locations and used unnatural mapping (*44*, *45*). Common issues found for active assistance systems regarded the information provided on how to operate systems, their limitations, and driver responsibilities (*53*). Accessing supporting information for ALC was deemed distracting (*29*), and accelerations operated by LKAS were in some cases rated as uncomfortable and jolty (*42*, *49*).

The scale was developed in response to the need to generate meaningful assessment ratings for a large number of assistance systems available on vehicles of different makes and models. In its current status, UTA-S represents a valuable methodology to be used for benchmarking purposes, and in the early stage of the development of assistance systems.

Acknowledgment

This project is supported by a research grant issued by the AAA Foundation for Traffic Research.

Author Contributions

The authors confirm contribution to the article as follows: study conception and design: FB; data collection: MMC, DG, and RG; analysis and interpretation of results: FB, DS, and JC; draft article preparation: FB. All authors reviewed the results and approved the final version of the article.

References

1. Mosquet, X., M. Andersen, and A. Arora. A Roadmap to Safer Driving Through Advanced Driver Assistance Systems. *Auto Tech Review*, Vol. 5, No. 7, 2016, pp. 20–25. https://www.bcgperspectives.com/content/articles/automotive-roadmap-to-safer-driving-through-advanced-driver-assistance-systems/
2. Stanton, N. A., and M. S. Young. Driver Behaviour with Adaptive Cruise Control. *Ergonomics*, Vol. 48, No. 10, 2005, pp. 1294–1313.
3. Vollrath, M., S. Schleicher, and C. Gelau. The Influence of Cruise Control and Adaptive Cruise Control on Driving Behaviour – A Driving Simulator Study. *Accident Analysis and Prevention*, Vol. 43, No. 3, 2011, pp. 1134–1139.
4. Biondi, F., R. Goethe, J. Cooper, and D. Strayer. Partial-autonomous Frenzy: Driving a Level-2 Vehicle on the Open Road. *Proc., 13th Conference of Human Computer Interaction International*, 2017. In International Conference on Engineering Psychology and Cognitive Ergonomics pp. 329–338. Orlando, FL
5. SAE. *Automated Driving. Levels of Driving Automation Are Defined in New SAE International Standard J3016*. 2014. SAE International.
6. Rossi, R., M. Gastaldi, F. Biondi, and C. Mulatti. Warning Sound to Affect Perceived Speed in Approaching Roundabout: Experiment With a Driving Simulator. *Procedia - Social and Behavioral Sciences*, Vol. 87, 2013, pp. 269–278.
7. Biondi, F., R. Rossi, M. Gastaldi, and C. Mulatti. Beeping ADAS: Reflexive Effect on Drivers' Behavior. *Transportation Research Part F: Traffic Psychology and Behaviour*, Vol. 25, 2014, pp. 27–33.
8. Adell, E., A. Várhelyi, and M. Hjälmdahl. Auditory and Haptic Systems for In-car Speed Management – A Comparative Real Life Study. *Transportation Research Part F: Traffic Psychology and Behaviour*, Vol. 11, No. 6, 2008, pp. 445–458.
9. Biondi, F., D. L. Strayer, R. Rossi, M. Gastaldi, and C. Mulatti. Advanced Driver Assistance Systems: Using Multimodal Redundant Warnings to Enhance Road Safety. *Applied Ergonomics*, Vol. 58, 2017, pp. 238–244.
10. Scott, J. J., and R. Gray. A Comparison of Tactile, Visual, and Auditory Warnings for Rear-End Collision Prevention in Simulated Driving. *Human Factors: The Journal of the Human Factors and Ergonomics Society*, Vol. 50, No. 2, 2008, pp. 264–275.

11. Biondi, F., M. Leo, M. Gastaldi, R. Rossi, and C. Mulatti. How to Drive Drivers Nuts: Effect of Auditory, Vibrotactile and Multimodal Warnings on Perceived Urgency, Annoyance and Acceptability. *Transportation Research Record: Journal of the Transportation Research Board*, 2018: In press.
12. Mehler, B., Reimer, B., & Dusek, J. A. (2011). *MIT AgeLab delayed digit recall task (n-back)*. Massachusetts Institute of Technology, Cambridge, MA, 1-23.
13. Biondi, F., J. Turrill, J. R. Coleman, J. M. Cooper, and D. L. Strayer. Cognitive Distraction Impairs Drivers' Anticipatory Glances: An On-road Study. *Proc., 8th International Driving Symposium on Human Factors in Driver Assessment, Training and Vehicle Design distractive*, 2015. Salt Lake City, UT. pp. 1–13.
14. Strayer, D., J. Cooper, J. Turrill, J. R. Coleman, and R. J. Hopman. *Measuring Cognitive Distraction in the Automobile III: A Comparison of Ten 2015 In-vehicle Information Systems*, AAA Foundation for Traffic Safety, Washington, D.C., 2015.
15. ESoP - European Statement of Principles on human-machine interface. *Official Journal of the European Union*, 2008. 1, 1–30.
16. Stevens, A., and S. Cynk. *Checklist for the Assessment of In-vehicle Information Systems*. Report MIS005. Transportation Research Laboratory, 2011.
17. Campbell, J. L., J. L. Brown, J. S. Graving, C. M. Richard, M. G. Lichty, T. Sanquist, L. P. Bacon, R. Woods, H. Li, D. N. Williams, J. F. Morgan, and G. Divekar. *Human Factors Design Guidance for Driver-Vehicle Interfaces (DVI)*. National Highway Traffic Safety Administration. 2016.
18. Alliance of Automobile Manufacturers. *Statement of Principles, Criteria and Verification Procedures on Driver Interactions with Advanced In Vehicle Information and Communication Systems*. Driver Focus-Telematics Working Group, Alliance of Automobile Manufactures. Washington, DC. 2006.
19. ISO.ISO 3958. *Road Vehicles, Passenger Cars - Driver Hand Control Reach*, 2nd ed. 1996. International Organization for Standardization.
20. ISO. ISO 4040. *Road Vehicles, Passenger Cars – Location of Hand Controls, Indicators and Tell-tales in Motor Vehicles*. 5th ed. 2009. International Organization for Standardization.
21. SAE. SAE J 287. *Driver Hand Control Reach*. 2007. Society for Automotive Engineers.
22. SAE. SAE J 1138. *Design Criteria - Driver Hand Controls Location for Passenger Cars, Multipurpose Passenger Vehicles, and Trucks (10 000 GVW and Under)*. 2009. Society for Automotive Engineers.
23. NHTSA. *Fatalities and Injuries in Motor Vehicle Backing Crashes*. Report to the House Committee on Transportation and Infrastructure and the Senate Committee on Commerce, Science, and Transportation. National Highway Traffic Safety Administration, Washington, D.C. 2009.
24. NHTSA. *On-Road Study of Drivers Use of Rearview Video Systems (ORSDURVS)*. DOT HS 811 024. National Highway Traffic Safety Administration, Washington, D.C. 2008.
25. IIHS. Rearview Camera Rule Aims to Reduce Backover Crashes. *Status Report*, Vol. 49, No. 4, 2014, pp. 1–8.
26. NHTSA. *Safety Technologies. Backup Cameras*. National Highway Traffic Safety Administration, Washington, D.C. 2017.
27NHTSA. Federal Motor Vehicle Safety Standards: Rear Visibility. RIN 2127-AK43. Vol. 79, No. 66, 2014. National Highway Traffic Safety Administration, Washington, D.C.
28. NHTSA. *Driver-Vehicle Interfaces for Advanced Crash Warning Systems: Research on Evaluation Methods and Warning Signals*. 2015. National Highway Traffic Safety Administration, Washington, D.C. 2015.
29. NHTSA. Visual-Manual NHTSA Driver Distraction Guidelines for In-Vehcile Electronic Devices. *Federal Register*, Vol. 77, No. 37. 2012. 11200-11250.
30. International Standard Organization. *ISO 16505:2015. Road Vehicles — Ergonomic and Performance Aspects of Camera Monitor Systems — Requirements and Test Procedures*. 2015. International Organization for Standardization.
31. Neurauter, M. L., R. E. Llaneras, B. Li, and C. Green. Issues Related to the Use and Design of a Backing Rear Cross Traffic Alert System. *SAE Technical Paper*, 2011. National Highway Traffic Safety Administration, Washington, D.C. 4(2011-01-0578), 462-466.
32NHTSA. Federal Motor Vehicle Safety Standards: Rear Visibility. RIN 2127-AK43. Vol. 79, No. 66, 2014. National Highway Traffic Safety Administration, Washington, D.C.
33. Mehler, B., B. Reimer, M. Lavallière, J. Dobres, and J. F. Coughlin. *Evaluating Technologies Relevant to the Enhancement of Driver Safety*. AAA Foundation for Traffic Safety, 2014.
34. IIHS. *Visibility of Children Behind 2010-13 Model Year Passenger Vehicles Using Glances, Mirrors, and Backup Cameras and Parking Sensors*. 2013. Insurance Institute for Highway Safety.
35. Biondi, F., R. Rossi, M. Gastaldi, and C. Mulatti. Beeping ADAS: Reflexive Effect on Drivers' Behavior. *Transportation Research Part F: Traffic Psychology and Behaviour*, Vol. 25, 2014, pp. 27–33.
36. Lipp, O. V., D. A. Siddle, and P. J. Dall. The Effect of Warning Stimulus Modality on Blink Startle Modification in Reaction Time Tasks. *Psychophysiology*, Vol. 37, No. 1, 2000, pp. 55–64.
37. Biondi, F., D. L. Strayer, R. Rossi, M. Gastaldi, and C. Mulatti. Advanced Driver Assistance Systems: Using Multimodal Redundant Warnings to Enhance Road Safety. *Applied Ergonomics*, Vol. 58, 2017, pp. 238–244.
38. Ho, C., N. Reed, and C. Spence. Assessing the Effectiveness of "Intuitive" Vibrotactile Warning Signals in Preventing Front-to-rear-end Collisions in a Driving Simulator. *Accident, Analysis and Prevention*, Vol. 38, No. 5, 2006, pp. 988–96, 238–244.
39. SAE. *SAE J2395. ITS In-Vehicle Message Priority*. 2002. Society for Automotive Engineers.
40. ISO. *ISO 16951. Road Vehicles — Ergonomic Aspects of Transport Information and Control Systems (TICS) — Procedures for Determining Priority of On-board Messages Presented to Drivers*. 2004. International Organization for Stndardization.
41. National Transportation Safety Board. *The Use of Forward Collision Avoidance Systems to Prevent and Mitigate Rear-End Crashes*. Special Investigation Report, No. NTSB/SIR-15/01, 2015. National Transportation Safety Board.
42. Sullivan, J. M., M. J. Flannagan, A. K. Pradhan, and S. Bao. *Literature Review of Behavioral Adaptations to Advanced Driver Assistance Systems*. AAA Foundation for Traffic Research. 2016.

43. SAE. *SAE J2400. Human Factors in Forward Collision Warning Systems: Operating Characteristics and User Interface Requirements*. 2003. Society of Automotive Engineers.
44. NHTSA. *Blind Spot Monitoring in Light Vehicles — System Performance*. 2014. National Highway Traffic Safety Administration.
45. SAE. *SAE J2802. Blind Spot Monitoring System (BSMS): Operating Characteristics and User Interface Standard*. 2016. Society of Automotive Engineers.
46. Koziol, J., V. Inman, M. Carter, J. Hitz, W. Najm, S. Chen, A. Lam, M. Penic, M. Jensen, M. Baker, M. Robinson, and C. Goodspeed. *Evaluation of the Intelligent Cruise Control System*. 1999. US Department of Transportation National, 1.
47. de Winter, J. C. F., C. M. Gorter, W. J. Schakel, and B. van Arem. Pleasure in Using Adaptive Cruise Control: A Questionnaire Study in the Netherland. *Traffic Injury Prevention*, Vol. 18, No. 2, 2016, pp. 1–15. 216-224.
48. NHTSA. *Use of Advanced In-vehicle Technology by Young and Older Early Adopters: Survey Results on Headlamp Systems*, DOT HS 810. 2008. National Highway Traffic Safety Administration.
49. IIHS. *Driver Trust in Five Driver Assistance Technologies Following Real-world Use in Four Production Vehicles*, 9588. 2016. Insurance Institute for Highway Safety.
50. Jermakian, J. S. Crash Avoidance Potential of Four Passenger Vehicle Technologies. *Accident Analysis and Prevention*, Vol. 43, No. 3, 2011, pp. 732–740.
51. Wilson, B. H., M. D. Stearns, J. Koopmann, and C. Y. Yang. *Evaluation of a Road-departure Crash Warning System*. DOT-HS-810-854, National Highway Traffic Safety Administration. Washington, D.C., 2007.
52. SAE. *SAE J2808. Lane Departure Warning Systems: Information for the Human Interface*. 2017. Society of Automotive Engineers
53. NHTSA. *Human Factors Evaluation of Level 2 and Level 3 Automated Driving Concepts. Human Factors Evaluation of Level 2 and Level 3 Automated Driving*. 2015. National Highway Traffic Safety Administration.
54. NHTSA. *Federal Automated Vehicles Policy*. 2016. National Highway Traffic Safety Administration.
55. IIHS. LATCH Ratings Improve. *Status Report*, Vol. 51, No. 7, 2016, 1-8.
56. ISO. *ISO 11270. Intelligent Transport Systems – Lane Keeping Assistance Systems (LKAS) – Performance Requirements and Test Procedures*. 2014. International Organization for Stndardization.
57. Navarro, J., F. Mars, and M. S. Young. Lateral Control Assistance in Car Driving: Classification, Review and Future Prospects. *IET Intelligent Transport Systems*, Vol. 5, No. 3, 2011, p. 207.
58. SAE. *SAE J3048. Driver-Vehicle Interface Considerations for Lane Keeping Assistance Systems*. 2015. Society of Automotive Engineers.

The Standing Committee on User Information Systems (AND20) peer-reviewed this paper (18-01558).

Article

TRR
JOURNAL OF THE TRANSPORTATION RESEARCH BOARD

Transportation Research Record
2018, Vol. 2672(37) 123–131

DOI: 10.1177/0361198118792128
journals.sagepub.com/home/trr
SAGE

Evaluation of Downstream Merge Behaviors Resulting from Driver Lane Choice: A Driving Simulator Study

Francis Tainter[1], Alyssa Ryan[1], Cole Fitzpatrick[1], Eleni Christofa[1], and Michael Knodler, Jr.[1]

Abstract
With the ever-increasing demand to add roadway capacity in a safe and efficient manner, the application of auxiliary through lanes (ATLs) at intersections has increased in recent years. Despite the potential capacity-related benefits, ATLs also introduce the potential for unique driver operation tactics, leading to challenging merge scenarios downstream of signalized intersections. Furthermore, the benefits of increasing capacity are only realized when drivers are making decisions that balance the utilization of the ATL and the adjacent continuous through lane. Moreover, balancing lane utilization upstream of an ATL signalized intersection may introduce an improvement in safety and operations in merging conflicts downstream. Previous research investigated the effectiveness of ATLs through microsimulation and field studies. This research employed two full-immersion driving simulation studies. The initial experiment was conducted using simulation technology to evaluate aspects of driver behavior associated with lane utilization at ATLs. Lane utilization of the ATL significantly increased when the driver was informed of the fundamental design concept, ultimately improving capacity. The second experiment was based upon the results of the initial experiment, coupled with ongoing research evaluating alternative merge signage. Specifically, the second experiment aimed to evaluate downstream merge signage that could mitigate some of the behaviors resulting from the less than optimal lane utilization for ATLs. With an emphasis on improving intersection performance, the results from this paper suggest that the implementation of alternative merging signage may lead to improved merging behavior.

With the ever-increasing demand to add roadway capacity in a safe and efficient manner, the application of auxiliary through lanes (ATLs) at intersections has increased in recent years. ATLs introduce the potential for unique driver operation tactics, leading to challenging merge scenarios downstream of signalized intersections. Focusing on these downstream merge scenarios, the application of alternative merge design has been investigated in recent literature, including various experimental signage for merging. Previous research investigated ATL intersections seeking to create better guidance in design and gain an understanding of anticipated driver operations through microsimulation (*1–3*). This current research endeavor extends the study of ATLs further through a driving simulation environment and seeks an understanding of real-world driving behavior in ATL intersections and experimental alternative merge scenarios.

Auxiliary Through Lane Design and Lane Utilization

An ATL may be added at signalized intersections with the intent to maximize the capacity of the intersection. Many studies have referenced the variance in geometric design of ATLs, including most notably the taper lengths both upstream and downstream of the intersection. These studies have been motivated from previous field studies that observed the utilization of the continuous through lanes (CTL) that are adjacent to the ATL. Lane utilization of CTLs and ATLs at intersections were found to be unbalanced at roughly 80% and 20%, respectively (*3*, *4*). This leads to the question of whether auxiliary lane design affects this unbalanced utilization. Tarawneh et al. suggest that there is a significant impact on intersection performance with the upstream length of the auxiliary lane, stating that longer auxiliary lanes produced a greater volume of through traffic as opposed to shorter auxiliary lanes (*5*). Alternatively, in the NCHRP Web-Only Document 178, research states that the length of the downstream ATL does not have a significant

[1]University of Massachusetts Amherst, Amherst, MA

Corresponding Author:
Address correspondence to Francis Tainter: ftainter@umass.edu

effect on the volume of vehicles traversing through the ATL upstream of the intersection (*2*).

Further research has analyzed the deceleration area approaching a taper lane entrance of an ATL. NCHRP Report 780 research led to a standardized distance required to allow lane changes, which was used to modify the American Association of State Highway and Transportation Officials Green Book (*1*). In addition to lane distance effects, there are many variables that may alter driver behavior when approaching an intersection with an ATL. These behaviors may vary depending on the surrounding area (rural, urban, etc.). According to Ring et al., it is possible to determine a fluctuating value of lane utilization from multiple sites located in two different cities, and concludes that there may be different outcomes for different location types (*6*). As a result, the geometric design of ATLs may have a significant impact on lane utilization at intersections (*2, 3, 6*).

Geometric Design of a Downstream Merge

An important aspect with the addition of ATLs at intersections is the merge that occurs downstream of the intersection. With the introduction of a lane drop, vehicles are forced into a single lane via a gradual taper. Most intersections involve the standard flow from right-to-left or right-lane to left-lane merge. These standard merges may be altered by increasing or decreasing the length of the taper. Altering the taper length affects the driver's judgment of merging from one lane to the other. According to Nevers et al., drivers must be aware when an ATL is ending downstream of an intersection, as this is when the appropriate merging behavior occurs (*3*). In fact, it has been found that drivers are more controlled in their ability to merge efficiently and safely when there is a greater length of taper (*7*). The tapers for ATLs are typically recommended to be longer than the standard lengths calculated using the formula provided in the *Manual on Uniform Traffic Control Devices* (MUTCD):

$$L = WS \tag{1}$$

where

L = length of the transition area (ft)

W = width of the lane (ft)

S = speed limit posted, 85th percentile speed, or calculated speed (mph)

The design of ATLs is a function of downstream merge design. Primarily, there are two varied approaches for downstream merge design, including both the traditional approach that is most commonly employed at intersections across the United States, as well as the alternative merge design that has become prevalent in recent literature.

Traditional Merge Design. As mentioned previously, the standard merge that exists most frequently in downstream lane-drops, particularly ATL intersections, is a right-lane to left-lane merge. The standard signage for this maneuver, as defined by the MUTCD, is represented by the W4-2 sign, also known as the "Lane Ends" sign (*8*). This warning sign depicts the standard taper, merging the right lane into the left lane. Although drivers perceive this sign with the intent to yield right-of-way to the drivers in the left lane, this pattern of merging does not adhere to any guidance or regulation (*9*).

Alternative Merge Design. Recent literature has investigated the implementation of alternative merge tactics, both in altering geometric design and merge signage. Specifically, the concept of "joint-merge" or "zipper-merge" has evolved over recent years as an alternative design of downstream merge geometry. This concept suggests the potential for adjacent vehicles yielding to one another as they approach a geometrically centralized merge. This design has been found in previous literature to reduce congestion during these merge scenarios, while having the potential to improve safety (*9, 10*). These joint-merges have been introduced in highway work zones and, as previously mentioned, applied with a dual-side taper method as compared to the standard single taper (*11*). Joint-merges are anticipated to create an even distribution of vehicles in each lane, as well as reduce the quantity of lane changes made by vehicles approaching a lane drop. Therefore, with this balance in lane distribution, drivers merge in a more coordinated, alternating pattern (*12*). Further, the alternative merge concept has been tested at signalized intersections and was found to improve overall traffic flow and safety at downstream merges (*9*). It is important to note that in this research by Feldblum et al., researchers experimented with downstream merge scenarios at intersections that did not involve ATLs (*9*). However, this research yields significant promise for the application of alternative merging at intersection ATLs.

Safety Concerns in Design Alternatives

Over 6 million vehicle crashes occur in the United States every year (*13*), which makes safety a major priority in transportation design. The implementation of ATLs must be designed to certain specifications, while remaining focused on maximizing safety. For example, the merging maneuvers downstream of intersections may lead to more crashes when compared with an intersection without an ATL. A previous study examined numerous ATL intersections and found that only 4.5 crashes (sideswipe and rear-end) were averaged over a given year (*3*). These crashes are hypothesized to occur in the downstream merge section. Improper merging maneuvers, according to the Highway Safety and Information System, are the second most dangerous maneuvers, behind only driving inattentively, on their list of 26 (*14*). As mentioned in the previous section, the joint-merge alternative design may have the potential to improve safety at these downstream merging scenarios (*9*). In addition, Wohlson et al. found that the alternative merge system provided an increase in driver mobility through the merge section, as well as an increase in safety (*11*). The incorporation of alternative merge designs, either with or without ATLs present, suggests a potential for improved roadway safety as well as an effective increase in intersection performance and operations.

Research Motivation and Objectives

This research investigates the underlying principles behind unbalanced lane utilization occurring at ATLs, with an emphasis on understanding the downstream merge impacts. Based in part upon hypothesized imbalances in lane utilization, there was also an interest in evaluating measures that, when implemented, result in greater balance of lane utilization. Specifically, the introduction of experimental alternative merge signage in a driving simulator environment allows for real-world driver comprehension and behavior characteristics to emerge in downstream merge scenarios. By researching the upstream lane utilization of intersections with two through lanes (and the anticipated downstream lane-drop), this research aims to improve intersection operations in addition to the safety of downstream merge scenarios.

Methodology

To evaluate the reasoning for unbalanced lane utilization upstream of ATL signalized intersections, two driving simulator experiments were employed to 1) understand the lane selection upstream of an intersection based on the introduction of an ATL, and 2) investigate the operational impacts of introducing alternative merge signage to drivers upstream of the intersection, before they reach a downstream lane-drop.

Participants

The two driving simulator experiments in this paper recruited participants from around the University of Massachusetts Amherst campus. The breakdown of age, sex, driving experience, and compensation for each study is explained in the following sections.

Experiment One: Auxiliary Through Lane Study. A total of 32 drivers (16 males, 16 females) were recruited and compensated to participate in this study. Each participant was compensated US$20 for their time. The ages of participants ranged from 18 to 62 years old with an average age of 23.6 years (SD = 7.9 years) and average driving experience of 7 years.

Experiment Two: Alternative Merge with Experimental Signage Study. An additional 24 participants (12 male, 12 female) were recruited and again compensated US$20 in the second experiment. The ages of participants ranged from 19 to 47 years old with an average age of 24 years (SD = 6.8 years) and average driving experience of 6 years.

Apparatus

Each of these driving simulator experiments were performed in the Arbella Insurance Human Performance Laboratory, located at the University of Massachusetts Amherst. A Realtime Technologies Inc. (RTI) driving simulator, depicted in Figure 1 used in the current study is a full-cab, fixed-base setup that includes a fully equipped 1996 Saturn sedan, with three screens subtending 135° horizontally. At a resolution of 1024 × 768 pixels and at a frequency of 60 Hz, the virtual environment is projected on each screen through a network of four advanced Realtime Technologies Inc. (RTI) simulator servers equipped with high-end, multimedia chips. The participant sits in the driver's seat and operates the controls, just as he or she would in a normal car. A Dolby surround sound system consisting of side speakers and two sub woofers located under the hood of the car provides realistic wind, road, and other vehicle noises with appropriate direction, intensity, and Doppler Effect Shift.

Scenarios and Experimental Design

The experimental design and focus of each driving simulator experiment is discussed in the following sections.

Figure 1. Driving simulator at University of Massachusetts Amherst.

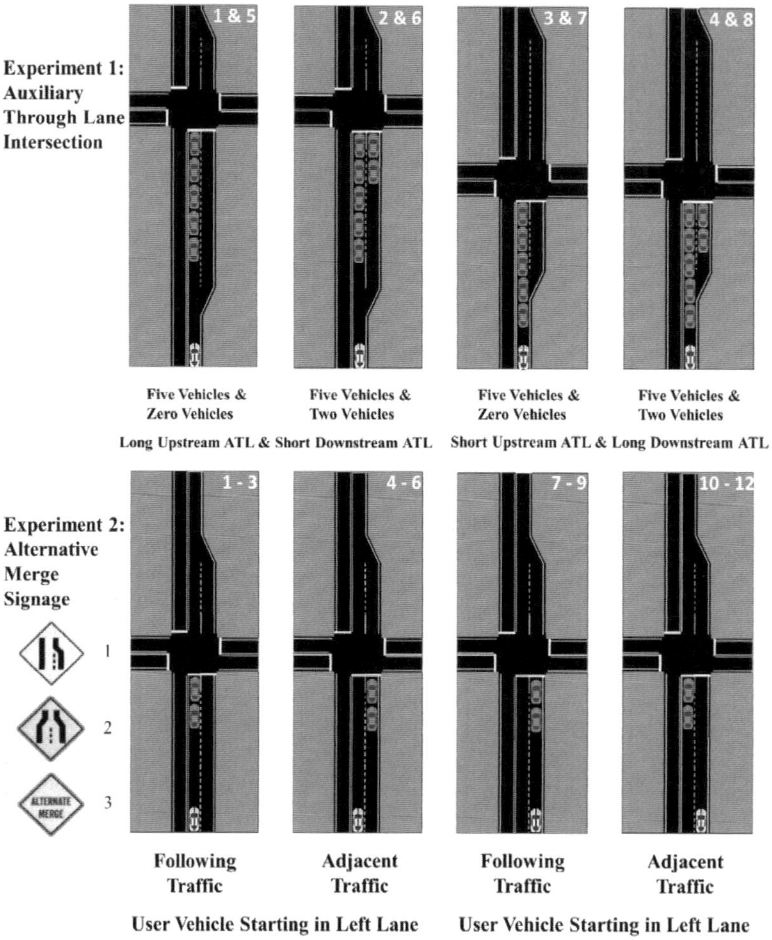

Figure 2. Driving simulator scenario breakdown for Experiment One (top) and Experiment Two (bottom).
Note: The white vehicles represent user vehicles and the gray vehicles represent scripted traffic.

Experiment One: Auxiliary Through Lane Study. In the first experiment, three factors were tested: ATL length, vehicle queues in CTL/ATL, and a familiarity concept. Within the simulator environment, participants were presented with combinations of long upstream and short downstream ATLs, or short upstream and long downstream ATLs; five vehicles present in the CTL with either zero or two vehicles in the ATL. Each of these combinations were presented to the participants before the researcher would step in and discuss, with brevity, the concept of engineering design in ATLs. The same combinations were run in similar order, thus providing a variable of familiarity for each participant. These options resulted in eight scenarios to be analyzed (Figure 2). It is important to note that only the geometric design and queue variables are displayed in this graphic, as the four scenarios were simply replicated to test the familiarity concept. Further, the density of roadside elements was reduced specifically at the merging points in each approach, providing less distraction to the participants and their driving actions. To eliminate any effects caused by the scenario order given to each participant, a Latin square was utilized to provide a balance of order. In this case, the Latin square was utilized for both the unfamiliar and familiar combinations of scenarios.

To better understand the operational impacts of ATLs, this study aimed to analyze the driver behavior associated with the introduction of ATLs at signalized intersections. In previous research, Tainter et al. completed a field study in Amherst, MA that ultimately laid the framework for the development of this experiment (*15*). The simulator environment developed for this experiment emulated this intersection from the field study. The results in this previous research found that a queue break point existed, meaning drivers were more likely to utilize the ATL once the queue reached a limit.

In addition, previous research found that the length of the ATL has an impact on lane utilization upstream of the intersection (*5*). These previously researched variables, including queue and lane design, were chosen for experimentation.

Experiment Two: Alternative Merge with Experimental Signage Study. In the second experiment, three factors were tested: initial lane positioning, traffic configuration, and experimental merge signage. Again, the simulator environment was used to convey combinations of left lane user vehicle starts or right lane user vehicle starts; user following traffic or user adjacent to traffic; and the standard merge sign, alternative merge sign, or word sign. These options resulted in 12 scenarios to be analyzed (Figure 2). Here, each of the three experimental merge signs was implemented on each of the four base scenarios. It is important to note that an overhead display of each experimental merge sign was presented at the beginning of the drive to convey the respective merge sign to be seen downstream of the intersection (Figure 3). The overhead displays were presented in an effort to eliminate any familiarity of scenarios that could affect the participants' actions. Further, the density of roadside elements was reduced specifically at the merging points in each approach, providing less distraction to the participants and their driving actions.

In order to understand the effects of downstream merging scenarios that are presented at lane-drops, a need exists to evaluate the upstream lane utilization. Previous research has studied the comprehension of experimental alternative merge signage. Ryan et al. developed a survey to analyze various alternative merge signage, in addition to the traditional merge sign (otherwise known as W4-2 in the MUTCD) (*16*). This previous research proposed further analyses on three signs, as seen in Figure 2. These experimental signs were well-received in previous research. In addition, these signs were originally tested to understand the driver's preferential lane choice, when provided with each of the signs at a downstream merge. As a result, three signs were selected for implementation in the second experiment: standard merge sign, alternative merge (geometry) sign, and word sign.

Figure 3. Screenshot of Experiment Two with traditional merge sign displayed overhead.

Procedure

For each of the two driving simulator experiments, participants completed a written consent form as well as pre-study questionnaire. The questionnaire collected demographic information of all participants in each study, as mentioned previously. After, participants were provided 5 min of practice in the driving simulator, enhancing their maneuvering capabilities of the vehicle before their experimental scenarios.

Experiment One: Auxiliary Through Lane Study. In the first experiment, each of the 32 participants completed a post-study questionnaire following their completion of the eight driving simulator scenarios. The questionnaire consisted of questions relating to ATL familiarity and opinions on geometric design, particularly the downstream merge design. Lastly, the final two questions focused on introducing the drivers to the joint-merge concept, mentioned previously in this paper. Each participant was given the opportunity to explain his or her thoughts on the existing design of ATLs comparatively with the alternative merge concept.

Experiment Two: Alternative Merge with Experimental Signage Study. In the second experiment, each of the 24 participants completed a computer-based static evaluation following the completion of their 12 driving simulator scenarios. This survey was developed, in part, through the research completed by Ryan et al (*16*). At the end of this evaluation, participants were allotted space to provide general feedback about the application of experimental alternative merge signs.

Results

Following the completion of the driving simulator experiments, an analysis of upstream lane utilization was conducted. The results from each experiment are presented in the subsequent sections.

Lane Utilization of Auxiliary Through Lane Intersections

Once all 32 participants completed the ATL driving simulator study, an analysis was performed on all eight scenarios. As previously mentioned, the scope of this research involved a combination of three key variables: ATL length, vehicle queues in CTL/ATL, and a familiarity concept.

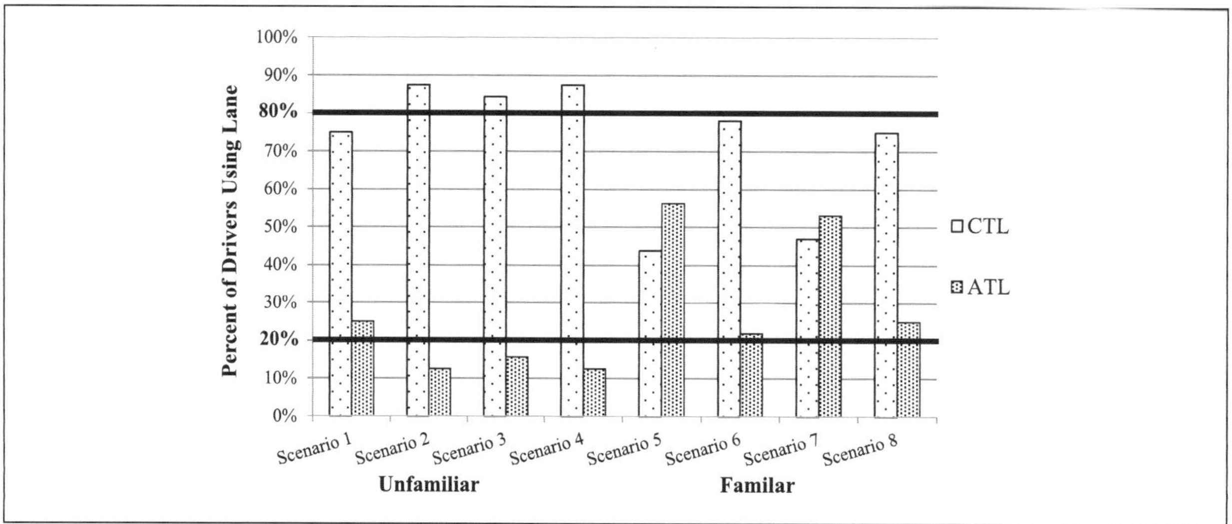

Figure 4. Comparison of lane utilization between ATL and CTL.
Note: The horizontal shaded regions above represent the expected 20% and 80% lane utilization, respectively, for ATLs and CTLs based on previous literature (*3*, *4*).

Table 1. Statistical Significance for Improving ATL Utilization

Variable	Option (ATL lane utilization)	Option (ATL lane utilization)	*P*-value
Vehicle queue in CTL/ATL	Five vehicles/two vehicles (33%)	Five vehicles/zero vehicles (35%)	0.87
Familiarity concept	Familiar (39%)	Unfamiliar (16%)	<0.01[a]
Auxiliary through lane length	Long upstream/short downstream lane (30%)	Short upstream/long downstream lane (27%)	0.44

[a]Statistical significance indicated by a p-value ≤ 0.05.

With the focus on upstream lane utilization at intersections with ATLs, the lane choice for each of the 32 participants was aggregated to display a percentage-based relationship (Figure 4). As mentioned previously, an analysis of driver familiarity with ATL intersections was conducted within this study. The results of this familiarity concept are displayed in Figure 4 below. Here, a significant difference in intersection operations appears as a function of the driver's familiarity of the ATL concept.

To test the statistical significance of the ATL length, vehicle queues in CTL/ATL, and familiarity concept, a collection of paired *t*-tests were performed (Table 1). Overall, each of the variables presented a unique result with respect to lane utilization of ATLs. The queue length variable presented to each driver did not yield significant results, meaning there was no significant impact from queue on ATL utilization. Conversely, the familiarity concept, introduced through this research, found a significant relation between the drivers' utilization of ATLs and their comprehension of the basic engineering design principles of ATLs. Most controversially, the geometric length of the ATL did not yield significant results on improving ATL utilization. These statistical results will be discussed in more detail later in the paper.

Comparison of Experimental Merge Signage

The second driving simulator experiment, as previously mentioned, was initiated to investigate the effects of upstream lane utilization based on the introduction of experimental merge signage downstream of the intersection. The variables taken into consideration in this experiment were: initial lane positioning, traffic configuration, and experimental merge signage.

The lane choices upstream for all 24 participants were aggregated and represented in terms of lane utilization versus experimental sign displayed (Figure 5). Here, the "Left Lane Start" and "Right Lane Start" represent the scenarios where the user vehicles were placed following scripted traffic upstream of the intersection. This graphic specifically represents a summation of the user vehicles, not including the scripted vehicles within the driving simulator scenario. It is important to note that the scenarios involving adjacent scripted traffic were left out of

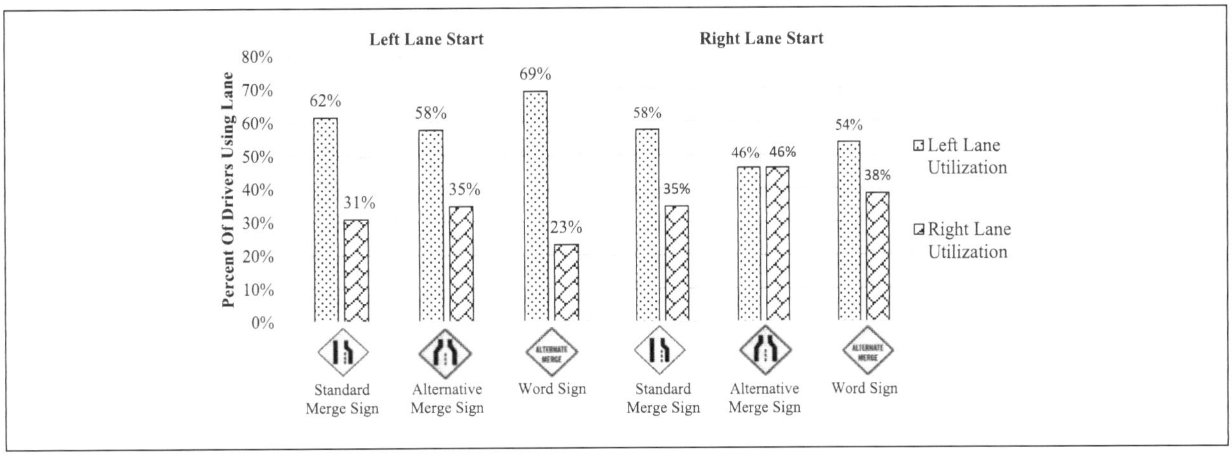

Figure 5. Comparison of lane utilization based on experimental merge signs displayed.

Table 2. Statistical Significance for Effecting Lane Utilization Upstream of Intersection

Variable	Option (right through lane utilization)	Option (right through lane utilization)	*P*-value
Lane positioning	Left lane start (18%)	Right lane start (33%)	0.04[a]
Traffic configuration	Adjacent traffic (6%)	Following traffic (42%)	<0.01[a]

[a]Statistical significance indicated by a *p*-value ≤ 0.05.

Figure 5, as these data were found to be insignificant to altering lane utilization.

Similar to the first experiment, a collection of paired *t*-tests was performed. These tests focused on understanding the significance in lane utilization based on traffic configuration and initial lane positioning. Each of these variables yielded significant results, as seen in Table 2. The lane positioning at the beginning of the scenario had a significant impact on the drivers' lane choice as they approached the signalized intersection. Similarly, the traffic configuration surrounding the vehicle (adjacent or following) also affected the lane utilization upstream of the intersection.

The significance of each sign impacting the lane utilization as drivers approached the intersection was inconclusive. These results will be discussed in more detail later in this paper.

Discussion

Based on the results of the two aforementioned driving simulator experiments, the following sections provide further analyses.

ATL Length and Familiarity vs. Lane Utilization

From the results displayed in Table 1, it is apparent that drivers are more comfortable with using the ATL if they have been previously exposed to its fundamental purpose. The lane choice data from scenarios 5 through 8 showed that drivers who were familiar with ATLs were drastically more likely to utilize the ATL than drivers unfamiliar with the concept. That being said, the results of lane utilization based on ATL length yielded the most conflicting results. Previous literature found that the length of the downstream ATL did not have an impact on the lane utilization upstream of the intersection (*2*). Although these results are consistent with this, the post-study questionnaire revealed that many of the drivers in this study appeared to dislike the rapid downstream merge that existed with a short downstream ATL length. Future research should further investigate the safety impacts of aggressive merging behavior at the downstream merge section, as correlated to the lane utilzation factors upstream of the intersection.

Lane Utilization Impact on Downstream Merging

The results provided in the first experiment in terms of unbalanced lane utilization at intersections with ATLs are consistent with previous literature (*3*, *4*). The lane utilization for the ATL, through all eight scenarios, averaged approximately 27%. It is important to note that this value skewed as a result of the significant impact of driver familiarity. Still, there is a unique merge dynamic that develops downstream of the intersection based on the unbalanced lane utilization. Therefore, a need to investigate alternative merge tactics became apparent in this experiment. The results of the post-study questionnaire found that participants were confused with the hypothetical concept of an alternative merge; however, some noted that with proper education, drivers could adapt to the new downstream merging maneuvers.

Potential for Improved Safety with Alternative Merge

According to recent literature, alternative merge tactics downstream of signalized intersections may lead to improvements in safety, as well as lower the impact of congested merging sections (*9*, *11*). The results from Experiment Two found that the participant's lane positioning at the beginning of each scenario had a significant impact on the lane choice at the signalized intersection. Further, the results in Figure 5 showed that the alternative merge sign had a slight impact on balancing lane utilization upstream of the intersection. Moreover, drivers that started their drives in the right lane while following traffic were likely to shift lanes and create a balanced lane split at the intersection. The experimental word sign did not yield any strong results; however, the results from the post-study static evaluation found that participants had a challenging time with word-based signs and preferred symbol signs instead. This finding was consistent with previous research on driver comprehension of alterative merge signage (*16*).

Limitations

Each of the two simulator experiments was conducted with 32 and 24 participants, repectively. An increased number of participants in both experiments could alter the outcome of the previously discussed results. In addition, the average age of participants for Experiment One and Two was 23.6 and 24 years, respectively. Many of the recruited participants were from the area surrounding the University of Massachusetts Amherst campus, therefore the average age was lower than preffered. Of note, age was not a variable that was controlled for in the experimental design because the focus was on infrastructure-related variables. The addition of older participants could affect the observed results and should be considered in any future experimentation.

Conclusion

This research employed an immersive driving simulator to better capture real-world driving behavior at ATL intersections and experimental merge scenarios. The first experiment sought to analyze the following key variables: ATL length, vehicle queues in CTL/ATL, and a familiarity concept. Although the familiarity concept yielded significant results in improving lane utilization balance upstream of ATL intersections, the ATL length did not have an impact on balancing lane utilization. The experimental merge concept was presented in the post-study questionnaire, and the feedback received resulted in an intrigue to employ this concept in a driving simulator environment.

The second experiment aimed to investigate the impact of experimental merge signage on the lane utilization upstream of signalized intersections. The variables analyzed in this study were: initial lane positioning, traffic configuration, and experimental merge signage. The traffic configuration at the start of the scenarios (i.e., following or adjacent) had a significant impact on the lane choice upstream of the intersection. Similarly, the vehicle lane positioning at the beginning of each scenario also had a significant impact on lane choice. The experimental merge signage, alternative merge sign, and word sign, did not have a significant impact on lane utilization upstream of the intersection; however, neither sign proved to cause any negative driving interaction or behavior throughout the study.

This research is one of the first studies to investigate alternative merge driver behavior in a dynamic driving environment. Future studies should focus on the safety impacts of alternative merge scenarios downstream of intersections.

Acknowledgments

This research was funded in part through a research project awarded by the Safety Research Using Simulation (SAFER SIM) University Transportation Center (UTC) at the University of Iowa. Funding for the UTC program is provided by the Office of Assistant Secretary for Research and Innovation (OST-R) of the United States Department of Transportation.

Author Contributions

Study conception and design: FT, AR, CF, EC, MK; data collection: FT, AR; analysis and interpretation of results: FT, AR, CF, EC, MK; draft manuscript preparation: FT, AR, CF, MK. All authors reviewed the results and approved the final version of the manuscript.

References

1. Fitzpatrick, K., B. A. Marcus, P. Dorothy, and E. S. Park. *NCHRP Report 780: Design Guidance for Intersection Auxiliary Lanes.* Transportation Research Board of the National Academies, Washington, D.C., 2014.
2. Kittleson & Associates, Inc. *NCHRP Web- Only Document 178: Assessment of Auxiliary Through Lanes at Signalized Intersections.* Transportation Research Board of the National Academies, Washington, D.C., 2011.

3. Nevers, B., S. Hermanus, Y. Merezczak, and Z. Clark. *NCHRP Report 707: Guidelines on the Use of Auxiliary Through Lanes at Signalized Intersections*. Transportation Research Board of the National Academies, Washington, D.C., 2011.
4. Bugg, Z., R. M. Nagui, and B. Schroeder. Lane Choice Model for Signalized Intersections with an Auxiliary Through Lane. *Journal of Transportation Engineering*, Vol. 139, No. 4, 2013, pp. 371–378. http://dx.doi.org/10.1061/(ASCE)TE.1943-5436.0000513.
5. Tarawneh, M. S., and T. M. Tarawneh. Effect on Utilization of Auxiliary Through Lanes of Downstream Right-Turn Volume. *Journal of Transportation Engineering*, Vol. 128, No. 5, 2002, pp. 458–464. https://doi.org/10.1061/(ASCE)0733-947X(2002)128:5(458).
6. Ring, J. B., and A. W. Sadek. Predicting Lane Utilization and Merge Behavior at Signalized Intersections with Auxiliary Lanes in Buffalo, NY. *Journal of Transportation Engineering*, Vol. 138, No. 9, 2012, pp. 1143–1150. https://doi.org/10.1061/(ASCE)TE.1943-5436.0000426.
7. Theiss, L., M. D. Finley, and G. L. Ullman. *Merging Taper Lengths for Short Duration Lane Closures*. Report 476990. Florida Department of Transportation, Tallahassee, FL, 2009.
8. *Manual on Uniform Traffic Control Devices*. FHWA, U.S. Department of Transportation, Washington, D.C., 2009.
9. Feldblum, E. G. *Alternative Merge Sign at Signalized Intersections*. Report No. CT-2233-F-05-4. Bureau of Engineering and Highway Operations Division of Research, Connecticut Department of Transportation, Newington, CT, 2005.
10. Grillo, L., T. Datta, and C. Hartner. Dynamic Late Lane Merge System at Freeway Construction Work Zones. *Transportation Research Record: Journal of the Transportation Research Board*, 2008. 2055: 3–10.
11. Wolshon, B., S. Ishak, and W. Idewu. *Design of Lane Merges at Rural Freeway Construction Work Zones*. Publication FHWA/LA.11/484. FHWA, U.S. Department of Transportation, Washington, D.C., 2012.
12. Idewu, W. *Development and Operational Analysis of Highway Alternating Merge Transition Zones*. Doctoral Dissertations, Louisiana State University, Baton Rouge, LA, 2009.
13. National Center for Statistics and Analysis. *2014 Motor Vehicle Crashes: Overview*. Publication Traffic Safety Facts Research Note. Report No. DOT HS 812 246. National Highway Traffic Safety Administration, Washington, D.C., 2016.
14. Council, F., D. Harkey, D. Nabors, A. Khattak, and Y. Mohamedshah. Examination of Fault, Unsafe Driving Acts, and Total Harm in Car-Truck Collisions. *Transportation Research Record: Journal of the Transportation Research Board*, 2003. 1830: 63–71.
15. Tainter, F., M. A. Knodler, C. Fitzpatrick, and E. Christofa. Analysis of Driver Performance and Operations at Intersection Auxiliary Through Lanes. Presented at Road Safety and Simulation International Conference, Orlando, FL, 2015.
16. Ryan, A., F. Tainter, C. Fitzpatrick, M. Page, E. Christofa, and M. A. Knodler, Jr. Alternative Merge Design Downstream of Intersections. Presented at Road Safety and Simulation International Conference, The Hague, Netherlands, 2017.

The Standing Committee on Simulation and Measurement of Vehicle and Operator Performance (AND30) peer-reviewed this paper (18-03610).

Article

TRR
JOURNAL OF THE TRANSPORTATION RESEARCH BOARD

Transportation Research Record
2018, Vol. 2672(37) 132–140

DOI: 10.1177/0361198118776519
journals.sagepub.com/home/trr
SAGE

Investigation of Time and Speed Perception using a Driving Simulator

Andronikos Keklikoglou[1], Cole D. Fitzpatrick[1], and Michael A. Knodler[1]

Abstract

Thirty-four drivers participated in a driving simulator experiment which investigated time and speed perception as it related to cognitive workload resulting from secondary tasks. Each participant drove the virtual drive twice, once with either an audio or a map task and again with no distractions as a control. Participants knew from a practice drive that they would be asked to estimate their speed and time duration of driving, thus this study used the prospective paradigm. Based on previous literature it was expected that there would be an underestimation of time and an overestimation of speed. The reverse occurred, participants overestimated time and underestimate their speed. This suggests that drivers may have found the drive unstimulating, despite the secondary tasks, and that the rural environmental may have impacted speed perception. In addition, a large group of participants, nine out of 34, crashed the virtual vehicle at a horizontal curve that was not problematic in previous simulator studies. When investigating these crashes further, it was found that drivers who crashed in the second drive had significantly worse time perception in the first drive as compared to drivers who did not crash in the second drive. This finding suggests that current time perception may be a predictor of future speed selection.

Distracted driving is defined as any activity that diverts attention from the primary driving task, including calling or texting, eating or drinking, talking to passengers in your vehicle, playing with the radio, or adjusting an entertainment or navigation system (*1*). Distracted driving is a major problem for roadway safety when the usage of mobile phones and other electronic devices becomes the primary task instead of paying attention to the road. It has been estimated that 660,000 drivers use electronic devices when driving every day (*1*). In 2015, 3,477 people lost their lives because of distracted driving and more than 391,000 were injured in the United States (*1*). Young drivers, 16–24 years old, were the largest age group reported as being distracted in fatal crashes. In research conducted by the National Highway Traffic Safety Administration (NHTSA), it was estimated that distracted driving is a contributing cause in 25 % of police-reported crashes.

Previous research has shown that people can be distracted in autonomous vehicles (AVs) if there are pre-alerts before the takeover occurs and the driver takes control of the vehicle (*2*). Without alerts or AVs, driving performance is degraded significantly by cell phone use (*3*). However, drivers might also become too distracted to the point that they are unable to meet the demands of the driving environment or counteract for the loss in attention to the primary driving task when manual driving becomes necessary. For example, distracted drivers have less time to reduce their speed or perform maneuvers caused by slower reaction times (*4*). In this current study, drivers' perception of time and speed will be investigated from a distracted driving perspective.

Cognitive load is defined in psychology as the total amount of mental effort being used in the working memory, which directly affects short-term decision making (*5*). Previous studies have shown that when cognitive load increases, the performance of driving tasks which require cognitive control degrades (*6*). In this study, we will observe how drivers react under varying levels of cognitive loads and in what way cognitive loads may or may not alter their perception of time and speed.

Most of the studies that have been conducted which focused on the perception of time used two approaches, the retrospective and the prospective paradigms. The retrospective paradigm approach is when study participants are unaware that they are going to be asked to estimate

[1]Department of Civil and Environmental Engineering, University of Massachusetts, Amherst, MA

Corresponding Author:
Address correspondence to Cole D. Fitzpatrick: cfitzpat@umass.edu

the size of an interval. In the prospective paradigm, participants are aware that they will be asked to estimate the length of an interval during the study (*7*). In the prospective paradigm, which is what will be used in this specific study, subjects tend to underestimate time if the cognitive load is high. Cognitive load has been identified as a main factor influencing interval duration estimates for both the retrospective and prospective paradigms, regardless of the paradigm chosen (*8*). By contrast, when a task has a low cognitive load, or is boring, previous research has shown that time durations are overestimated (*9*). It is important to note that previous studies involving this simulator found that participants, with no distractions, perceived their travel speed to be approximately 5 mph higher in the driving simulator but actual speeds observed in the field closely matched the speeds observed in simulated environments which replicated the field drive (*10*, *11*).

Although there is significant research on time perception as the cognitive load varies, there is limited research on how speed perception varies with cognitive load. There is a need to determine how drivers perceive time and speed when distracted, versus when they are solely focused on the driving task. More specifically, there is a need to quantify how different types of distractions, such as texting or navigating, affect drivers' perception of time and speed. This is critical because with the implementation of automation, drivers will often need to quickly re-engage in the driving task and speed and time perception could influence subsequent speed selection. The research presented here investigates the extent to which drivers overestimate and the speeds and times under varied levels of distraction.

Methods

The driving simulator in the Arbella Insurance Human Performance Laboratory at UMass Amherst was used to conduct a simulation experiment with a previously-developed environment (*12*) which was altered to decrease the visual cognitive workload outside of the car so that drivers could concentrate on in-vehicle secondary tasks (distractions). A user-friendly environment was developed to accurately examine how participants perceived time and speed under varying cognitive workloads. This section details the methods that were used to address the objectives of this study.

Participants

Thirty-four licensed drivers, 19 years and older (19 males and 15 females) from the greater Amherst, MA area were recruited to participate in this study. During the recruitment stage, it was advertised that participants would be paid $20 compensation for their time and that people who experienced motion sickness should not participate. Participants were selected based on the need for a wide variety of age groups to participate in the study. As younger drivers more frequently text while driving than older drivers, the majority of participants were in the 18–24 age group. Moreover, a larger number of male participants were not able to complete some of the drives, so extra male participants were recruited in an effort to have a similar number of completed experiments for both males and females.

The experiment consisted of four different groups (Table 1), all of which drove the same virtual scenario twice, once with no distraction and once with either an audio or navigating distraction. Most of the participants completed all of the drives, without any issues. One participant did not complete their last drive, because of technical difficulties with the simulator that resulted in a partial dataset. Nine participants did not complete one of their drives, because of high speeds along turns which led to a "crash" of the virtual vehicle. These crashes also resulted in partial datasets. The reason for these crashes along with their implications is discussed in the section on Crashes. A comparison of participant demographics by group is shown in Table 1.

Before the virtual practice drive, participants completed a questionnaire which evaluated their aggressive driving tendencies. Participants were asked to rate each question as either "Never," "Rarely," "Sometimes," or "Often." The questionnaire included 13 actions such as "Tailgate others to force move" and "Deliberately prevent other from passing." By assigning a value of 1–4 for Never to Often, a mean aggressiveness score could be computed for each participant and thus each group. The mean scores of the different groups were statistically

Table 1. Participant Demographics

Group: drive 1/drive 2	Male total (complete)	Female total (complete)	Driver age in years mean ± SD	Driving experience in years mean ± SD
Control/map	4 (3)	3 (3)	25.3 ± 4.3	9.0 ± 4.4
Map/control	5 (3)	4 (3)	21.2 ± 3.2	3.1 ± 1.9
Control/audio	4 (4)	3 (2)	26.9 ± 5.5	8.7 ± 7.3
Audio/control	6 (3)	5 (3)	27.2 ± 10.7	9.2 ± 7.9

Figure 1. Driving simulator at UMass Amherst.

similar for the four groups this fact, coupled with the balancing of age and sex, indicates that the driving tendencies of the four groups were likely similar.

Apparatus

A Realtime Technologies Inc. (RTI) driving simulator, depicted in Figure 1, was used in the current study and is a full-cab, fixed-base, setup that includes a fully equipped 1996 Saturn sedan, with three screens subtending 135º horizontally. At a resolution of 1024 × 768 pixels and at a frequency of 60 Hz, the virtual environment is projected on each screen through a network of four advanced RTI simulator servers equipped with high-end, multimedia chips. The participant sits in the driver's seat and operates the controls, just as he or she would in a normal car. A Dolby surround system consisting of side speakers and two sub woofers located under the hood of the car provides realistic wind, road and other vehicle noises with appropriate direction, intensity and Doppler effect shift.

Scenarios and Experimental Design

The entire drive consisted of a rural two-lane roadway with no posted speed limit and contained two signalized intersections, Figure 2. At these intersections, drivers had to make an unprotected left turn with vehicles traversing the intersection in the opposite direction.

There were three left horizontal curves and three right horizontal curves. Each curve had a length of 157 m and radius of 100 m. Lanes were 3.66 m wide (12 ft) with a 0.30 m shoulder (1 ft). There were no significant roadside objects or hazards. Individual vehicles were scripted in the oncoming direction to travel at 35 mph. This individual scripting allowed the oncoming traffic to be consistent for every participant.

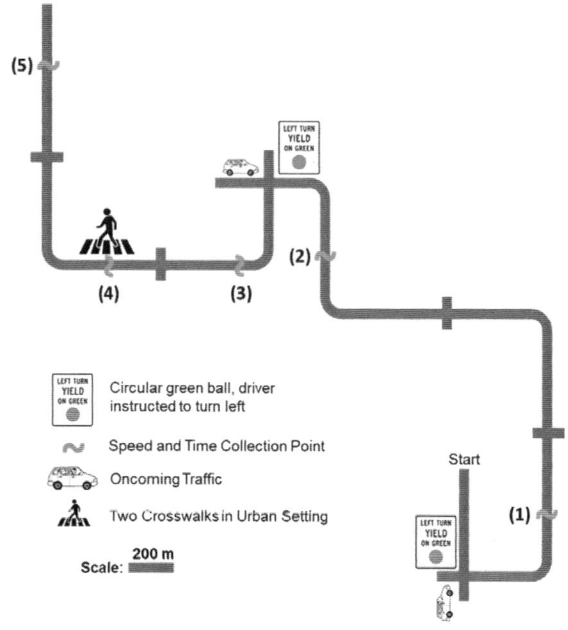

Figure 2. Schematic of virtual drive depicting various elements participants encountered.

"Pass with Care" signs were placed along tangent segments throughout the drive and served as a cue to the experimenter to play the audio which asked participants, "How fast are you currently going? How long have you been driving?" Participants left all watches and phones outside of the vehicle and the car clock and speedometer were obscured so that participants had no way to definitively measure their time or speed. Each experimental drive lasted 9–12 minutes. A full layout of the virtual drive is depicted in Figure 2.

Procedures

The objectives of this study were addressed by subjecting participants to two types of distractions, an audio distraction task or a visual distraction task. By using both distraction types, the amount, and type, of cognitive load used on each task could be examined. First, each participant provided informed consent, filled out a demographic and driving aggressiveness questionnaire then completed a short practice drive to get used to the features and controls of the driving simulator. In the practice drive, the distraction tasks and questions were presented and explained so that participants were comfortable with the tasks and questions during the experimental drive. Next, participants were randomly assigned with a distraction task to complete throughout a drive, either an audio or visual distraction task. The participants also had to complete the same drive without the distraction task. The two drives occurred in a random order so that the ordering of the drives did not produce a confounding effect. The three different drive types were the following:

1. Audio Distraction Task (Audio Task): Participants were asked to verbally respond to a sentence, played through the speakers, by stating the subject and object of the sentence, as well as whether or not the sentence made sense. Example: The kid threw the ball. Subject: the kid, object: the ball, sentence makes sense.
2. Visual Distraction Task (Map Task): Participants were asked through the speakers, to locate a street on a map while driving. They had to verbally respond with the coordinates of where the street name appeared on the map.
3. Control: In this task, participants were only asked to perceive the time intervals since the start of the drive and the speed at each checkpoint. The control drive did not include distractions.

Results and Discussion

The current driving simulator study examines how drivers' perception of time and speed is impacted by secondary tasks and varying workload levels. A hybrid between/within-subjects experimental design was utilized in which each participant did a control drive with no secondary task and a second drive with either an audio or map task. The order of the drives was counter-balanced, so half the participants did the control drive first and the distraction task second and half did the distraction task drive first. All statistical tests conducted were unpaired two-sample Student's t-tests using the software package Minitab. All error bars represent 95 % confidence intervals, a statistically significant difference at 90 % ($p < 0.10$) is denoted by (*) and a statistically significant difference at 95 % ($p < 0.05$) is denoted by (**).

Actual Speeds and Speed Perception

As depicted previously in Figure 2, actual speeds and times along with perceived times and speeds were collected at five fixed points spaced roughly even throughout the drive. Figure 3 displays the mean speeds at these five fixed points for the three drive types. Within the groups when comparing checkpoints, checkpoint three and four had statistically lower mean speeds than checkpoints one, two, and five for both the control drive and audio task, but not the map task. Looking back at Figure 2, lower speeds at checkpoint three were likely

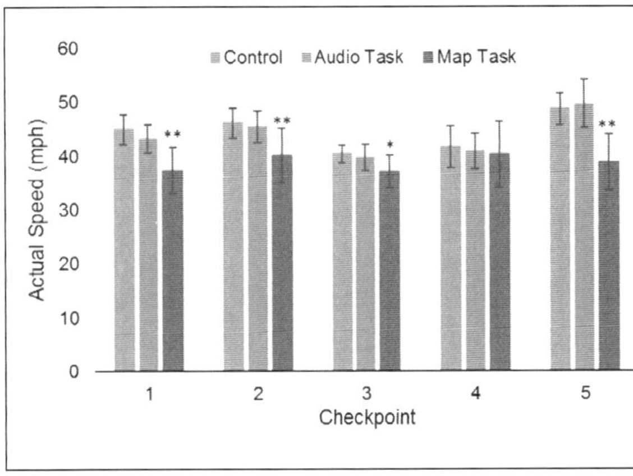

Figure 3. Mean actual speeds of participants for three drive types at five checkpoints.

because of its proximity to the right horizontal curve and lower speeds at checkpoint four were likely because of the urban stretch of roadway with crosswalks. By comparison, the other three checkpoints occurred on straight rural environments. Comparing the groups, there were no statistically significant differences in actual speeds between the audio task and control drives. However, speeds at the following checkpoints for the map task were significantly lower than the control drive: checkpoint one, control (M = 45.0, SD = 7.8), map task (M = 37.3, SD = 8.4); t(27) = 3.10, p = 0.004; checkpoint two, control (M = 46.2, SD = 7.0), map task (M = 40.2, SD = 9.5); $t(24)$ = 2.21, p = 0.037; checkpoint three, control (M = 40.4, SD = 4.4), map task (M = 37.2, SD = 6.0); $t(24)$ = 1.92, p = 0.066; checkpoint five, control (M = 48.7, SD = 8.2), map task (M = 38.8, SD = 8.2); $t(24)$ = 1.92, p = 0.002. The varying degrees of freedom in the unpaired two-sample t-test are because of the varying sample size at the five checkpoints as a result of the "crashes" which happened at various points throughout the drive. The cause of these "crashes" and the insights they provide on experimental design are discussed later in this section.

Table 2 displays the sample sizes at each checkpoint in the brackets, along with the mean actual and perceived speeds. The sample size for the control group was much higher than the audio or map tasks because all subjects drove the control drive, whereas the subjects drove with either the map or audio task. The mean actual speeds are an early indicator that the map task was a larger distraction than the audio task and resulted in a higher cognitive workload. Subjects did not have to reduce speeds to safely complete the audio task, but speed reduction was necessary for the map task.

At the five fixed points where speed was captured, participants were asked how fast they thought they were going. Without having a speedometer, the vibrations of

Table 2. Mean Actual and Perceived Speeds and Sample Size at Each Checkpoint for the Three Groups

Checkpoint	Mean speed in mph (sample size at checkpoint) mean perceived speed								
	Control			Audio task			Map task		
1	45.0	(34)	37.2	43.3	(18)	37.8	37.3	(16)	31.7
2	46.2	(29)	41.2	45.5	(18)	40.7	40.2	(16)	36.1
3	40.4	(28)	38.6	39.8	(17)	38.7	37.2	(16)	34.1
4	41.7	(27)	32.9	40.9	(17)	33.6	40.3	(16)	34.2
5	48.7	(26)	42.3	49.6	(17)	39.3	38.8	(13)	34.9

Table 3. Differences in Speed Perception and Statistical Comparisons with Checkpoint 3

Checkpoint	Mean difference between actual and perceived speed in mph			p-value of difference with checkpoint #3 (difference in parentheses)		
	Control	Audio task	Map task	Control	Audio task	Map task
1	7.8	5.5	5.6	0.003 (56)*	0.130 (31)	0.386 (29)
2	5.0	4.8	4.1	0.126 (54)	0.198 (31)	0.775 (28)
3	1.8	1.0	3.1	N/A	N/A	N/A
4	8.8	7.3	6.1	0.002 (52)*	0.032 (30)*	0.383 (29)
5	6.3	10.3	3.9	0.143 (38)	0.012 (25)*	0.834 (23)

*Significance at 95% confidence ($p < 0.05$).

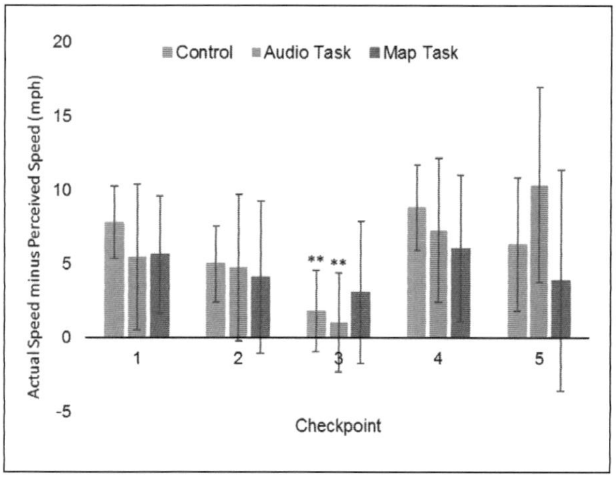

Figure 4. Mean differences between actual and perceived speeds for the three drive types.

pavement conditions from a real vehicle, or the familiarity of their own vehicle, this question was very difficult for participants. Figure 4 shows the difference between actual speeds and perceived speeds at the five checkpoints across the three groups. When combining all groups and all checkpoints, subjects perceived their speed 5.6 mph lower than their actual speeds. This finding is counter to the results found in a previous study which found speed perception to be ~5 mph higher than actual simulator speeds (*7*). However, in this study, the simulated rural environment provided minimal visual flow, whereas in the previous study a more stimulating suburban environment was utilized. This open environment was likely the cause of the lower perceived speeds as previous research on roadside vegetation and clear zone size showed that as the density of roadside vegetation, and thus the level of visual flow increases, perceived speeds exceed actual speeds (*9*).

When comparing speed perception by drive type at each checkpoint, there were no statistically significant differences. The largest difference occurred at the last checkpoint between the audio task and map task, however because of the smaller sample size of the distraction drives by the fifth checkpoint ($n = 17$ for audio task and $n = 14$ for map task), this difference was not statistically significant. Statistically significant differences were present within the groups between the different checkpoints. For all three groups, subjects had more accurate speed perception at checkpoint three than at any other checkpoint with significant differences occurring with the control and audio task groups. For this reason, the four other checkpoints were compared to checkpoint three to evaluate differences. Table 3 shows the mean differences in speed perception at each checkpoint along with statistical comparisons of checkpoints three and the other four checkpoints.

The reason for the improved speed perception at checkpoint three was likely because of the lower actual speeds at checkpoint three. As discussed previously, these significantly lower speeds only existed in the control and

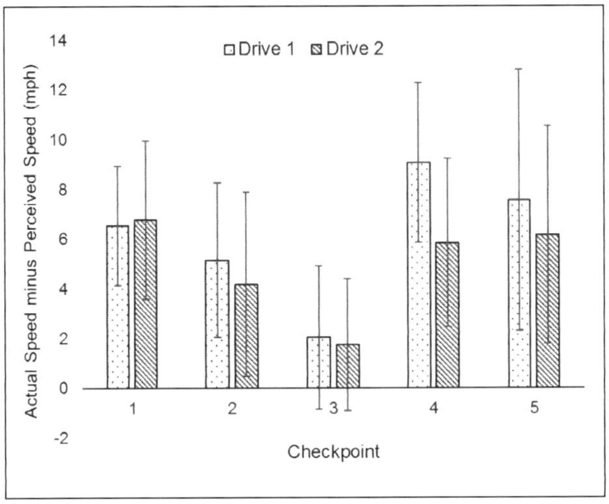

Figure 5. Mean differences between actual and perceived speeds for drive one and two.

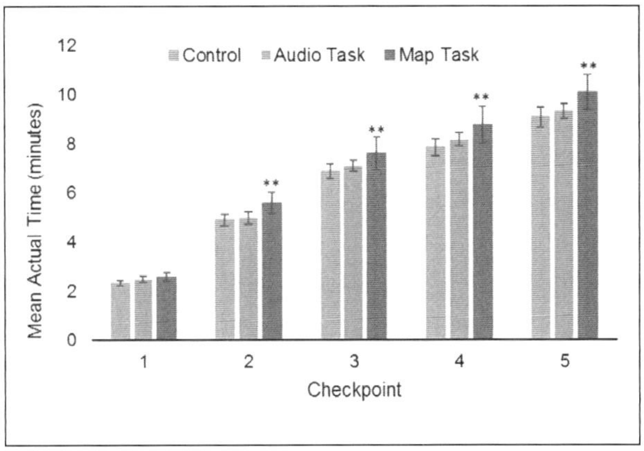

Figure 6. Mean actual times at each checkpoint for the three groups.

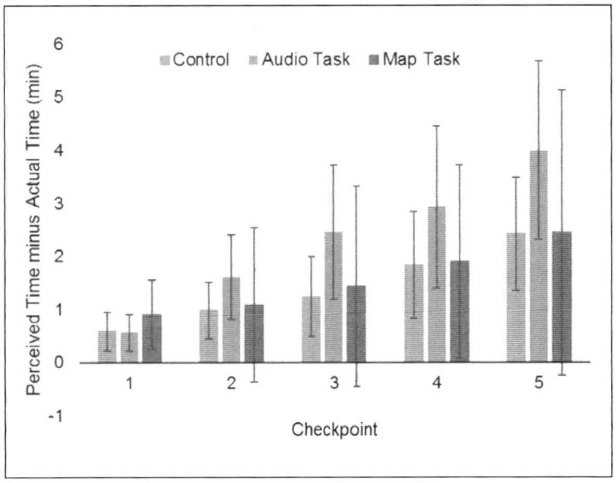

Figure 7. Mean differences in perceived time and actual time for the three drive types.

audio task groups and those were the same groups which had significantly better speed perception. This result indicates that speed perception is easier at lower speeds because of the smaller margin of error. Although checkpoint four also had significantly lower actual speeds, there was not the same improvement in speed perception as seen at checkpoint three. This is likely because of the roadway environment, checkpoint three was in the same rural environment as the other checkpoints but checkpoint four was in an urban stretch of the drive with crosswalks and parked vehicles. Although participants did slow down at checkpoint four, the checkpoint occurred toward the end of the urban section so their actual speed was not as low as they had perceived over the past minute of urban driving.

Figure 5 shows the mean speed perception based on drive order. Although there appears to be a slight improvement in speed perception between the first and second drive, specifically in checkpoints four and five, these differences were not statistically significant (p = 0.162 and 0.679, respectively). There were no differences in actual speeds based on drive order.

Time Perception

The second focus of this study was time perception, participants were asked at the same five checkpoints how long they thought they had been driving. The clock in the vehicle was obscured and watches/phones were left outside of the vehicle. Figure 6 displays the mean actual times of the three groups at each checkpoint. These times obviously correlate with the vehicle speeds. There are no significant differences between the control and audio task but subjects with the map task took 68 seconds longer to reach the fifth checkpoint than subjects in the control drive, a statistically significant result (p = 0.013).

In the introduction, previous research by Sweller was cited which mentioned that subjects in the prospective paradigm tended to underestimate time and Engström et al. found that when the cognitive load was low, the task was "boring" and subjects overestimated time (*5*, *6*). Figure 7 shows the difference in speed perception at the five checkpoints between the three groups. Despite using the prospective paradigm, in which subjects knew they would be asked about time duration, all three groups overestimated the time interval, possibly indicating that the driving simulator task was found to be boring or that the cognitive workload was low despite the secondary tasks. Although a trend shows that time perception with the audio task was worse than with the map task or the control, these differences were not statistically

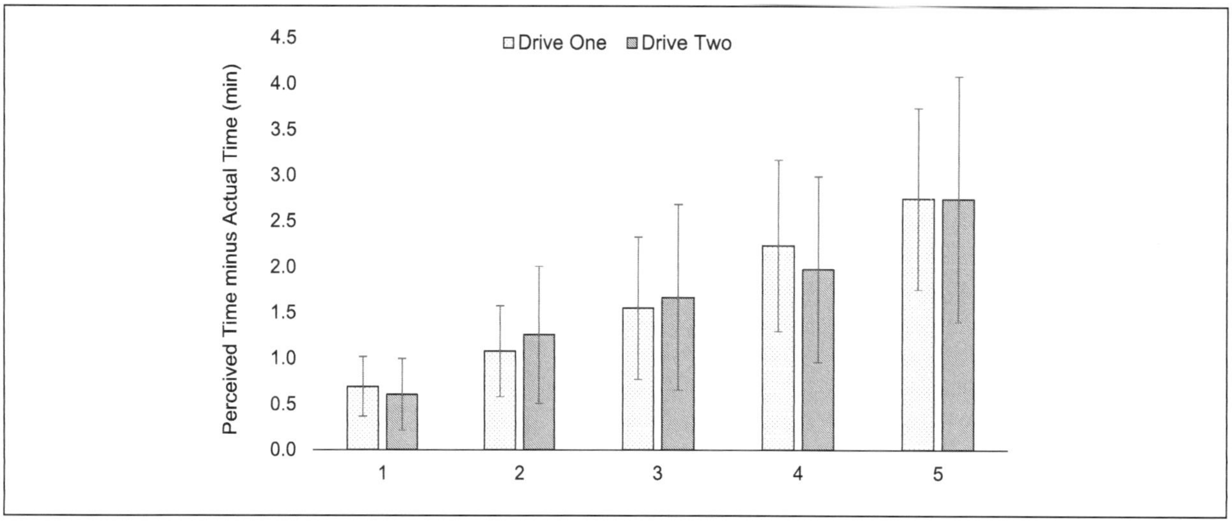

Figure 8. Mean differences in perceived time and actual time for drive one versus drive two.

significant. This difference could have been the result of an individual subject who poorly perceived time during the experiment and did the audio task but not the map task.

Figure 8 examines the difference in time perception between the first drive and the second drive. There were no significant changes in time perception between the drives, this was the expected result as participants did not receive any feedback on the accuracy of their time perception between drive one and drive two.

Crashes

As mentioned previously, nine out of the 34 drivers crashed the virtual vehicle, ending the drive and resulting in a partial dataset. Six subjects crashed in the second drive, two crashed during the first drive, and one crashed in both the first and second drive. All of these crashes occurred on a right horizontal curve. The same roadway geometry was used by Fitzpatrick et al. when examining roadside design and no crashes occurred (*13*). However, during that study, the curve warning signs had an advisory speed and participants could see their speedometer. While there were curve warning signs in the current study, there were no advisory speed limits (so that speed perception was not influenced) and the speedometer was obscured. When looking at the last recorded speed of crashing drivers compared to speeds of non-crashing drivers, it is clear that these crashes occurred because of excessive speeds: last speed recorded of crashing drivers, (M = 54.8, SD = 11.3), all speeds of non-crashing drivers (M = 41.6, SD = 7.6); $t(9) = -3.67, p = 0.005$).

Figure 9 further investigates the issue of crashing drivers by comparing the perceived time differences during

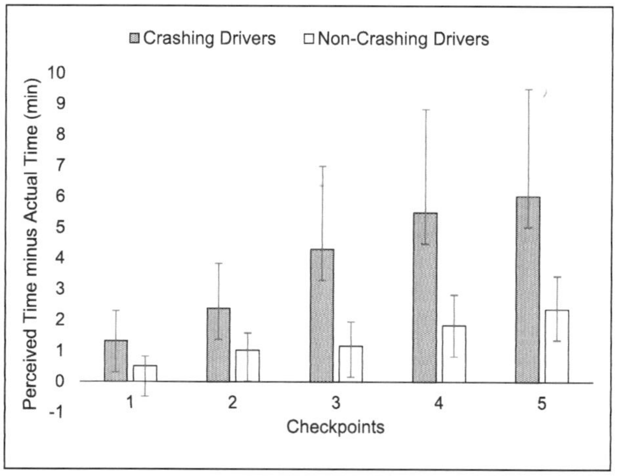

Figure 9. Mean differences in perceived and actual times during the first drive for drivers who crashed in the second drive versus drivers who did not crash.

the first drive of crashing drivers who crashed in the second drive versus drivers who did not crash in the second drive. Despite the extremely small sample of drivers who crashed in the second drive and not in the first ($n = 6$), the difference in time perception with drivers who did not crash at all was statistically significant at 90 % for checkpoints three, four, and five. This shows that drivers who perceived a longer time interval during the first drive, or were bored, were more likely to crash during the second drive. A previous study showed that boredom can be induced by low or constant speeds, which might lead to frustration and speeding (*14*). This is an interesting finding as it suggests that current time perception may be used as a predictor for future speed selection.

Limitations and Future Work

The limited sample size was a significant limitation to this study. The high number of crashes because of excessive speeds around corners was unanticipated and severely limited the power of the statistical analyses. A future study should be conducted which alters the roadway geometry so that the horizontal curves are more safely navigated at higher speeds. The overestimation of time in the prospective paradigm was unexpected based on previous literature, a future study could use a more stimulating roadway environment to see if the roadway environment significantly affects the cognitive stimulation and thus influences time perception. A more stimulating environment may also lead to more accurate speed perception, making the distinction worth exploring. Additionally, assigning participants to one secondary task or the other may obfuscate the results because of individual behavioral variances. A future study should focus on a single secondary task so that definitive conclusions can be drawn about the influence of distraction on time and speed perception. Another future study could examine whether time and speed perception is an individual skill or is simply influenced by the situation. Finally, a future study should further investigate our finding that current time perception is a predictor of future speed selection. If this finding was confirmed, safety could be improved through early interventions when unsafe speed selections are predicted.

Conclusion

Thirty-four drivers participated in a driving simulator experiment which investigated time and speed perception as it related to cognitive workload resulting from secondary tasks. Each participant drove the virtual drive twice, once with either an audio or a map task and again with no distractions as a control. Participants knew from a practice drive that they would be asked to estimate their speed and time duration of driving, thus this study used the prospective paradigm. Based on previous literature it was expected that there would be an underestimation of time and an overestimation of speed. The reverse occurred, participants overestimated time and underestimate their speed. This suggests that drivers may have found the drive unstimulating, despite the secondary tasks, and that the rural environmental may have impacted speed perception. Additionally, a large group of participants, nine out of 34, crashed the virtual vehicle at a horizontal curve that was not problematic in previous simulator studies. When investigating these crashes further, it was found that drivers who crashed in the second drive had significantly worse time perception in the first drive as compared to drivers who did not crash in the second drive. This finding suggests that current time perception may be a predictor of future speed selection.

Author Contributions

Study conception and design was carried out by Andronikos Keklikoglou, Cole Fitzpatrick, and Michael Knodler. Data collection was performed by Andronikos Keklikoglou and Cole Fitzpatrick. Analysis and interpretation of results was carried out by Andronikos Keklikoglou, Cole Fitzpatrick, and Michael Knodler. Draft manuscript preparation was carried out by Andronikos Keklikoglou and Cole Fitzpatrick. The final manuscript was prepared by Andronikos Keklikoglou, Cole Fitzpatrick, and Michael Knodler. All authors reviewed the results and approved the final version of the manuscript

References

1. National Center for Statistics and Analysis. *Pedestrians: 2013 Data*. Traffic Safety Facts. Report No. DOT HS 812 124. National Highway Traffic Safety Administration, Washington, D.C. http://www-nrd.nhtsa.dot.gov/Pubs/812124.pdf. Accessed November 4, 2015.
2. van der Heiden, R. M. A., S. T. Iqbal, and C. P. Janssen. Priming Drivers before Handover in Semi-Autonomous Cars. *Proc., IGCHI Conference on Human Factors in Computing Systems*, Denver, Colorado, ACM, New York, 2017, pp. 382–404.
3. Caird, J. K., C. R. Willness, P. Steel, and C. Scialfa. A Meta-Analysis of the Effects of Cell Phones on Driver Performance. *Accident Analysis & Prevention*, Vol. 40, No. 4, 2008, pp. 1282–1293.
4. Haigney, D., R. Taylor, and S. Westerman. Concurrent Mobile (Cellular) Phone Use and Driving Performance: Task Demand Characteristics and Compensatory Processes. *Transportation Research Part F: Traffic Psychology and Behaviour*, Vol. 3, No. 3, 2000, pp. 113–121.
5. Sweller, J. Cognitive Load During Problem Solving: Effects on Learning. *Cognitive Science*, Vol. 12, No. 2, 1988, pp. 257–285.
6. Engström, J., G. Markkula, T. Victor, and N. Merat. Effects of Cognitive Load on Driving Performance: The Cognitive Control Hypothesis. *Human Factors*, 2017, Vol. 59, No. 5, pp. 734–764.
7. Block, R. A. *Cognitive Models of Psychological Time*. L. Erlbaum Associates, Hillsdale, N.J., 1990.
8. Block, R. A., and D. Zakay. Prospective and Retrospective Duration Judgments: A Meta-Analytic Review. *Psychonomic Bulletin & Review*, Vol. 4, No. 2, 1997, pp. 184–197.
9. Sucala, M., S. Stefan, A. Szentagotai-Tatar, and D. David. Time Flies When You Expect to Have Fun. An Experimental Investigation of the Relationship Between Expectancies and the Perception of Time Progression. *Cognition Brain Behavior*, Vol. 14, 2010, pp. 231–241.
10. Hurwitz, D. S., M. A. Knodler, and D. Dulaski. Speed Perception Fidelity in a Driving Simulator Environment. *Proc., Driving Simulation Conference 2005 North America*, Driving Simulator Conference of North America, Orlando, Fla., 2005, pp. 343–352.

11. Hurwitz, D. S. *Static and Dynamic Evaluation of the Driver Speed Perception and Selection Process*. University of Massachusetts Amherst, 2006.
12. Fitzpatrick, C. D., S. Samuel, and M. A. Knodler. The Use of a Driving Simulator to Determine How Time Pressures Impact Driver Aggressiveness. *Accident Analysis and Prevention*, Vol. 108, 2017, pp. 131–138.
13. Fitzpatrick, C. D., S. Samuel, and M. A. Knodler. Evaluating the Effect of Vegetation and Clear Zone Width on Driver Behavior Using a Driving Simulator. *Transportation Research Part F: Traffic Psychology and Behaviour*, Vol. 42, 2016, pp. 80–89.
14. Steinberger, F., A. Moeller, and R. Schroeter. The Antecedents, Experience, and Coping Strategies of Driver Boredom in Young Adult Males. *Journal of Safety Research*, Vol. 59, 2016, pp. 69–82.

The Standing Committee on Simulation and Measurement of Vehicle and Operator Performance (AND30) peer-reviewed this paper (18-01281).

TRR
JOURNAL OF THE TRANSPORTATION RESEARCH BOARD

Article

Transportation Research Record
2018, Vol. 2672(37) 141–152

DOI: 10.1177/0361198118790372
journals.sagepub.com/home/trr
SAGE

Multi-Level Driver Workload Prediction using Machine Learning and Off-the-Shelf Sensors

Paul van Gent[1], Timo Melman[2], Haneen Farah[1], Nicole van Nes[3], and Bart van Arem[1]

Abstract

The present study aims to add to the literature on driver workload prediction using machine learning methods. The main aim is to develop workload prediction on a multi-level basis, rather than a binary high/low distinction as often found in literature. The presented approach relies on measures that can be obtained unobtrusively in the driving environment with off-the-shelf sensors, and on machine learning methods that can be implemented in low-power embedded systems. Two simulator studies were performed, one inducing workload using realistic driving conditions, and one inducing workload with a relatively demanding lane-keeping task. Individual and group-based machine learning models were trained on both datasets and evaluated. For the group-based models the generalizing capability, that is the performance when predicting data from previously unseen individuals, was also assessed. Results show that multi-level workload prediction on the individual and group level works well, achieving high correct rates and accuracy scores. Generalizing between individuals proved difficult using realistic driving conditions but worked well in the highly demanding lane-keeping task. Reasons for this discrepancy are discussed as well as future research directions.

Research into driver workload has been conducted for at least three decades (*1*, *2*). Recently, research efforts have shifted to using powerful machine learning (ML) methods, giving promising results (*3*, *4*). ML methods have been used for other driver-related classification problems, such as driver distraction (*5*), driver interruptibility (*6*) or driver identification (*7*). The present study aims to fill the gaps in the existing research on predicting driver workload using ML methods in several ways, as will be explained in the next paragraphs.

First, ML studies into predicting driver workload often focus on a binary classification problem (high workload versus low workload). A more fine-grained prediction of workload may be desirable to enable adaptive interfaces for in-vehicle advice systems (IVIS), systems that may simplify their content (*8*), or driver assistance systems that may incrementally increase their level of support based on the level of driver workload. The experiments described in this paper attempt to predict workload on 7- and 10-point workload scales.

Second, studies to date often use intrusive sensors or measure variables (e.g., electroencephalogram, EEG) that are not practical in the driving environment [see for example (*3*, *4*)]. Additionally, it is unknown how well results obtained by the high-grade intrusive sensors used in experiments translate to low-cost sensors. This work uses low-cost sensors that can be integrated into the real-world driving environment, and uses measures that can be obtained non-intrusively. This is important, because especially low-cost sensors are likely to be integrated into the driving environment in real-world applications.

Lastly, the models generated in most studies are not generally publicly available for use by the research community. The models developed in this study will be made available for scientific use after publication of results (https://github.com/paulvangentcom).

[1]Faculty of Civil Engineering and Geosciences, Delft University of Technology, Delft, Netherlands
[2]Faculty of Mechanical, Maritime and Materials Engineering, Delft University of Technology, Delft, Netherlands
[3]SWOV—Stichting Wetenschappelijk Onderzoek Verkeersveiligheid, The Hague, Netherlands

Corresponding Author:
Address correspondence to Paul van Gent: P.vanGent@tudelft.nl

Research Objectives

The previous section outlined the main research gaps and ways to add to the present literature. This led to the formulation of three criteria for predicting driver workload in the present work. The main goal is to develop a workload algorithm that (*a*) has usable accuracy when predicting multiple workload levels, while generalizing among individuals; (*b*) uses data that can be measured with available low-cost sensors that can be integrated into the driving environment; and (*c*) is implementable on embedded hardware (for example in a smart steering wheel).

The first criterion (*a*), predicting workload at a higher resolution than the binary low/high found in previous literature, while generalizing among individuals, is addressed in the experimental design and data analysis presented in subsequent sections.

The second criterion (*b*) entails using sensor inputs from readily available, low-cost sensors that are easy to implement in the driving environment. By using low-cost sensors, which are likely to present more noise in the signal compared with high-end sensors, results will give a better reflection of real-world performance compared with studies using high-end sensors. Apart from having been used successfully in other workload prediction studies, selected variables should be measurable non-intrusively in the driving environment. This led to the selection of heart rate, skin response, blink rate and several performance measures [for an overview of the selection process, see (*9*)]. This criterion ensures that any results are directly applicable to in-car settings at a low cost, and that results obtained are likely to translate well to real-world applications.

Criterion (*c*), ensuring the model is implementable on an embedded system, means it must be efficient both in memory use and in computational requirements. Two ML algorithms were selected that can satisfy this criterion: "Random Forest" and "Support Vector Machine" algorithms. Random Forests (*10*) are computationally efficient (*11*) but can have a large memory footprint. Solutions have been proposed that allow embedded implementations while maintaining performance (*12*), making this a suitable algorithm to use. Support Vector Machine (*13*) implementations can suffer from computational complexity, as well as high memory footprint for more complex models. Methods have been proposed, however, that achieve remarkable efficiency increases without sacrificing performance (*14*, *15*), making SVMs also a suitable candidate algorithm.

Two experiments were conducted to evaluate the feasibility of the previously defined criteria. First, a simulator experiment was performed, in which workload was induced using realistic driving situations. Results of this experiment were explored further using a dataset obtained from another driving simulator experiment that induced workload with a demanding lane-keeping task. At the end of this paper, the results of both experiments are discussed and future steps are outlined.

Estimating Workload in a Realistic Driving Scenario Study

To assess the feasibility of predicting driver workload in realistic driving settings, a simulator study was performed. The main goal was to evaluate the prediction of multi-level driver workload in realistic driving conditions.

Methods

Equipment. The study was performed in a fixed-base, medium-fidelity driving simulator. A dashboard mockup with three 4K-displays (resolution 4096 * 2160 px) provided roughly 180° vision. Actuators consisted of a Fanatec steering wheel and pedals, and a custom blinker control. The simulation ran in Unity3D. The simulated vehicle had an automatic gearbox and a top speed of 165 km/h. Figure 1*a* illustrates the setup.

Physiological data were recorded at 100 Hz, using low-cost sensors powered by an Atmel ATMega328p embedded processor board. Heart rate was recorded using a photoplethysmographic (PPG) method (*16*) at the left index finger. Skin response was recorded at the middle and ring finger of the same hand (see Figure 1*b*). Additionally, blink data were recorded using a GoPro HERO + camera on the dashboard, running at 1080p at 30 Hz. Simulator data were logged at 50 Hz.

Simulator Scenarios. Two scenarios were created in Unity3D, one scenario with situations likely to induce high workload ("high workload" scenario) and one with situations that are not likely to induce high workload ("low workload" scenario). Road geometry was based on a part of the Cooperative-ITS (C-ITS) corridor in the Netherlands: the A67, a two-lane highway between Eindhoven and Venlo with a speed limit of 130 km/h. Three weather conditions were designed for each scenario: clear weather, and two levels of fog with visibility of approximately 150 m ("light fog") and below 25 m ("heavy fog"). This gave a total of six scenarios.

To accurately design the road geometry, CAD drawings of the road segments were secured from the open data program of the Dutch government (https://data.overheid.nl). Using Autodesk 3DS Max, the data in the CAD files were converted to 3D models and textured. The surrounding terrain was generated using height map data obtained from the Microsoft Bing Maps API (https://www.bingmapsportal.com/). Canals and wooded

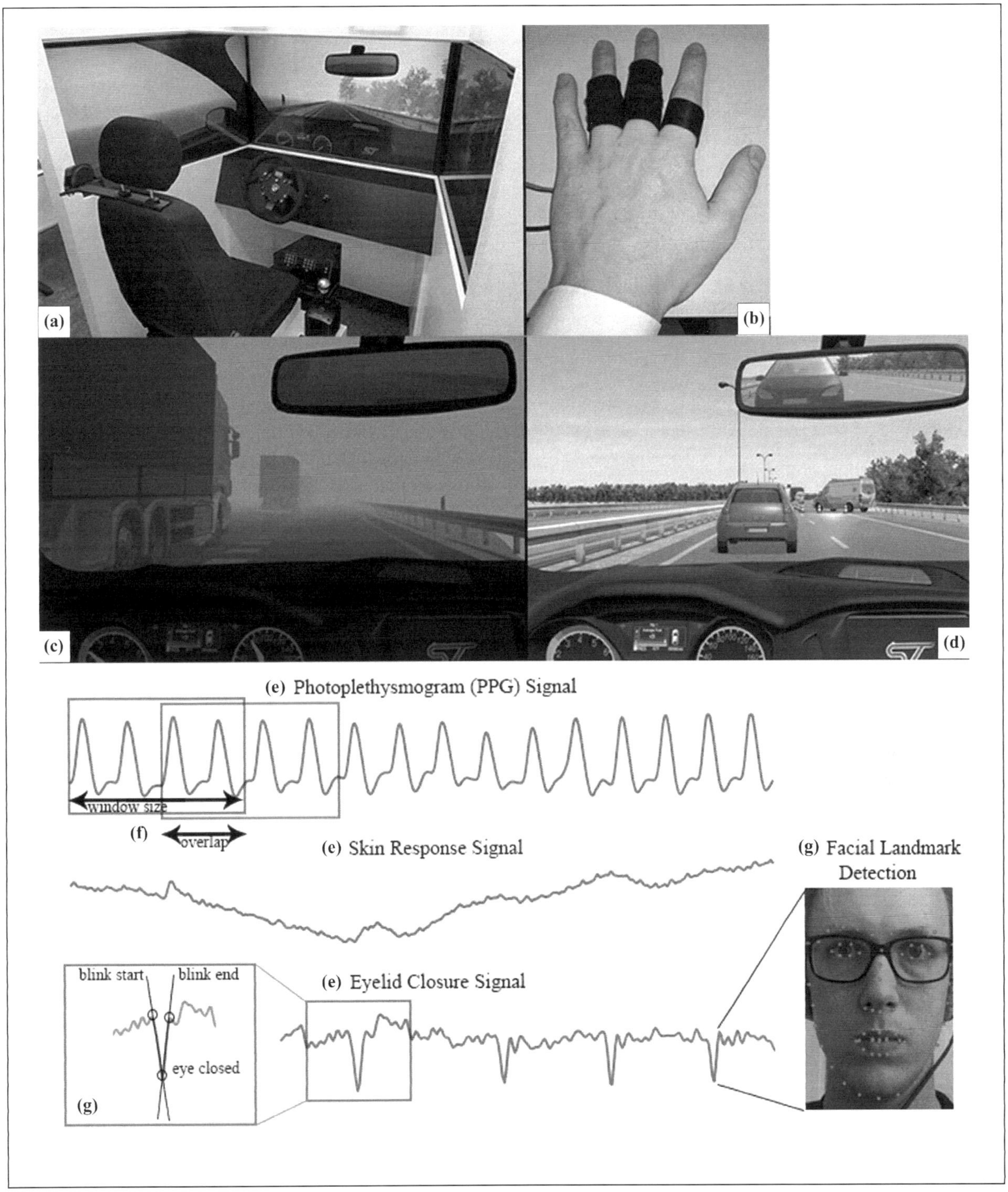

Figure 1. Figure showing (*a*) the simulator setup, (*b*) physiological sensors, (*c*) the merging of a platoon of trucks in dense fog, (*d*) the accident site at the end of the "high workload" scenario, (e) examples of the raw signal data, (*f*) the concepts of window size and overlap factor, and (*g*) an example of the facial landmark detection and the resulting process of analysing the blink rate signal.

areas were extracted automatically from satellite imagery, and adjusted by hand if necessary. The location, shape, and content of traffic signs were inferred from Google Street View, designed in 3DS Max and manually placed at the corresponding locations in the scenario.

The "high workload" scenario was 15.9 km in length, and ran between Eindhoven and Someren. Participants would encounter several workload-inducing "events" spread out across the scenario. After accelerating across an on-ramp, the first event was encountered: participants had to merge into a dense platoon of trucks (4–5 m headway, Figure 1*c*), a maneuver shown to increase workload on the driver (*17*). The second event was encountered 2 km downstream and consisted of a segment of slow-moving traffic in the right lane, designed to nudge the participants to drive in the left lane. While passing the slow-moving traffic, an ambulance approached from behind exhibiting auditory and visual signals, traveling at the maximum speed legally allowed in the Netherlands (170 km/h, a maximum of 40 km/h difference from other traffic). This placed the participant in the demanding situation of quickly having to find a gap in the much slower moving lane to the right and perform a merging maneuver. The third event was a game of "20 questions" (*18*), intended to simulate an engaging (phone) conversation. By asking at most 20 polar (yes/no) questions, participants had to guess which animal, object, or person the experimenter had in mind. The final event came near the end of the scenario. The right lane was closed off because of an accident, with slow-moving (<15 km/h) traffic in the left lane (Figure 1*d*). The 20 questions game was played until the accident site was reached. If participants finished early, the game was restarted with a different subject. After this, participants took the next exit and stopped the car.

The "low workload" scenario consisted of self-paced driving in light traffic for 20.5 km. The simulated road was a replica of the A67 road between Someren and Venlo. There were no events. Participants drove until reaching a designated exit, where they stopped the car.

Experimental Procedure. Approval for the study was obtained from the ethics committee at Delft University of Technology. Participants drove the six scenarios spread out over three separate days, each day driving one randomly assigned "high workload" and one "low workload" scenario. This approach was taken because physiological measures can vary from day to day, as well as to avoid a fatigue effect from asking participants to drive six 10–15-min scenarios consecutively.

In the "high workload" scenario, participants were asked to rate their experienced mental effort and task difficulty on a seven-point scale after each event, leading to six workload data points per run. In the "low workload" scenario, the questions were asked at fixed positions in the scenario, leading to four workload data points per run. The exact questions were "*How much mental effort did the driving task take in the last few moments, on a scale of 1–7?*" and "*How difficult was the driving task in the last few moments, on a scale of 1–7?*". Scale labels ranged from very low/easy, to very high/difficult, and were explained to participants before the experiment started. Note that the study did not use a standardized workload scale such as the NASA Task Load Index (TLX) or Rating Scale Mental Effort (RSME), because the intent was to keep interaction time with and demands on the driver to a minimum.

Participants who registered for the experiment received a copy of the informed consent. It was signed and brought to the first session. After being seated in the simulator, a relaxation period of 3 min was given to the participants. This was to allow the physiological measures of each participant to return to their baseline. Sensors were attached, after which the signal quality was checked. A physiological baseline was recorded first. After the baseline, it was briefly explained to the participant that there would follow a drive on a segment of the A67 highway. Participants were instructed to drive at their own pace, but not exceed the speed limit as indicated on roadside signs. If a participant was unfamiliar with "20 questions," a test round was played to familiarize them with the game.

Data Analysis

Participants were asked to rate their mental effort and driving task difficulty on a seven-point scale. Because querying the driver might influence workload, the "high workload" scenario was constructed in such a way that there was at least 1 min of driving between each two events, to allow signals to return to baseline. The data recorded between two events were not used in the analysis. In the case of the "low workload" scenario, 1 min of data following each question was excluded from the analysis.

Preprocessing of Physiological Data. An algorithm was developed to extract the most commonly used features from the measured heart rate signal (*19*, *20*), using a sliding window approach (see Figure 1*f*). The output measures are divided into time-domain (*21*) and frequency-domain measures (*22*). In the time domain, the measures included are BPM (beats per minute), IBI (interbeat interval), MAD (median absolute deviation of intervals between heart beats), SDNN (standard deviation of intervals between heart beats), RMSSD (root mean square of successive differences between neighboring heart beat intervals), SDSD (standard deviation of successive differences between neighboring heart beat intervals), and the pNN50 and pNN20 (proportion of differences between successive heart beats greater than 50 ms and 20 ms, respectively). In the frequency domain, included measures are LF (the low frequency band: 0.04–0.15 Hz), which is related to short-term blood pressure variation, and HF (the high frequency band: 0.16–0.5 Hz), which reflects breathing rate, and the LF/HF ratio, a measure of sympathetic–parasympathetic balance (*22, 23*).

Skin response consists of a tonic and phasic component (*24*). Tonic represents the long-term, slow variation in the signal, indicative of general psycho-physiological arousal (*25*). Phasic reflects relatively quick responses to discrete external stimuli, occurring generally 1–3 s after stimulus onset (*25*). Power in the frequency spectrum of skin response between 0.03 Hz and 0.5 Hz has been linked to short-term workload changes (*26*). The mean, max-min difference, MAD (median absolute difference), and 0.03–0.5 Hz frequency spectrum were extracted from the GSR (galvanic skin response) signal, using the same window approach as for heart rate. Frequency spectra were extracted using a trapezoidal integration of the area under corresponding frequency bands in the power spectrum.

Blink data were detected offline from recorded video data. An algorithm was developed to extract blink number, blink duration, and interblink interval. It functioned by detecting 68 "facial landmarks" (*27*), then calculating eyelid distance for each frame. Blinks were detected in the resulting signal by finding large slopes, then finding the lowest point of reversal. The process is displayed visually in Figure 1*g*.

Driver Performance Data. Performance measures reflect how the control the driver exerts over the vehicle varies across conditions. The study included steering wheel angle, steering wheel reversals, speed, variation in lateral and longitudinal position, and headway and time to collision when available (for more information, see [*9*]).

Generating Machine Learning Sets. Machine learning sets were generated from the raw data and labeled based on self-report data, by varying window size and overlap factor. Window size refers to how much data is used for the calculation of features, and overlap factor refers to how much data any window W_i shares with the previous window W_{i-1}. Both concepts are visualized in Figure 1*f*. Window sizes of 5, 10 and 30 s, and overlap factors of 0% and 50% were used, leading to a total of six sets.

Model Development and Evaluation. Two different ML algorithms were used: a Random Forest Regressor (RFR), and a Support Vector Machines Regressor (SVR). The RFR creates an ensemble (forest) of regression trees in which each tree is trained on a random subset of the features. They have been used in, for example, (*28*). Support Vector Machines function by mapping the data to a higher dimensional space, and solving an optimization problem to identify a set of hyperplanes that separate the training data into classes. They have been used in for, example, (*5, 7*). With the SVR, the Polynomial kernel [SVR(poly)], and the Radial Basis Function kernel [SVR(rbf)] were evaluated. Algorithms that were used are taken from the SciKit-Learn repository (*29*).

The resulting models were evaluated using several metrics. Model error was evaluated using mean absolute error (AE_μ) and median absolute error ($AE_{\mu 1/2}$), both measures of the accuracy of the predictions. The coefficient of determination (R^2) was also computed as a goodness-of-fit measure. Performance for class-based predictions was also evaluated, expressed as correct rate.

Results

Participants. Nineteen participants took part in the experiment. Data from one participant were excluded because of a failure to understand some tasks, caused by a language barrier. This left 18 participants, of whom 12 were males and six were females. The average age was 34.56 years (SD 10.09). Of the 18 participants, 12 owned a car and reported using it three to four times a week on average, and traveling between 2,500 and 15,000 km annually. All participants held a valid driver's license. No simulator sickness severe enough to terminate a driving session was reported. Reported mental effort and perceived difficulty correlated with weather conditions and with scenario type independently and in line with expectations, although no interaction effect was present (*9*).

Individual Models. The training and testing sets for the individual models were generated by dividing the dataset of each driver into training and testing sets with an 80% to 20% split ratio, respectively. This split ratio was chosen to ensure sufficient training data, as individual datasets were relatively small.

The results indicated that the models functioned well, with the RFR outperforming the SVR. For all individual models with a window size of 5 s and overlap of 0%, the AE_μ was 0.343, the $AE_{\mu 1/2}$ was 0.129, R^2 was 0.679, and Correct Rate (CR) was 76.30% when predicting discrete classes, and 93.80% when miss-by-one errors were allowed (CR +/−1). This indicated that, on average, predictions were off by 0.343, and that half the predictions had an error of less than 0.129, from a total scale of seven classes. See Table 1 for an overview of all results. Model performance increased with a larger overlap

Table 1. Performance Metrics RFR Models

Window size	5 s		10 s		30 s	
Overlap factor	0%	50%	0%	50%	0%	50%
Individual model						
AE_{μ}	0.343	0.219	0.431	0.280	0.613	0.492
$AE_{\mu 1/2}$	0.129	0.565	0.296	0.109	0.490	0.291
R^2	0.679	0.8716	0.590	0.794	0.071	0.306
CR	76.30%	85.21%	67.88%	80.77%	49.68%	60.82%
CR +/− 1	93.80%	97.61%	92.93%	96.13%	85.81%	89.55%
Group model						
AE_{μ}	0.605	0.455	0744	0.553	0.898	0.801
$AE_{\mu 1/2}$	0.406	0.250	0.565	0.344	0.628	0.652
R^2	0.661	0.774	0.564	0.709	0.372	0.504
CR	57.40%	69.57%	46.12%	62.48%	40.47%	43.82%
CR +/− 1	90.60%	93.81%	87.02%	91.42%	80.60%	84.56%
Generalizing model						
AE_{μ}	1.522	1.536	1.457	1.519	1.375	1.424
$AE_{\mu 1/2}$	1.163	1.201	1.199	1.253	1.174	1.230
R^2	−0.538	−0.623	−0.460	−0.602	−0.299	−0.396
CR	20.07%	20.05%	19.81%	20.21%	20.21%	20.47%
CR +/− 1	55.18%	55.19%	55.46%	54.94%	57.21%	55.89%

Note: The table displays the mean (μ) and median ($\mu 1/2$) absolute error metrics, the coefficient of determination (R^2), the correct rate (CR), and the miss-by-one correct rate (CR +/− 1).

factor. This was expected, because a larger overlap creates a larger training set to fit the model to, and because a larger overlap factor indicates more shared variance between adjacent samples. Interestingly, an inverse relationship between window size and model performance was observed, contrary to what has been reported previously (*3*). Miss-by-one errors indicate predictions that are "almost correct," and still contain enough information about the true workload states. For example, if workload is predicted as "6" but the true value is "7," the information in the prediction is still useful: in either case workload is on the high end.

Group Models. The second step was to estimate the model performance within the entire group. The dataset containing data from all drivers was split into training and testing sets with a 60% to 40% split ratio. Because the size of the group dataset was much larger compared with the individual dataset, a more stringent split ratio could be chosen while maintaining a sufficiently large training set.

Results indicated group models performed well. The AE_{μ} for the model with window size 5 s and 0% overlap was 0.605, the $AE_{\mu 1/2}$ 0.406, R^2 0.661, CR 57.40%, and CR + /−1 90.60%.

Generalizing Group Models. The last step was to assess how models would perform in a realistic setting, such as a setting in which workload from an unknown driver is predicted based on data from a pool of other drivers. To achieve this, data were sampled using a k-fold approach, with $k = N_{participants}$. For every k_i, the training set consisted of all data except the held out participant k_i. Workload for participant k_i was then predicted and model performance evaluated. This method simulated how the trained models would perform when predicting data from previously unseen individuals. This obtained performance measure reflects real-world settings, in which it is impractical for models to be trained on all possible drivers and generalizing power is thus preferable.

Results showed that models did not perform well when generalizing to unknown drivers. The AE_{μ} for all individual models with window size 5 s and 0% overlap was 1.522, $AE_{\mu 1/2}$ was 1.163, R^2 was −0.538, CR 20.07%, and CR + /−1 55.18%. The strongly negative coefficient of determination suggests unsatisfactory performance (the mean of the data is a better predictor than the trained model). The relatively low (though above chance level, not satisfactory) absolute error rates given R^2 are explained by a class imbalance in the dataset, in which two classes (workload levels 1 and 2) dominate. To assess whether this was a possible cause for the poor performance of the models, data were resampled using SMOTE (Synthetic Minority Over-Sampling Technique) (*30*). This had little discernible effect on the model performance, and it was concluded that low performance was not caused by the class imbalance in the dataset. It was also observed that R^2 increases slightly with increasing window size, in accordance with earlier studies (*3*) and contrary to the individual and group models in the present study.

Conclusion

The results of this study showed that predicting self-reported workload in a simulated realistic environment was possible at the individual and group level, but proved difficult when generalizing to unknown drivers. Several causes can be identified. The simulated scenarios might not have induced sufficient workload to be measurable with performance or physiological measures. Indeed, most participants indicated that driving in the simulator felt very different from actual driving, and was not that difficult at all. Because a self-report measure was used, which is a subjective measure, it is possible that different participants had biased response tendencies. Lastly, it might also be the case that different physiological response patterns to workload exist, in which case the sample size of 18 could have been too small to account for all occurring patterns.

This raises the question of whether workload prediction is at all possible on non-binary scales, while generalizing across drivers. To further explore this possibility, a dataset from a study with a lane-keeping task was obtained. This study and the results are discussed in the next section.

Estimating Workload in a Forced-Pace Simulator Study

A dataset was re-used from a previously executed study by Melman et al. (*31*) to further assess multi-level workload prediction in drivers. The study featured a challenging lane-keeping task, which had the potential to induce higher workload than the previous study. The same physiological and performance measurements were used as in the previously described simulator study.

Method

Equipment. The study was performed in a fixed-base driving simulator at the faculty of Aerospace Engineering, Delft University of Technology. The simulator consisted of a mockup dashboard with three LCD projectors (BenQ W1080ST 1080p) that provided roughly 180° vision. The simulated vehicle had an automatic gearbox and a top speed of 210 km/h.

Physiological data were logged using a biosignalsPlux wireless hub at 1000 Hz. Heart rate was recorded using three pregelled Ag/AgCl (silver/silver chloride) electrodes at the heart's v3-node. Skin response was measured using the same pregelled electrodes, placed inside the palm and on the wrist of both hands. Simulator data were logged at 100 Hz.

Scenarios. The scenarios used to induce workload in drivers each consisted of a 25-km long, single-lane road. The road was divided into four 6-km sections of different lane width (3.6 m, 2.8 m, 2.4 m, 2.0 m). Each section had seven curves, five with an inner radius of 750 m and two with a 500-m radius. Transitions between sections of different width always took place in a 750-m radius curve, and were preceded by a road sign indicating a narrowing road. The four sections were identical, with the exception that the curves of segments 2 and 4 four were mirrored with respect to sections 1 and 3.

Cones were placed 8 m apart on the road markings on both sides of the road. The main task was to hit as few cones as possible. A cone hit was indicated to the participant visually by a red dot on the side of the car where the cone was hit, and by a loud auditory beep. Extra difficulty in lane keeping was induced by a perturbation added to the vehicle's lateral motion. This perturbation was an unpredictable multi-sine signal with five frequencies between 0.067 Hz and 0.25 Hz, with a maximum summed amplitude of 1,000 N. Without the perturbation, lane keeping (especially on straight segments) was not considered challenging enough. The width of the simulated vehicle was 1.8 m.

Three runs were driven with the aim of inducing different levels of workload: a self-paced run and two forced-pace runs of 90 km/h and 130 km/h. In the self-paced run, participants had full longitudinal control over the car and could drive at their own pace. In the forced-pace conditions, however, the car's speed was automated and kept constant at 90 km/h and 130 km/h. This would push participants into curves at high speeds, with the goal of raising their workload significantly. The three runs were presented to the participants in randomized order.

Procedure. Participants read and signed an informed consent form, informing them of the purpose and procedure of the study. Participants were instructed that the main task was to minimize the total number of cone hits. Furthermore, participants were informed that during the experiment, a beep would sound every 20 s. At the sounding of this beep, participants were asked to verbally answer the question "*From 0 to 10, how much effort does the current driving task take you?*" with 0 being "no effort," 5 being "moderate effort," and 10 being "a lot of effort."

Before the experiment started, participants were familiarized with the simulator and the procedure by driving two 3.7-km trial runs. The first trial run was self-paced, the second was forced-pace with a speed of 110 km/h.

Table 2. Performance Metrics RFR Models

Window size	5 s		10 s		30 s	
Overlap factor	0%	50%	0%	50%	0%	50%
Individual model						
AE_{μ}	1.046	0.823	1.213	0.853	1.127	0.870
$AE_{\mu 1/2}$	0.662	0.511	0.833	0.518	0.959	0.694
R^2	0.635	0.763	0.600	0.675	0.561	0.735
CR	40.74%	50.31%	33.93%	45.83%	20.83%	40.28%
CR +/−1	77.31%	84.34%	70.83%	81.94%	65.83%	81.48%
Group model						
AE_{μ}	0.904	0.730	0.984	0.808	1.084	0.876
$AE_{\mu 1/2}$	0.638	0.482	0.722	0.546	0.792	0.663
R^2	0.774	0.830	0.740	0.802	0.718	0.811
CR	41.61%	51.30%	35.12%	46.44%	34.22%	37.87%
CR +/−1	82.30%	88.18%	80.32%	85.88%	73.21%	82.41%
Generalizing model						
AE_{μ}	1.878	1.988	1.988	1.989	1.809	1.717
$AE_{\mu 1/2}$	1.831	1.844	1.718	1.741	1.680	1.568
R^2	0.118	0.079	0.196	0.177	0.411	0.433
CR	14.09%	13.45%	12.62%	13.44%	15.72%	15.21%
CR +/−1	41.92%	40.70%	44.15%	42.29%	47.16%	46.32%

Note: The table displays the mean (μ) and median ($\mu 1/2$) absolute error metrics, the coefficient of determination (R2), the correct rate (CR), and the miss-by-one correct rate (CR +/−1).

After the trial run, any question the participant had was answered. The electrodes were attached, and a 1-min baseline was recorded.

Analysis

Participants rated their mental effort on a scale of 0–10, every 20 s. This rating was annotated by the experimenter and added to the dataset. What data were logged, data preprocessing, ML set generation, model development, and evaluation are identical to what has been described in the previous study.

Results

Participants. In total 24 participants took part in the experiment (17 male, seven female). The average age was 24.6 years (SD 2.4). Participants reported driving multiple times a week (11 participants), at least once a month (7 participants), or less than one month (6 participants). All participants held a valid driving license. Reported mental effort was sensitive to the lane width variations, although with respect to speed only to the 130 km/h forced-pace condition (*31*).

Individual Models. As in the previous study, training and testing sets for the individual models were generated by dividing the dataset into two stratified sets. More data per participant were collected than in the previous experiment, so data were split with the more stringent 60% to 40% split ratio.

Results were similar to the previous study, and indicated that the models performed well, with RFR outperforming SVR. An inverse relationship between model performance and overlap factor was observed, as well as increasing performance with increasing overlap factors, both as in the previous experiment. For all individual models with a window size of 5 s and overlap of 0%, the AE_{μ} was 1.046, the $AE_{\mu 1/2}$ 0.662, R^2 0.635, CR 40.74%, and CR +/−1 77.31%. The relatively larger absolute errors, compared with individual models in the previous study, might have resulted from the wider workload scale, the different nature of the driving task, or the more frequent reporting of mental workload. More information is displayed in Table 2.

Group Models. To evaluate performance at the group level, data were split with a 60% to 40% split ratio. Results indicated group models attained high performance. For the model with window size 5 s and 0% overlap, the AE_{μ} was 0.904, the $AE_{\mu 1/2}$ 0.638, R^2 0.774, CR 41.61%, and CR +/−1 82.30%. Table 2 displays the full results. Performance increased with larger overlap factors, and again a (weak) inverse relationship between performance and window size was observed.

Generalizing Group Models. Model performance when generalizing to unknown individuals was then assessed, which had not performed well in the first simulator experiment. Data sampling methods were identical to the previous study.

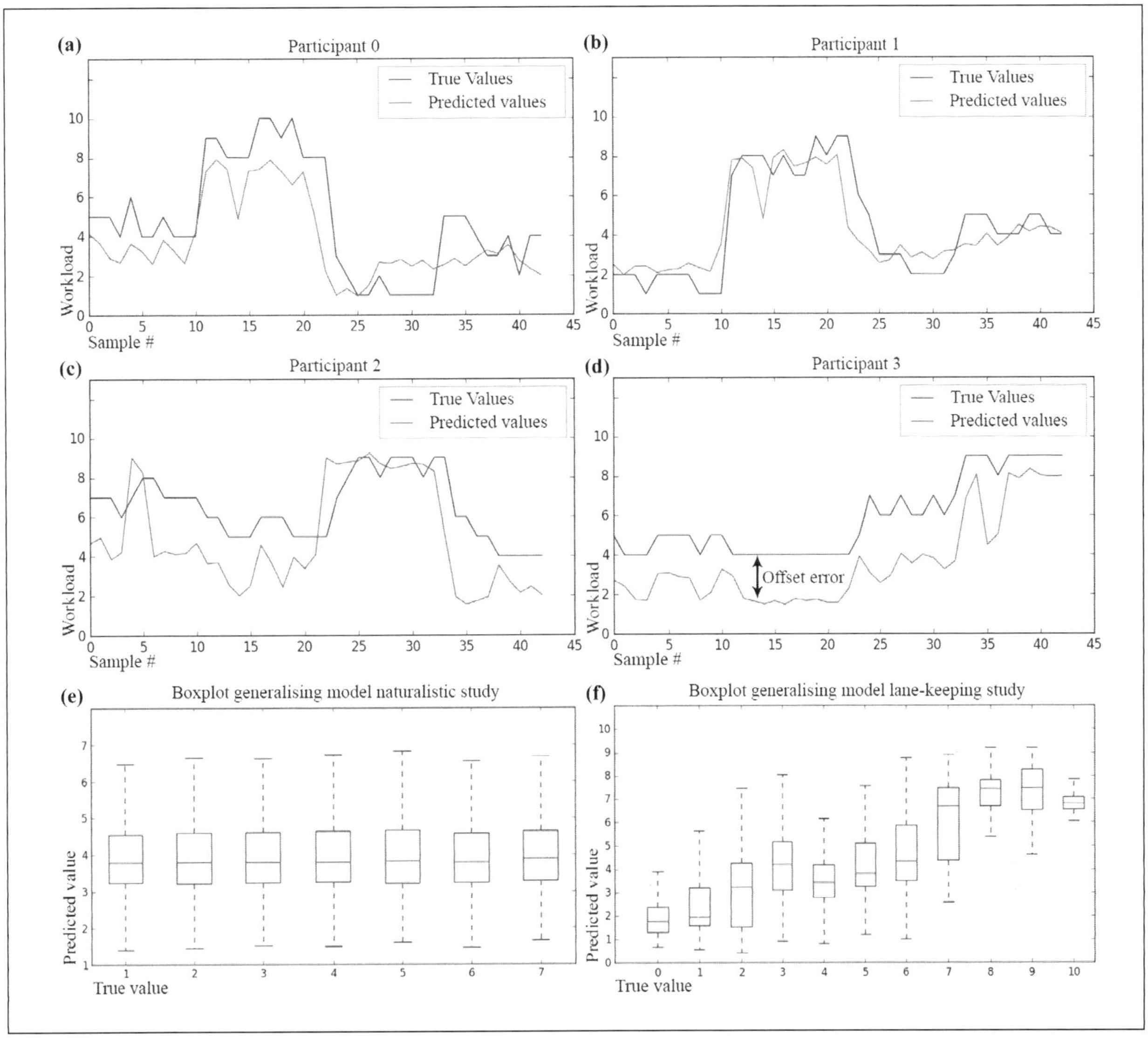

Figure 2. The top four windows (*a–d*) show plots from the first four participants, indicating that the models performed well, with the relatively large absolute errors likely to have resulted from individual scaling problems in the predictions. These offset errors are indicated in (*d*): the general trend is predicted well but there is a constant offset error. The last two windows (*e–f*) show box plots, further exploring the generalizing models from both studies.

Results indicated models performed moderately well. For the best performing model with window size 30 s and 50% overlap, the AE_{μ} was 1.717, the $AE_{\mu 1/2}$ 1.568, R^2 0.433, CR 15.21%, CR +/−1 46.32%. Although model absolute error was relatively large, the coefficient of determination indicated a moderate relationship between model and data. Figure 2 displays the predicted and true values for the first four participants. Individual model performance varied, with workload being predicted well for some participants, whereas for others it showed a correct trend but with a constant offset error. These offset errors inflated the absolute error rates and deflated the predictive accuracy despite good model performance. In general, a decreased performance with increased overlap factor was observed (except for the largest window size of 30 s), as well as increased performance with increased window size. The effect is similar to results for the model generalization step in the previous study, but more pronounced. The effect also corresponds with what has been reported before (*3*).

Conclusion

The results of this study show similarities with the previous study for individual and group-based models. In addition, this second experiment shows that, when predicting multi-level workload (11 classes), generalizing performance was satisfactory, although still with room for improvement.

This study seems to indicate that indeed non-binary workload prediction that generalizes to unknown individuals is possible using ML methods. Although models generalizing between individuals showed variations in performance based on which individual's workload was being predicted, including constant offset errors in several participants, overall performance was promising.

Overall Conclusions and Discussion

The present study tried to model driver workload using ML techniques that can run on embedded systems, with data collected from low-cost sensors. Results showed that individual models and within-group models functioned well in both a realistic driving setting and an artificial lane-keeping task setting. When generalizing to unknown drivers, only the lane-keeping study produced usable results. As displayed in Figure 2, *e* and *f*, in the first study the generalized model learns to predict values around the mean to optimize accuracy, whereas in the second study the model learns to predict based on the reported workload.

Because the data gathered in the study are time-series human physiological and performance data, they are likely to exhibit strong autocorrelation from one sample to the next. This might be a potential explanation for the higher performance in the individual and group models in both studies because, with random sampling, shared variance between samples from the training set and the prediction set might bias the classifier toward a higher accuracy. To better assess performance, training cases were included for which the models had to generalize to unknown individuals. These give a more accurate indication of performance, as with this approach there is no shared variance between training set (all participants minus participant k) and the testing set (participant k). As such, only the generalizing training case offers a reliable index of performance. This is an important distinction, because it shows that although using ML to predict driver workload can lead to promising results, care must be taken when interpreting the results. Without care in selecting the sampling techniques used, model performance might be inflated.

Possible reasons for the discrepancy in generalizing performance between the two studies could include the workload induced in the realistic settings being too low to be reflected in the physiological or performance signals, or the workload induced by artificial tasks being more easily measurable than that induced by more realistic tasks. It could also be that different physiological response patterns to workload might exist and that the sample in the first study was either too small or contained too much individual variation.

Possible limitations of the present study are that a self-report measure was employed as ground truth of the experienced mental workload of the drivers. The study did not employ standardized workload scales such as NASA TLX, to keep interaction time and demand with the driver to a minimum. However, this may have contributed to lower model performance through participant response tendencies, and leaves some doubt as to the degree to which the data captures workload. In addition to this, the study did not look at compensatory behavior drivers might employ to manage their workload, such as reducing speed in complex or demanding situations.

Future directions are planned. These include feature space normalization of the dataset to attempt to reduce the offset errors observed in some individuals, as well as exploring additional feature extraction methods. After this, on-road testing is planned to explore model performance in real-world driving settings. Lastly, development of an embedded variant of the model is planned.

Author Contributions

The authors confirm contribution to the paper as follows: study conception and design: Paul van Gent, Timo Melman, Haneen Farah, Nicole van Nes, Bart van Arem; data collection: Paul van Gent, Timo Melman; analysis and interpretation of results: Paul van Gent, Haneen Farah, Nicole van Nes, Bart van Arem; draft manuscript preparation: Paul van Gent, Haneen Farah, Nicole van Nes, Bart van Arem. All authors reviewed the results and approved the final version of the manuscript.

References

1. de Waard, D. *The Measurement of Drivers' Mental Workload.* Drukkerij Haasbeek, Alphen aan den Rijn, 1996.
2. Aasman, J., G. Mulder, and L. J. M. Mulder. Operator Effort and the Measurement of Heart-Rate Variability. *Human Factors*, Vol. 29, No. 2, 1987, pp. 161–170.
3. Solovey, E. T., M. Zec, E. A. Garcia Perez, B. Reimer, and B. Mehler. Classifying Driver Workload Using Physiological and Driving Performance Data. *Proc., 32nd Annual ACM Conference on Human Factors in Computing Systems—CHI '14*, ACM, New York, NY, 2014, pp. 4057–4066. https://doi.org/10.1145/2556288.2557068.
4. Jarvis, J., F. Putze, D. Heger, and T. Schultz. Multimodal Person Independent Recognition of Workload Related Biosignal Patterns. *Proc., 13th International Conference on Multimodal Interfaces—ICMI '11*, ACM, New York, NY, 2011, p. 205. https://doi.org/10.1145/2070481.2070516.
5. Liang, Y., M. L. Reyes, and J. D. Lee. Real-Time Detection of Driver Cognitive Distraction Using Support Vector Machines. *IEEE Transactions on Intelligent Transportation Systems*, Vol. 8, No. 2, 2007, pp. 340–350. https://doi.org/10.1109/TITS.2007.895298.

6. Kim, S., J. Chun, and A. K. Dey. Sensors Know When to Interrupt You in the Car. *Proc., 33rd Annual ACM Conference on Human Factors in Computing Systems—CHI '15*, ACM, New York, NY, 2015, pp. 487–496. https://doi.org/10.1145/2702123.2702409.
7. Moreira-Matias, L., and H. Farah. On Developing a Driver Identification Methodology Using In-Vehicle Data Recorders. *IEEE Transactions on Intelligent Transportation Systems*, Vol. 18, No. 9, 2017, pp. 2387–2396. https://doi.org/10.1109/TITS.2016.2639361.
8. Birrel, S., M. Young, N. Staton, and P. Jennings. Using Adaptive Interfaces to Encourage Smart Driving and Their Effect on Driver Workload. *Proc., AHFE 2016 International Conference on Human Factors in Transportation*, Orlando, FL, 2016, p. 764. https://doi.org/10.1007/978-3-319-41682-3.
9. van Gent, P., H. Farah, N. Van Nes, and B. Van Arem. Towards Real-Time, Nonintrusive Estimation of Driver Workload: A Simulator Study. *Proc., Road Safety and Simulation 2017 Conference*, The Hague, Netherlands, 2017.
10. Breiman, L. Random Forests. *Machine learning*, Vol. 45, No. 1, 2001, pp. 5–32. https://doi.org/10.1023/A:1010933404324.
11. Sventnik, V., A. Liaw, C. Tong, J. C. Culberson, R. P. Sheridan, and B. P. Feuston. Random Forest: A Tool for Classification and Regression in Compound Classification and QSAR Modeling. *Journal of Chemical Information and Computer Science*, Vol. 43, No. 6, 2003, pp. 1947–1958. https://doi.org/10.1016/j.rse.2008.02.011.
12. Mishina, Y., R. Murata, Y. Yamauchi, T. Yamashita, and H. Fujiyoshi. Boosted Random Forest. *IEICE Transactions on Information and Systems*, Vol. E98, No. D, 2015, pp. 1630–1636.
13. Cortes, C., and V. Vapnik. Support-Vector Networks. *Machine Learning*, Vol. 20, No. 3, 1995, pp. 273–297. https://doi.org/10.1023/A:1022627411411.
14. Theocharides, T., and S. Member. Embedded Hardware-Efficient Real-Time Vector Machines. *IEEE Transactions on Neural Networks and Learning Systems*, Vol. 27, No. 1, 2016, pp. 99–112. https://doi.org/10.1109/TNNLS.2015.2428738.
15. Bajaj, N., G. T. C. Chiu, and J. P. Allebach. Reduction of Memory Footprint and Computation Time for Embedded Support Vector Machine (SVM) by Kernel Expansion and Consolidation. *Proc., IEEE International Workshop on Machine Learning for Signal Processing, MLSP*, Reims, France, IEEE, New York, NY, 2014. https://doi.org/10.1109/MLSP.2014.6958875.
16. Jae Baek, H., H. Bit Lee, J. Soo Kim, J. Min Choi, K. Keun Kim, and K. Suk Park. Nonintrusive Biological Signal Monitoring in a Car to Evaluate a Driver's Stress and Health State. *Telemedicine and e-HEALTH*, Vol. 15, No. 2, 2009, pp. 182–189.
17. de Waard, D., A. Kruizinga, and K. A. Brookhuis. The Consequences of an Increase in Heavy Goods Vehicles for Passenger Car Drivers' Mental Workload and Behaviour: A Simulator Study. *Accident Analysis and Prevention*, Vol. 40, No. 2, 2008, pp. 818–828. https://doi.org/10.1016/j.aap.2007.09.029.
18. Kun, A. L., A. Shyrokov, and P. A. Heeman. Interactions between Human–Human Multi-Threaded Dialogues and Driving. *Personal and Ubiquitous Computing*, Vol. 17, No. 5, 2013, pp. 825–834. https://doi.org/10.1007/s00779-012-0518-1.
19. van Gent, P. Analyzing a Discrete Heart Rate Signal Using Python. *A Tech Blog About Fun Things With Python and Embedded Electronics*. http://www.paulvangent.com/2016/03/15/analyzing-a-discrete-heart-rate-signal-using-python-part-1/.
20. van Gent, P. Python Heart Rate Analysis Toolkit. *GitHub Repository*. https://github.com/paulvangentcom/heartrate_analysis_python.
21. Reimer, B., B. Donmez, M. Lavallière, B. Mehler, J. F. Coughlin, and N. Teasdale. Impact of Age and Cognitive Demand on Lane Choice and Changing under Actual Highway Conditions. *Accident Analysis and Prevention*, Vol. 52, 2013, pp. 125–132. https://doi.org/10.1016/j.aap.2012.12.008.
22. Montano, N., A. Porta, C. Cogliati, G. Costantino, E. Tobaldini, K. R. Casali, and F. Iellamo. Heart Rate Variability Explored in the Frequency Domain: A Tool to Investigate the Link between Heart and Behavior. *Neuroscience and Biobehavioral Reviews*, Vol. 33, No. 2, 2009, pp. 71–80. https://doi.org/10.1016/j.neubiorev.2008.07.006.
23. Billman, G. E. Heart Rate Variability—A Historical Perspective. *Frontiers in Physiology*, Vol. 2 2011, pp. 1–13. https://doi.org/10.3389/fphys.2011.00086.
24. Lim, C. L., C. Rennie, R. J. Barry, H. Bahramali, I. Lazzaro, B. Manor, and E. Gordon. Decomposing Skin Conductance into Tonic and Phasic Components. *International Journal of Psychophysiology*, Vol. 25, No. 2, 1997, pp. 97–109. https://doi.org/10.1016/S0167-8760(96)00713-1.
25. Seitz, M., T. J. Daun, A. Zimmermann, and M. Lienkamp. Measurement of Electrodermal Activity to Evaluate the Impact of Environmental Complexity on Driver Workload. *Proc., FISITA 2012 World Automotive Congress*, Lecture Notes in Electrical Engineering, Vol. 200, Springer, Berlin, Heidelberg, Germany, 2012, pp. 245–256. https://doi.org/10.1007/978-3-642-33741-3.
26. Shimomura, Y., T. Yoda, K. Sugiura, A. Horiguchi, K. Iwanaga, and T. Katsuura. Use of Frequency Domain Analysis of Skin Conductance for Evaluation of Mental Workload. *Journal of Physiological Anthropology*, Vol. 27, No. 4, 2008, pp. 173–177. https://doi.org/10.2114/jpa2.27.173.
27. Köstinger, M., P. Wohlhart, P. M. Roth, and H. Bischof. Annotated Facial Landmarks in the Wild: A Large-Scale, Real-World Database for Facial Landmark Localization. *Proc., IEEE International Conference on Computer Vision*, Barcelona, Spain, IEEE, New York, NY, 2011, pp. 2144–2151. https://doi.org/10.1109/ICCVW.2011.6130513.
28. Miyaji, M., M. Danno, H. Kawanaka, and K. Oguri. Driver's Cognitive Distraction Detection Using Adaboost on Pattern Recognition Basis. *Proc., 2008 IEEE International Conference on Vehicular Electronics and Safety, ICVES 2008*, Columbus, OH, IEEE, New York, NY, 2008, pp. 51–56. https://doi.org/10.1109/ICVES.2008.4640853.

29. Pedregosa, F., G. Varoquaux, A. Gramfort, V. Michel, B. Thirion, O. Grisel, M. Blondel, P. Prettenhofer, R. Weiss, V. Dubourg, J. Vanderplas, A. Passos, D. Cournapeau, M. Brucher, M. Perrot, and É. Duchesnay. Scikit-Learn: Machine Learning in Python. *Journal of Machine Learning Research*, Vol. 12, 2012, pp. 2825–2830. https://doi.org/10.1007/s13398-014-0173-7.2.
30. Chawla, N. V., K. W. Bowyer, L. O. Hall, and W. P. Kegelmeyer. SMOTE: Synthetic Minority Over-Sampling Technique. *Journal of Artificial Intelligence Research*, Vol. 16, 2002, pp. 321–357. https://doi.org/10.1613/jair.953.
31. Melman, T., D. A. Abbink, M. M. van Paassen, E. R. de Boer, and J. C. F. Winter. What Determines Drivers' Speed? A Replication of Three Behavioural Adaptation Experiments in a Single Driving Simulator Study. *Ergonomics*, Vol. 61, No. 7, 2018, pp. 966–987. https://doi.org/10.1080/00140139.2018.1426790.

The Standing Committee on Simulation and Measurement of Vehicle and Operator Performance (AND30) peer-reviewed this paper (18-02628).

Article

TRR
JOURNAL OF THE TRANSPORTATION RESEARCH BOARD

Transportation Research Record
2018, Vol. 2672(37) 153–163

DOI: 10.1177/0361198118792131
journals.sagepub.com/home/trr
SAGE

Models of Human Decision-Making as Tools for Estimating and Optimizing Impacts of Vehicle Automation

Gustav Markkula[1], Richard Romano[1], Ruth Madigan[1], Charles W. Fox[1], Oscar T. Giles[1], and Natasha Merat[1]

Abstract

With the development of increasingly automated vehicles (AVs) comes the increasingly difficult challenge of comprehensively validating these for acceptable, and ideally beneficial, impacts on the transport system. There is a growing consensus that virtual testing, where simulated AVs are deployed in simulated traffic, will be key for cost-effective testing and optimization. The least mature model components in such simulations are those generating the behavior of human agents in or around the AVs. In this paper, human models and virtual testing applications are presented for two example scenarios: (i) a human pedestrian deciding whether to cross a street in front of an approaching automated vehicle, with or without external human–machine interface elements, and (ii) an AV handing over control to a human driver in a critical rear-end situation. These scenarios have received much recent research attention, yet simulation-ready human behavior models are lacking. They are discussed here in the context of existing models of perceptual decision-making, situational awareness, and traffic interactions. It is argued that the human behavior in question might be usefully conceptualized as a number of interrelated decision processes, not all of which are necessarily directly associated with externally observable behavior. The results show that models based on this type of framework can reproduce qualitative patterns of behavior reported in the literature for the two addressed scenarios, and it is demonstrated how computer simulations based on the models, once these have been properly validated, could allow prediction and optimization of AV impacts on traffic flow and traffic safety.

In recent years, rapid technological progress has been made on automated vehicles (AVs); vehicles capable of taking over increasing shares of the driving task, with large hoped-for benefits in terms of increased mobility, improved traffic safety, reduced environmental footprint, and economic growth (*1*). As with any technology, it is important to subject AVs to sufficient testing and validation to ensure safe and effective operation. This is a daunting task, however, considering the essentially infinite variations of possible traffic situations that an AV needs to handle (*2*).

It is increasingly assumed that one important approach in the toolkit for AV validation will be *virtual testing* methods, where a computer simulation of an AV, integrating mathematical models of vehicle and sensors with the actual self-driving algorithms, can encounter a rich variety of simulated traffic situations (*3, 4*). These traffic situations can for some purposes be replayed from large static databases of actual real-traffic recordings, but if the objectives of a virtual test depends on some form of *interaction* between the AV and surrounding road users—e.g., by means of external human–machine interfaces (*5*)—or interaction between the AV and its onboard operator, the involved human agents also need to be simulated.

Human road user models do exist, and are currently being actively developed and applied to AV testing (*3, 6*), but so far mainly on the relatively coarse-grained level of *traffic microsimulation*, for which the trajectory of each road user (car, truck, pedestrian, etc.) is directly generated from equations of motion taking into account positions and velocities of surrounding road users. Though this level of granularity might be adequate for some purposes, it seems likely that in many cases, the outcome of interactions between humans and AVs will hinge on finer details and time dynamics of human perception, scene interpretation, and decision making (*7, 8*).

[1]Institute for Transport Studies, University of Leeds, Leeds, UK

Corresponding Author:
Address correspondence to Gustav Markkula: g.markkula@leeds.ac.uk

Capturing such phenomena accurately in virtual simulation is of course very challenging, but recent literature suggests some possibly fruitful directions: there is by now ample neuroscientific support for the idea that perceptual decision-making in typical laboratory tasks is underpinned by noisy neural *evidence accumulation*, to a decision threshold at which the overt response is initiated (*9*). Interestingly, such models have also been proven useful for explaining human driver reaction times in responding to stimuli in traffic; discrete stimuli such as brake lights (*10*) but importantly also graded, dynamic stimuli like the *visual looming* of approaching road users or collision threats (*11, 12*).

Here, this line of modeling will be taken one step further, and applied to the more complex decision-making situations that tend to arise in the AV context, where the emphasis of modeling naturally shifts away from low-level decisions like "more or less braking?" toward a higher-level assessment of the situation at hand – what is often referred to as *situational awareness* (*8, 13*). The basic modeling idea to be explored here is that situational awareness can be modeled as a number of interrelated perceptual decisions about the world, in which each decision is modeled as an accumulation process, but where these processes are also interconnected to influence each other in an excitatory or inhibitory fashion. This is reminiscent of connectionist- and activation-dynamics-type models of cognition (*14, 15*). However, to our knowledge, the present paper is the first time several decision-making processes of an evidence accumulator type have been connected in a network arrangement as a means of modeling complex, multi-stage decision processes.

After a first section below describing the two AV–human interaction scenarios addressed here, three consecutive sections will pursue the three main objectives of the paper: (i) to suggest models of human decision-making in the targeted scenarios, as examples of the modeling approach sketched above; (ii) to illustrate that these models can qualitatively reproduce non-trivial behavior patterns reported in the literature. These prior findings thus serve as a first high-level data set, to which model parameters are manually fitted; (iii) to illustrate how models at this level of detail can be useful in virtual testing of AVs. It is not an aim here to suggest final model formulations, or to perform full model validation and tuning; these will be matters for future work.

Targeted AV–Human Interaction Scenarios

Figure 1 shows the two traffic scenarios considered here. The pedestrian interaction scenario in panel (*a*) is based on the test track study by Schneemann and Gohl (*16*). They considered the question of how an AV should behave at a zebra crossing (crosswalk), by studying the interactive behavior of human drivers and pedestrians, as a function of initial car speed v_0 and *time to collision* (TTC; or alternatively in this scenario *time to arrival*) at the moment when the pedestrian was instructed to step up to the crossing, prompting decelerations from the (uninstructed) drivers. Here, what will be modeled is the pedestrian's decision to cross or yield as a function of car movements, as well as of more direct indications from the car (from either the driver or the AV, as the case may be) about having seen the pedestrian, and about its possible intentions to yield.

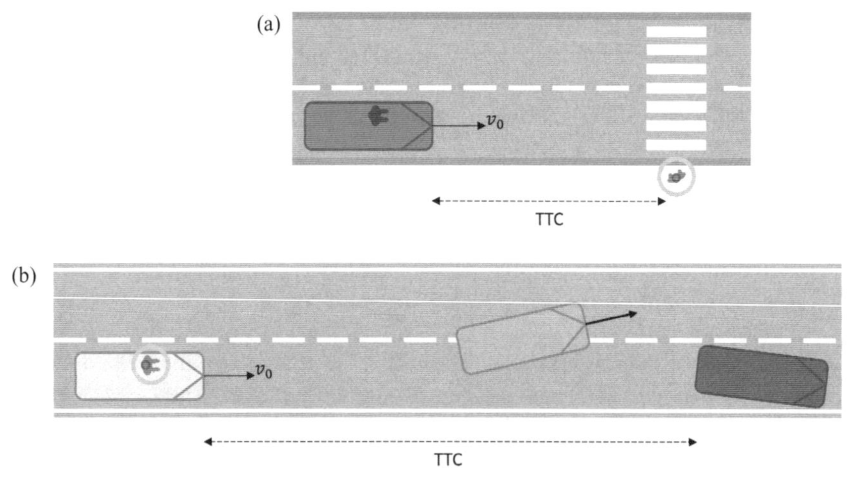

Figure 1. Schematic illustrations of the scenarios considered in this paper. The modeled human is indicated with a circle. TTC is short for time to collision: (*a*) pedestrian interaction scenario, and (*b*) safety-critical take-over scenario.

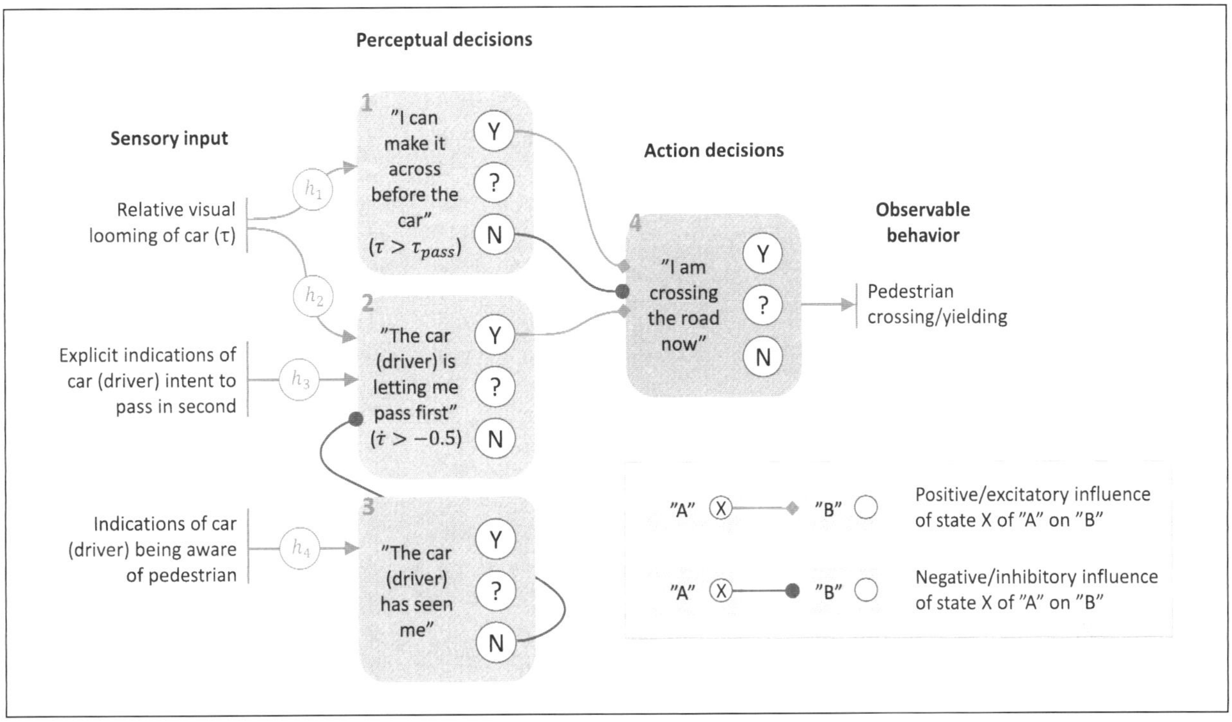

Figure 2. Schematic representation of the model for the pedestrian interaction scenario.

The safety-critical take-over scenario in panel (*b*) is based on the simulator study by Gold et al., in which drivers either manually drove a vehicle or were driven by an AV at $v_0 = 120$ km/h (75 mph), when a lead vehicle changed lane to reveal a stationary vehicle, at a TTC of either 5 s or 7 s, in the AV case prompting a *take-over request* to the driver (*17*). Similar types of take-over scenarios have recently been subject to intense research efforts (*7, 18*). What will be modeled here is the human's decisions, after the obstacle appears, on where to look and how to maneuver the car.

Models

Model of Pedestrian Crossing Decisions

The basic hypothesis behind the pedestrian model, illustrated in Figure 2, is that the pedestrian makes the action decision to cross based on either or both of two perceptual decisions: (i) a perceptual decision that it is possible to pass the road before the car arrives, made by observing the visual looming quantity $\tau = \theta/\dot{\theta}$, with θ the projected angle of the car on the pedestrian's retina. τ is a close approximation of TTC if assuming constant speeds (*19*), and is assumed to be compared to a threshold τ_{pass}. (ii) A perceptual decision that the car (driver or AV, but here "car" will be used for short) intends to yield, but not if the actual car movements clearly suggest that it is still unsafe to cross ($\tau < \tau_{pass}$). To make the decision about whether or not the car intends to yield, the pedestrian is assumed to monitor the quantity $\dot{\tau} = \mathrm{d}\tau/\mathrm{d}t$, which is ≥ -0.5 if the car stops at or before the zebra crossing (*19*), but also explicit communication acts (e.g., headlight flashes). However, if the pedestrian decides that the car has *not* seen him/her (e.g., based on driver head orientation), this is taken as evidence that the car is *not* intending to yield.

The remainder of this section provides a full mathematical specification of the model, but is not crucial for understanding the rest of this paper: Figure 2 shows four interconnected *decision units*, and at any point in time, the state of the model is described by the vector of unit *activations*, $\boldsymbol{A}(t) = [A_1(t) \dots A_4(t)]^T$, where each A_i is a two-sided accumulator, bounded at plus and minus one, implying "yes" and "no" decisions, respectively, and zero implying maximum uncertainty. This type of accumulator is often referred to as a "drift-diffusion model for two-choice decisions" or similar (*20*), although interconnecting several of these or simulating them beyond a threshold being reached is possibly novel. At each new simulation time step, the A_i are updated as follows:

$$\frac{d}{dt}\boldsymbol{A}(t) = -\frac{1}{T}\boldsymbol{A}(t) + \boldsymbol{W}_Y\sigma_Y(\boldsymbol{A}(t)) + \boldsymbol{W}_?\sigma_?(\boldsymbol{A}(t)) + \boldsymbol{W}_N\sigma_N(\boldsymbol{A}(t)) + \boldsymbol{F}(t) + \boldsymbol{\nu}(t) \quad (1)$$

where

T is a decay time constant,
$\boldsymbol{W}$ are matrices of connections between decision units,
σ are activation functions,
$\boldsymbol{F}(t)$ is a vector of external inputs, and
$\boldsymbol{\nu}(t)$ is a vector of noise, accounting for between-trial behavioral variability due to for example variations in brain activity, modeled here as Gaussian noise with standard deviation $\Sigma\sqrt{\Delta t}$, where Σ is a parameter and Δt is the simulation time step (*9*). In this first model implementation, to simplify the manual tuning process, all elements in $\boldsymbol{W}$ were kept either at zero (no connection) or ± 1 (excitatory and inhibitory connections), rather than being freely tuned. Also for simplicity, the activation functions were kept piecewise linear: $\sigma_Y(A) = 0$ and 1 for $A \leq 0$ and $A \geq 1$, respectively, and linearly increasing in between, and $\sigma_N(A) = \sigma_Y(-A)$, such that the two functions respond maximally when A indicates "yes" and "no", respectively. Further, $\sigma_?(A) = 1 - \sigma_Y(A) - \sigma_N(A)$, that is, zero at $A = \pm 1$ and linearly increasing from both sides to a maximum of one at $A = 0$. The external inputs to the decision units were:

- $F_1(t) = h_1(\tau(t)) = k_1(\tau(t) - \tau_{pass})$, corresponding to the perceptual decision about whether the time margin for crossing is currently above $\tau_{pass} = 3$ s.
- $F_2(t) = h_2(\tau(t)) + h_3(t)$, where $h_2(\tau(t)) = k_2(\dot{\tau}(t) - \dot{\tau}_{stop})$, corresponding to the perceptual decision about whether the car is decelerating to stop before the zebra crossing. Here $\dot{\tau}_{stop} = -0.55$ was used, to accumulate positive evidence also in the limit case $\dot{\tau} = -0.5$. The input $h_3(t)$ was set to one to model the case for which the car is providing some communicative act interpreted by the pedestrian as indicating an intention to yield, e.g., flashing the headlights, zero otherwise.
- $F_3(t) = h_4(t)$ was set to one to model the car or its driver doing something that the pedestrian takes as evidence of the car seeing the pedestrian, e.g., eye contact or related head pose in the case of a human driver, or some external human-machine interface (HMI) in the case of an AV, or minus one if the car is doing something that the pedestrian takes as evidence of the opposite, e.g., a car driver is looking somewhere else.
- $F_4(t) = 0$.

Manual tuning indicated satisfactory model behavior for $T = 2$s; $\Sigma = 0.5$; $k_1 = 0.5$; $k_2 = 4$.

Model of Driver Behavior during Safety-Critical Take-Over

The model for the take-over scenario, illustrated in Figure 3, is slightly more complex. It does not terminate at the first action decision like the pedestrian model; instead each new action affects how the scenario continues playing out over time. For reasons of space, all connections shown in Figure 3 will not be discussed individually here, but the main hypotheses behind the model are as follows: driver braking is assumed to function as proposed in Svärd et al., as discrete increases of brake pedal position applied each time a discrepancy between visual looming, quantified as τ^{-1}, and predicted visual looming τ_p^{-1}, updated after each new brake adjustment, accumulates to threshold (*12*). The decision to change lane can come about in the model in two ways: an early detection of catching up with the lead vehicle (which does not in itself prompt braking) or a later realization that attempted braking is not solving the conflict, and both of these are also modulated by visual looming directly. However, the lane changing decision is in all cases inhibited—as in made less probable—if the driver does not know that it is safe to change lane. Gaze sampling is assumed to be driven to a large extent by the need to resolve uncertainties in the perceptual decisions (*21*), and the current gaze target determines what sensory input the model receives (the "switches" in Figure 3). In this first version of the model, there is no direct account of the process of actual control take-over; as a first approximation it will be investigated to what extent the effects of taking over from the AV can be accounted for as a simple time delay, combined with lower initial awareness of the adjacent lane.

Again, the rest of this section gives mathematical model details: Just as in the pedestrian model, Equation 1 governs activations in the network. However, decisions between competing actions about where to look (units 4 and 5) or how to maneuver (units 6 and 7), are here modeled as "races" (*20*), bounded downward by zero instead of minus one, and each time an action threshold $A_i = 1$ is reached, the race in question is restarted by resetting the competing accumulators to zero. The exception is the "increase braking" action, which has its accumulator reset to 0.7, to reflect an increased brake readiness after the first brake application (*12*). Furthermore, before the first brake application, Decision Unit 2 is inactive. The looming-related inputs to the network, all set to zero if the driver's eyes are off the road, were defined as:

- $F_1(t) = k_1\tau^{-1}(t)$, modeling the perceptual decision of whether the vehicle ahead is coming closer.
- $F_2(t) = -k_2\left(\tau^{-1}(t) - \tau_p^{-1}(t)\right)$, modeling the perceptual decision of whether braking is solving the conflict.
- $F_6(t) = k_3\left(\tau^{-1}(t) - \tau_p^{-1}(t) - \tau_B^{-1}\right)$, modeling the decision to increase braking (*12*).
- $F_7(t) = k_4(\tau^{-1}(t) - \tau_S^{-1}) - \min(0, k_5(\tau^{-1}(t) - \tau_{NS}^{-1}))$, modeling a positive looming contribution to the

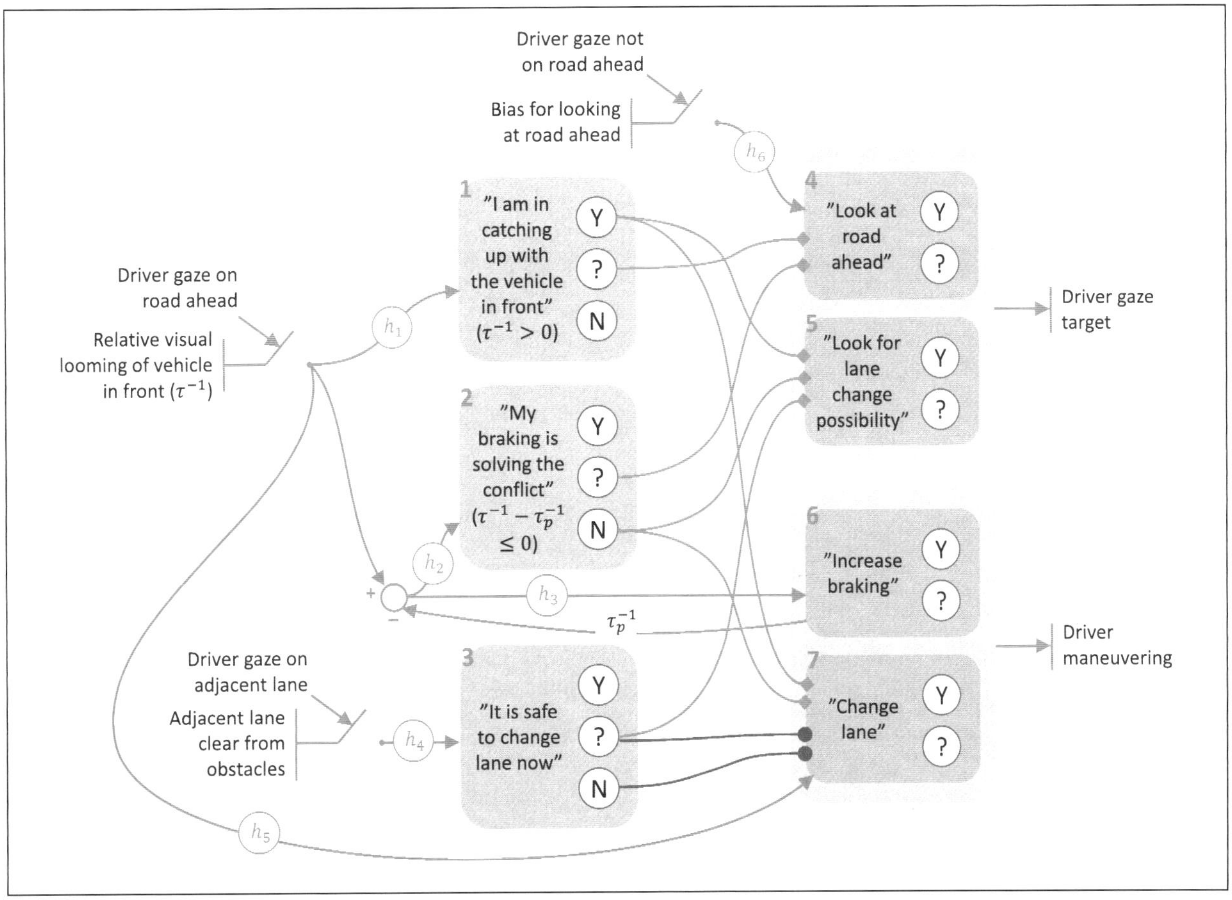

Figure 3. Schematic representation of the model for the safety-critical take-over scenario.

decision to change lane and a negative contribution to the same decision coming into play once $\tau^{-1}>\tau_{NS}^{-1}$, where $\tau_{NS}^{-1}>\tau_S^{-1}$.

The remaining inputs were defined as:

- $F_3(t) = k_6$ and $F_4(t) = k_7$, both if the driver is directing their gaze toward the adjacent lane (whether by mirror checks, shoulder checks, etc., is not defined in the model), zero otherwise.
- $F_5(t) = 0$.

Manual tuning indicated satisfactory model behavior for $T = 2\text{s}; \Sigma = 0.3; k_1 = k_2 = 20; k_3 = k_5 = 6;\ \ k_4 = 3;$ $k_6 = 1.5; k_7 = 1.2;\ \tau_B^{-1} = \tau_S^{-1} = 0.1\text{s}^{-1}; \tau_{NS}^{-1} = 0.5\text{s}^{-1}$.

Model Behavior – Reproducing Findings from Literature

Reproducing Pedestrian Crossing Behavior

Figure 4 shows example simulations with the pedestrian model, all at $v_0 =\ 50\,\text{km/h}$ (31 mph) and with TTC = 4 s at the moment when the pedestrian steps up to the zebra crossing, at which point the simulated car also initiates a deceleration to stop exactly at the zebra crossing. Panel (*a*) shows example simulations both with and without noise, whereas panels (*b*) and (*c*) show distributions for 10,000 simulations with noise.

In panel (*a*), note in the simulation without noise how the pedestrian comes close to a quick decision to cross while τ is above the threshold $\tau_{pass} = 3$ s, but as the margin $\tau - \tau_{pass}$ is small, the evidence accumulation is so slow that τ falls below τ_{pass} before a complete crossing decision has formed. From that point, despite picking up some signs of car deceleration, the pedestrian still judges from τ that crossing might be risky. Thus, as further illustrated by the delay time distribution in panel (*b*), one qualitative human behavior result that the model reproduces is the finding reported in Schneemann and Gohl, in which 20% of pedestrians waited until the car came to a complete stop before crossing (*16*). In the model simulations, the remaining $\approx 80\%$ are spread out in time with a wide peak in the time region while τ is still above τ_{pass}, in which the model can make the decision to cross even if it is unsure of the car's intentions. This aligns well

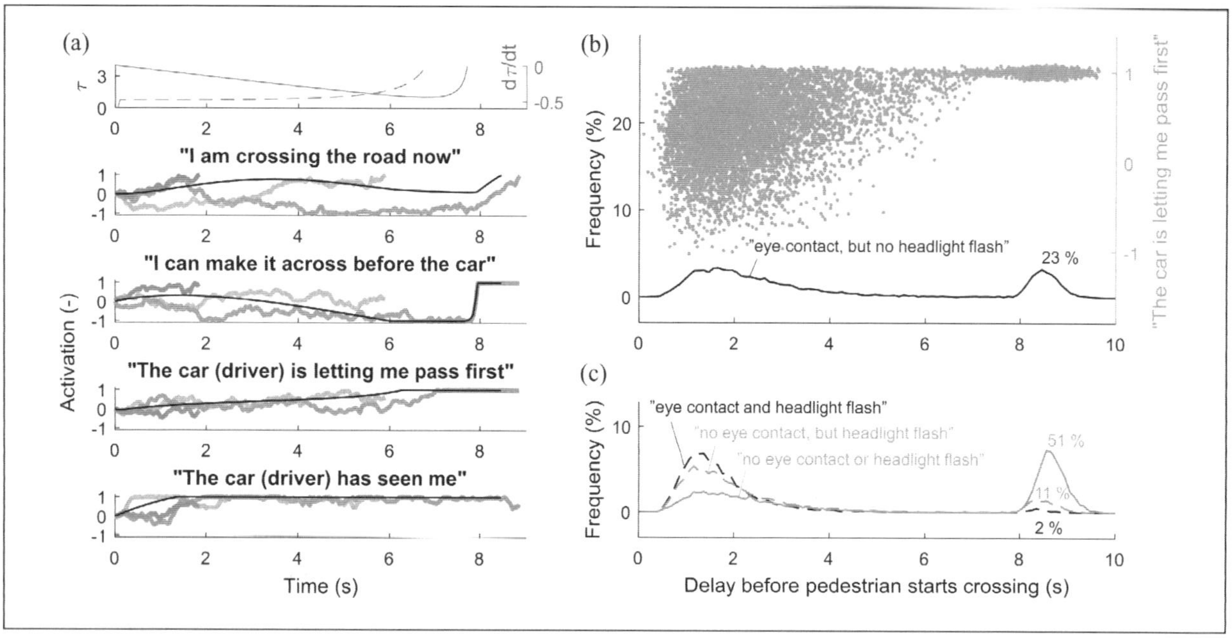

Figure 4. Example simulations of the pedestrian model: (*a*) visual looming input (top panel) and resulting model activations. The black activation traces show a model simulation without noise, and the lighter colored activation traces show three example simulations with noise; (*b*) the black line shows the probability distribution of delay times for the model pedestrian's decision to cross the road; the percentage indicates the fraction of simulations where the pedestrian waited until complete vehicle standstill. The light brown dots show the activation of the "the car is letting me pass first" unit at the time of crossing decision, as a function of the decision delay time, and (*c*) distribution of delay times for variations of the same scenario.

with another finding in Schneemann and Gohl, in which 25% of pedestrians stated that they made the decision to cross without having seen the car's deceleration (*16*). Exactly what activation A of the deceleration-detecting perceptual decision would yield a positive response to such a question is not defined in the model, but as a rough comparison, 25% of the light brown dots in panel (*b*) lie below $A = 0.3$.

The scenario in Figure 4*b* assumes that the pedestrian gets indications of being seen by the driver (e.g., eye contact) but no explicit communication about yielding intentions, whereas panel (*c*) shows variations in this interactive behavior. In the best case, in which the pedestrian gets indications both of the car being aware of him/her, and of the car's intention to stop, the model pedestrian almost never waits until the car is stationary (dotted black line), whereas more than half of the model pedestrians wait that long in the worst case for which neither of those indications are present (solid gray line).

Reproducing Safety-Critical Take-Over Behavior

Figure 5*a* shows four example time histories generated by the take-over driver model, all with TTC = 4 s and intermediate ($A = 0.5$) initial awareness of the adjacent lane being empty. These simulations differ only in the initial seed of the random noise in the decision processes, but it is notable that the model can nevertheless produce a range of qualitatively different sequences of behavior, both with and without one or more long or short glances away from the road ahead, and both with and without some degree of braking before a lane change is initiated (terminating the simulation).

Figure 5*b* shows the model-predicted frequency of different avoidance maneuvers, as a function of TTC at obstacle appearance, for a case with intermediate ($A = 0.5$) initial awareness of the adjacent lane being empty, as a possible emulation of a typical manual driving case. The obtained model distributions, and specifically the frequencies marked by vertical lines in the figure, align well with the pattern of driver responses reported in Gold et al. for the manual driving condition (*17*): all human drivers (8/8; 100%) responded with steering in the TTC = 7-s condition, whereas in the TTC = 5-s condition, some combined braking and steering was also observed (1/5 drivers; 20%).

Figure 5*c* shows that without any initial awareness of the adjacent lane ($A = 0$), the probability of braking increases. This was also reported in Gold et al. for the AV condition (*17*). Specifically, the model frequencies at the initial TTCs marked with vertical lines (4 s and 3.5 s) align qualitatively with the human responses in Gold et al. for the experimental conditions with take-over request at 7 s and 5 s TTC (*17*): a near 50/50 division

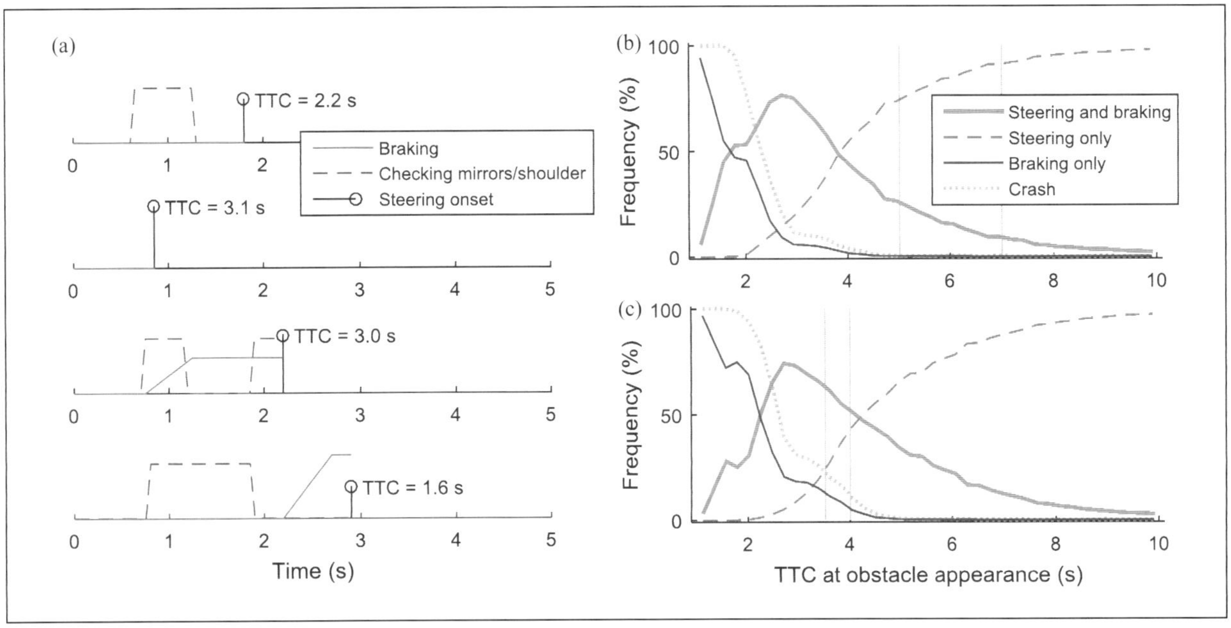

Figure 5. Example simulations of the hand-over model: (*a*) typical model behavior time histories, all for the same initial condition; (*b*) frequency of avoidance maneuvers performed by the model when run in a way so as to emulate manual driving (intermediate initial awareness of the adjacent lane), and resulting crash frequency, as a function of TTC at the time of obstacle appearance. The vertical lines indicate points of correspondence with Gold et al. (*17*), discussed in the text; and (*c*) as panel (*b*), but emulating a transition out of automated driving (with zero initial awareness of the adjacent lane).

between steering only and combined steering/braking was observed in the 7-s condition, whereas in the 5-s condition combined steering/braking rose to more clearly being the most common maneuver, followed by steering only, and some observations of pure braking. In other words, overall, the model simulations suggest the possibility that the take-over process in this scenario can be accounted for as a combination of reduced situational awareness and an urgency-dependent time delay, longer for the less critical scenario (7s – 4s = 3s, compared with 5s – 3.5s = 1.5s) . This aligns with the report in Gold et al. that hands-on-wheel and eyes-on-road delay times were longer in the less critical 7-s take-over scenario (*17*).

Figure 5, *b* and *c* also show simulated crash frequencies. The model's braking alone was typically not enough to avoid collision, whereas in Gold et al. seemingly no collisions occurred (*17*). This may indicate some problem with the tuning of the braking model. If the model applied steering avoidance, avoidance was considered successful as long as the time margin before collision, given speed and acceleration at the time of steering initiation, was above 1 s.

Model Applications in Virtual Testing and Tuning

This section will give examples of how the proposed models could be put to applied use in virtual, testing of AVs. It should be emphasized that as the models have yet to be fully validated and parameterized, the actual results given below are preliminary at best; especially for the transition scenario in which the model is both more complex and showed signs of possible limitations in its braking behavior. The main aim here is to illustrate the *type* of results that can be obtained.

Optimizing AV Traffic Flow at a Pedestrian Crossing

If an AV intends to yield to a pedestrian, how should it behave so as to help the pedestrian make the crossing decision as quickly as possible? To answer this question, it was assumed that once the pedestrian starts walking, the AV adapts acceleration to pass behind the pedestrian, and then accelerates at 1 m/s^2 back up to the initial 50 km/h (31 mph). The time lost in the interaction was then calculated by comparing the distance traveled before regaining the initial speed, to the distance that would have been traveled in the same amount of time at constant speed, that is, as if there had been no pedestrian.

Figure 6 shows, across a range of pedestrian appearance TTCs, (i) the model's predicted effect on time loss, of applying more than the deceleration needed to stop exactly at the zebra crossing, that is, rising above the curved edge of the gray shaded area in each panel, and (ii) the effect on time loss of external AV–human communication. Panel (*a*) shows a "minimal

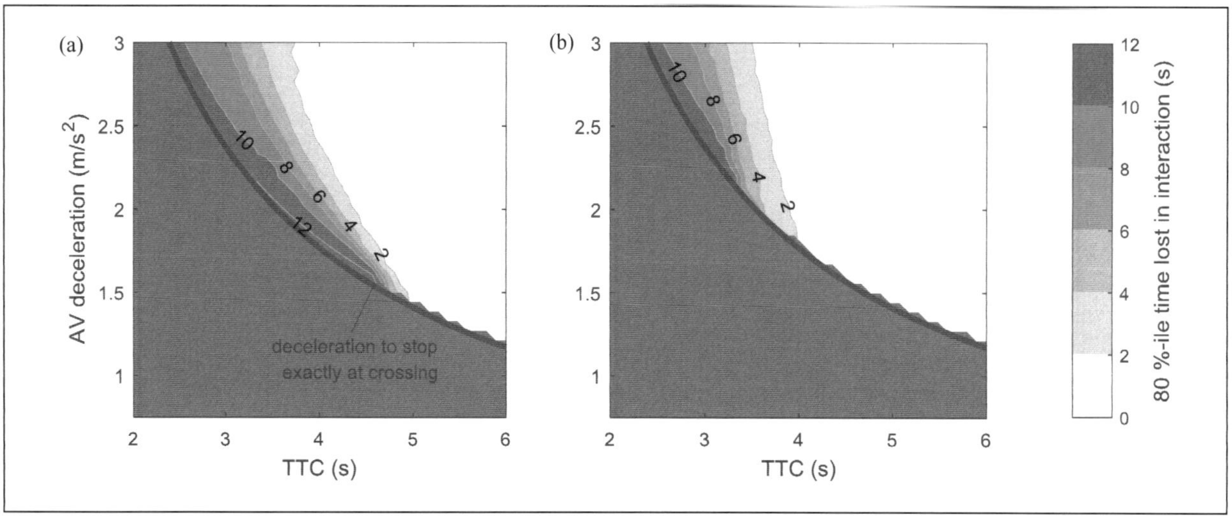

Figure 6. Model estimation of time lost for the AV in the pedestrian interaction, as a function of TTC at time of pedestrian presentation, and AV deceleration applied at the same moment: (*a*) no explicit communication from AV, and (*b*) AV explicitly communicates pedestrian awareness and yielding intention.

communication" case in which the AV does not give any explicit external signs of being aware of the pedestrian (i.e., nothing to replace eye contact) nor any signs, besides deceleration, of its intentions to yield, and panel (*b*) is a case in which the AV does show such explicit external signs.

Perhaps counter to what one might expect, the model predicts that increased deceleration can save considerable time. This is especially true for the AV without external communication means, whereas for some TTCs (e.g., TTC = 4.5 s) an extra 0.5 m/s^2 of deceleration can change a 12-s time loss to near zero. The reason this happens in the model is that the pedestrian's judgment of whether the car will stop before the crossing (i.e., of whether $\dot{\tau} > -0.5$) is difficult when the margin is narrow (i.e., when $\dot{\tau} \approx -0.5$), and the exaggerated deceleration makes this perceptual decision much easier and quicker.

The model predicts that time savings can also be achieved by letting the AV communicate its situational awareness and intentions (assuming that this can be achieved by some external HMI that is understandable to the pedestrian). The benefits of exaggerated decelerations are still present, but less pronounced, except for the lowest TTCs, where it starts being questionable whether the AV should yield at all to a waiting pedestrian, considering the comfort and safety of AV occupants and any following vehicles.

Minimizing Crash Risk in Time-Critical AV Hand-Overs

It has been suggested that detriments in performance during AV–human hand-overs are the result of a loss of situational awareness (*8*), with a recommendation that designers need to give in-depth consideration as to how best to facilitiate the process of bringing drivers back "into the loop" during transitions from automation (*18*). Exactly how to do so is of course a non-trivial matter, but here it is examined what benefit it might have in the studied scenario *if* a solution could be devised ensuring that the driver quickly knows, without having to look away from the road ahead, that it is possible to change lane, emulated here by initializing the "it is safe to change lane now" decision at full certainty. Whether any such benefit might interact, according to the model, with emergency braking applied by the AV itself is also examined here.

As shown in Figure 7, the model predicts a range of initial TTCs between 1.5 s and 3.5 s (remember that, factoring in the take-over process, this might correspond to a considerably higher obstacle appearance TTC) for which benefits of increased situational awareness may exist. For example, even if the AV decelerates with 6 m/s^2 in a TTC = 2 s scenario, immediate driver awareness of the feasibility of steering avoidance changes the model-predicted crash risk from about 80% to about 20%. In fact, according to the model, this low crash risk remains nearly unchanged even if the AV does not apply any emergency braking at all, which seems beneficial from a perspective of minimizing the risk of being struck from behind.

Discussion and Conclusion

This paper has presented first draft versions of novel models of human decision-making and behavior in two

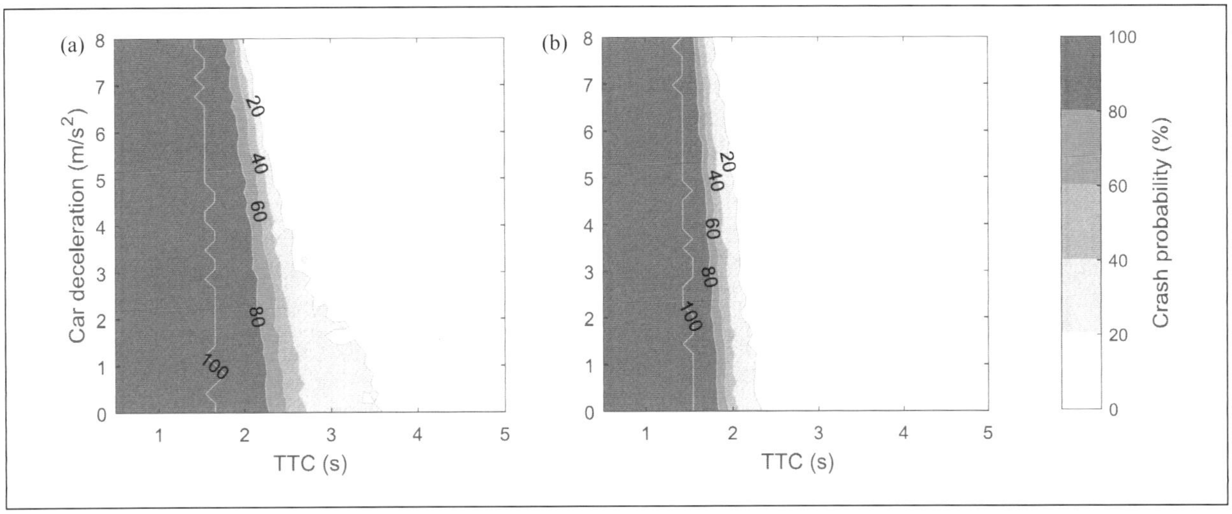

Figure 7. Model estimated effect of driver awareness of possibility to change lane, and AV emergency braking, on crash risk in the studied take-over scenario: (*a*) no driver awareness of adjacent lane, and (*b*) full driver awareness of adjacent lane.

scenarios relevant to development and testing of AVs. Both models seem to offer useful developments beyond what is currently available in the literature for these scenarios.

Pedestrian road-crossing has been modeled here as an interrelated set of perceptual decisions about oncoming vehicle movement and intentions. The model has illustrated how a car that applies exactly the right amount of deceleration for stopping might make it difficult for a waiting pedestrian to decide on crossing before the car has come to a complete stop, thus explaining empirical observations of such late pedestrian crossing decisions (in Schneemann and Gohl [*16*], but anecdotally also in many people's everyday experience). In contrast, existing microsimulation pedestrian models tend to emphasize the question of *whether* a pedestrian accepts a certain time gap between passing cars (*22*), and less on *when* the pedestrian accepts it, as a function of exact movement and communicative acts of a vehicle, making the model proposed here a potentially useful addition to the toolkit for virtual testing and tuning of AV behavior.

With regards to the other scenario targeted here, focusing on human avoidance maneuvering close to a potential rear-end crash, existing driver models for such situations typically decide on braking or steering, or both, by drawing from fixed, situation-independent probability distributions (*23*), whereas it is well known that human avoidance responses vary significantly with situation urgency (*16, 24*), in ways similar to what is exhibited by the model proposed here. The present work also provides some first steps toward modeling the actual process of taking over from an AV, suggesting that, at least in the presently studied scenario, the take-over could be thought of as an urgency-dependent time delay, combined with a low initial situational awareness.

Indeed, a striking feature of both models is that they provide concrete operationalizations of the concept of situational awareness in the given scenarios, in terms of neurobiologically plausible models of perceptual decisions. This may be a useful complement to existing computational models of situational awareness (e.g., [*13*]). Overall, the results presented here seem promising for the general modeling approach of interconnecting several perception and action decision accumulators.

Related but different models have been developed from connectionist and dynamical systems perspectives (*14, 25*), as well as in the field of naturalistic decision-making (*15*). One useful next step would be to benchmark the present modeling framework against these existing ones, and ideally also against more conceptually different alternatives based on cognitive architectures such as ACT-R (*26*), or from robotics (*27*).

An obvious next step will be to properly test and parameterize the models on detailed data sets of human behavior in the studied scenarios. If such validation tests are successful, the (so far very preliminary) virtual testing results presented here suggest that models at this level of detail, capable of predicting full AV–human interaction sequences unfolding in dynamic traffic scenarios, could be of great use when testing, validating, and optimizing future automated transport systems.

Acknowledgments

This work is part of the interACT project, funded by the European Union's Horizon 2020 research and innovation program under grant agreement 723395.

Author Contributions

The authors confirm contribution to the paper as follows: development of models and model simulation results: GM; interpretation of results: all authors; draft manuscript preparation: GM. All authors reviewed the results and approved the final version of the manuscript.

References

1. Fagnant, D. J., and K. Kockelman. Preparing a Nation for Autonomous Vehicles: Opportunities, Barriers and Policy Recommendations. *Transportation Research Part A: Policy and Practice*, Vol. 77, 2015, pp. 167–181. http://dx.doi.org/10.1016/j.tra.2015.04.003.
2. Kalra, N., and S. M. Paddock. Driving to Safety: How Many Miles of Driving Would it Take to Demonstrate Autonomous Vehicle Reliability? *Transportation Research Part A: Policy and Practice*, Vol. 94, 2016, pp. 182–193. http://dx.doi.org/10.1016/j.tra.2016.09.010.
3. Chen, B., D. Zhao, and H. Peng. Evaluation of Automated Vehicles Encountering Pedestrians at Unsignalized Crossings. *Proc., 2017 IEEE Intelligent Vehicles Symposium (IV)*, IEEE, New York, 2017.
4. Waymo. *On the Road to Fully Self Driving: Waymo Safety Report*. 2017. https://waymo.com/safetyreport/. Accessed October 16, 2017.
5. Clamann, M., M. Aubert, and M. L. Cummings. Evaluation of Vehicle-to-Pedestrian Communication Displays for Autonomous Vehicles. Presented at 96th Annual Meeting of the Transportation Research Board, Washington, D.C., 2017.
6. Kesting, A., M. Treiber, M. Schönhof, and D. Helbing. Adaptive Cruise Control Design for Active Congestion Avoidance. *Transportation Research Part C: Emerging Technologies*, Vol. 16, No. 6, 2008, pp. 668–683. http://dx.doi.org/10.1016/j.trc.2007.12.004.
7. Zeeb, K., A. Buchner, and M. Schrauf. What Determines the Take-Over Time? An Integrated Model Approach of Driver Take-Over after Automated Driving. *Accident Analysis and Prevention*, Vol. 78, 2015, pp. 212–221. http://dx.doi.org/10.1016/j.aap.2015.02.023.
8. Kaber, D. B., and M. R. Endsley. The Effects of Level of Automation and Adaptive Automation on Human Performance, Situation Awareness and Workload in a Dynamic Control Task. *Theoretical Issues in Ergonomics Science*, Vol. 5, No. 2, 2004, pp. 113–153. http://dx.doi.org/10.1080/1463922021000054335.
9. Purcell, B. A., R. P. Heitz, J. Y. Cohen, J. D. Schall, G. D. Logan, and T. J. Palmeri. Neurally Constrained Modeling of Perceptual Decision Making. *Psychological Review*, Vol. 117, No. 4, 2010, pp. 1113–1143.
10. Ratcliff, R., and D. Strayer. Modeling Simple Driving Tasks with a One-Boundary Diffusion Model. *Psychonomic Bulletin & Review*, Vol. 21, No. 3, 2014, pp. 577–589. http://dx.doi.org/10.3758/s13423-013-0541-x.
11. Markkula, G., J. Engström, J. Lodin, J. Bärgman, and T. Victor. A Farewell to Brake Reaction Times? Kinematics-Dependent Brake Response in Naturalistic Rear-End Emergencies. *Accident Analysis & Prevention*, Vol. 95, 2016, pp. 209–226.
12. Svärd, M., G. Markkula, J. Engström, J. Bärgman, and F. Granum. A Quantitative Driver Model of Pre-Crash Brake Onset and Control. *Proc., Human Factors and Ergonomics Society 61st Annual Meeting*, Austin, Tex., SAGE Publications, Vol. 61, No. 1, 2017, pp. 339–343.
13. McCarley, J. S., C. D. Wickens, J. Goh, and W. J. Horrey. A Computational Model of Attention/Situation Awareness. *Proc., Human Factors and Ergonomics Society 46th Annual Meeting*, Vol. 46, No. 17, 2002, pp. 1669–1673.
14. Roe, R. M., J. R. Busemeyer, and J. T. Townsend. Multialternative Decision Field Theory: A Dynamic Connectionist Model of Decision Making. *Psychological Review*, Vol. 108, No. 2, 2001, pp. 370–392.
15. Patterson, R., L. Fournier, B. Pierce, M. Winterbottom, and L. Tripp. Modeling the Dynamics of Recognition-Primed Decision Making. *Proc., 9th International Conference on Naturalistic Decision Making*, British Computer Society, London, UK, 2009, pp. 113–120.
16. Schneemann, F., and I. Gohl. Analyzing Driver-Pedestrian Interaction at Crosswalk: A Contribution to Autonomous Driving in Urban Environments. *Proc., 2016 IEEE Intelligent Vehicles Symposium*, Gothenburg, Sweden, IEEE, New York, 2016.
17. Gold, C., D. Damböck, L. Lorenz, and K. Bengler. "Take Over!" How Long Does it Take to Get the Driver Back into the Loop? *Proc., Human Factors and Ergonomics Society 57th Annual Meeting*, Sage Publications, 2013.
18. Merat, N., A. H. Jamson, F. C. Lai, M. Daly, and O. M. Carsten. Transition to Manual: Driver Behaviour when Resuming Control from a Highly Automated Vehicle. *Transportation Research Part F: Psychology and Behaviour*, Vol. 27, 2014, pp. 274–282. http://dx.doi.org/10.1016/j.trf.2014.09.005.
19. Lee, D. N. A Theory of Visual Control of Braking Based on Information About Time-to-Collision. *Perception*, Vol. 5, No. 4, 1976, pp. 437–459.
20. Bogacz, R., E. Brown, J. Moehlis, P. Holmes, and J. D. Cohen. The Physics of Optimal Decision Making : A Formal Analysis of Models of Performance in Two-Alternative Forced-Choice Tasks. *Psychological Review*, Vol. 113, No. 4, 2006, pp. 700–765. http://dx.doi.org/10.1037/0033-295X.113.4.700.
21. Friston, K., R. A. Adams, L. Perrinet, and M. Breakspear. Perceptions as Hypotheses: Saccades as Experiments. *Frontiers in Psychology*, Vol. 3, 2012, p. 151. http://dx.doi.org/10.3389/fpsyg.2012.00151.
22. Lu, L., G. Ren, W. Wang, C. Y. Chan, and J. Wang. A Cellular Automaton Simulation Model for Pedestrian and Vehicle Interaction Behaviors at Unsignalized Mid-Block Crosswalks. *Accident Analysis and Prevention*, Vol. 95, 2016, pp. 425–437.
23. Van Auken, R. M., J. W. Zelner, D. P. Chiang, J. Kelly, J. Y. Silberling, and R. Dai. *Advanced Crash Avoidance Technologies (ACAT) Program - Final Report of the Honda-DRI Team*. Publication DOT HS 811 454. USDOT/NHTSA, Washington, D.C., 2011.

24. Lee, S. E., E. Llaneras, S. G. Klauer, and J. Sudweeks. *Analyses of Rear-End Crashes and Near-Crashes in the 100-Car Naturalistic Driving Study to Support Rear-Signaling Countermeasure Development*. Publication DOT HS 810 846. USDOT/NHTSA, Washington, D.C., 2007.
25. Pellecchia, A., C. Igel, J. Edelbrunner, and G. Schoner. Making Driver Modeling Attractive. *IEEE Intelligent Systems*, Vol. 20, No. 2, 2005, pp. 8–12. http://dx.doi.org/10.1109/MIS.2005.31.
26. Keller, J., C. Lebiere, C. R. Shay, and K. Latorella. Cockpit System Situational Awareness Modeling Tool. In *Human Performance, Situation Awareness and Automation: Current Research and Trends: HPSAA II* (D. A. Vincenzi, M. Mouloua, and P. A. Hancock, eds.), Psychology Press, pp. 66–71.
27. Bai, H., S. Cai, N. Ye, D. Hsu, and W. S. Lee. Intention-Aware Online POMDP Planning for Autonomous Driving in a Crowd. *Proc., 2015 IEEE International Conference on Robotics and Automation*, Seattle, Wash., IEEE, New York, 2015.

The Standing Committee on Simulation and Measurement of Vehicle and Operator Performance (AND30) peer-reviewed this paper (18-03744).

The research presented here reflects the authors' views and the European Commission is not responsible for any use that may be made of the information herein.

Article

TRR
JOURNAL OF THE TRANSPORTATION RESEARCH BOARD

Transportation Research Record
2018, Vol. 2672(37) 164–171

DOI: 10.1177/0361198118798729
journals.sagepub.com/home/trr
SAGE

Assessing Driving Simulator Validity: A Comparison of Multi-Modal Smartphone Interactions across Simulated and Field Environments

Thomas McWilliams[1], Nathan Ward[2], Bruce Mehler[1], and Bryan Reimer[1]

Abstract

The use of a driving simulator as a tool to evaluate secondary task performance elicits the question of simulator validity. After upgrading an existing driving simulator from a medium-fidelity to a high-fidelity configuration with a new software environment, a study was run to benchmark this simulator against previously published highway-driving data. A primary goal was to assess relative and absolute validity in a simulated highway environment. Data from 72 participants who performed manual and voice-based contact dialing tasks with a center-stack-mounted smartphone in either the driving simulator or driving on the highway in one of two vehicles is considered. This analysis compared secondary task demand between the simulator and on-road vehicles by primarily considering driver off-road glance behavior. Mean total eyes-off-road time, mean single-glance duration, and the number of long off-road glances showed similar patterns relative to the manual versus voice-based tasks in the simulator and the two on-road vehicles. A driving performance metric, percentage change of standard deviation of velocity, showed differing results between the simulator and on-road vehicles. It is concluded that these data make a strong argument for relative validity, and in some cases absolute validity, for this simulator for studying glance behavior associated with in-vehicle devices under a highway-driving scenario.

Driving simulators have been used as an alternative to on-road driving for experimental purposes since as early as the 1930s (*1*). Driving simulators offer better environmental control and raise fewer safety concerns than real-world driving. Due to technological leaps and the development of personal computers, driving simulators can be relatively easily implemented and are used for research around the world. Driving simulators can be generally divided into three groups: low fidelity, medium fidelity, and high fidelity (*2*). The fidelity of a driving simulator refers to how realistic an experience it can provide. Low-fidelity simulators are the most basic, often consisting of a computer monitor with a portable steering wheel and pedals placed at a desk. Medium-fidelity simulators typically employ a larger screen, or several screens, to encompass a larger portion of the participant's field of view. These types of simulators often also have a partial vehicle cab with a driver's seat, instrument cluster, and center stack with which the driver can interact. High-fidelity simulators generally use multiple, large screens to support a broader range of a driver's field of view, often use full vehicle cabs, and can provide physical motion feedback through actuators mounted to the base of the simulator. There is evidence of the benefits of using high-fidelity simulators over lower fidelity, for example in driver training (*3*), but the level of fidelity needed often depends on the research question at hand. Although high-fidelity simulators may be more attractive in relation to their greater face-validity, they can require a sizable amount of space and resources to build and operate, whereas low-fidelity simulators can be relatively inexpensive and can be set up in small spaces. The present report considers data collected from a high-fidelity simulator and compares it with a subset of previously collected data (see publications *4, 5*) from two on-road vehicles under actual highway-driving conditions.

At face value, the simulator in question is nearly identical to a commercially available vehicle; however, it is

[1]MIT Age Lab, Cambridge, MA
[2]Tufts University, Medford, MA

Corresponding Author:
Address correspondence to Thomas McWilliams: tmcwill@mit.edu

crucial to go beyond physical appearances to compare human performance in the simulator to that on the road. The physical validity of a driving simulator refers to the similarity of the simulator to an actual vehicle. This is often seen though a correspondence between the layout of components in the vehicle and the simulator, and similarities in the operation and dynamics of both environments (*6, 7*). Not surprisingly, high-fidelity simulators are often seen to have higher physical validity than low- and medium-fidelity simulators (*6*). The simulator in question arguably has physical validity in the sense that the driver sits in an actual Volkswagen New Beetle cab and interacts with a full sized steering wheel and the original vehicle accelerator and brake pedals. The simulated external environments presented to the drivers of presently available simulators are not able to provide the same level of physical validity. The world may be presented in a realistic manner, and interactions may be similar in the real world, but the environment is highly controlled by researchers and the environment does not present any substantive risk to the driver.

Putting physical validity aside, one must also evaluate whether or not a simulator produces behavioral results that are representative of those observed in the real world. Arguably, the more important type of validity is behavioral validity (*7*). There are two main aspects of behavioral validity: absolute and relative. The extent to which numerical values obtained in the simulator and in real-world driving are consistent is known as absolute validity. For example, if eye-glance behaviors (e.g., amount of time looking off-road during a specific task) are quantitatively the same in a simulator and on road, then one may consider the simulator to have absolute validity. Relative validity refers to the degree to which the direction and magnitude of behavior observed in the simulator and real-world driving correspond. For example, if two different tasks are performed in both a simulator and on road, and participants spend less time looking off-road during one task versus the other task in both environments, the simulator would have relative validity.

The present study was motivated by a major upgrade of an existing medium-fidelity simulator to a high-fidelity configuration that included the implementation of an entirely new simulation software package. The previous medium-fidelity configuration had been validated against on-road behavior in a number of ways. For example, physiological arousal in response to increasing levels of cognitive demand was evaluated in the simulator and on road using a structured working memory task (*8*). Though the absolute levels of heart rate and skin conductance were found to be higher under actual highway-driving conditions (thus not demonstrating absolute validity), the relative change in arousal with each level of demand was identical across the two environments (supporting relative validity). In a separate effort (*6*), a number of metrics (initial response time, mean task duration, and a range of glance measures) during interactions with three different configurations of a destination entry task, mapped in a very similar manner in both the simulator and during interaction with same interfaces on-road, showing good relative validity and, in some cases, absolute validity as well. Findings for driving performance metrics were somewhat mixed, as has been observed in other studies (*7, 9*). Other aspects of validity, considering the relationship between self-reported behavior and behavior observed driving the simulator have also been previously evaluated (*10*).

Of primary concern with the new simulator configuration, was the extent to which it could be considered a valid tool for continuing an ongoing program evaluating the demands on drivers associated with interacting with various in-vehicle devices and user interfaces as an analog for on-highway behavior. Of particular interest in this regard are glance behaviors that the National Highway Traffic Safety Administration (NHTSA) has utilized for visual demand assessment in driving simulators (*11*). Off-road glances are a frequently used metric to measure the amount of demand a device places on the driver (*2, 6*). There has been significant research on the connection between glance behavior and crash risk (*12, 13*), as well as the issuance of guidelines by NHTSA on glance behavior during secondary task performance while driving (*14*). In this context, the present paper focuses on the validation of the new driving simulator configuration for assessing the visual demand of manual and voice-based in-vehicle secondary tasks.

Methods

Participants

Data from a total of 72 participants was considered. A sample of 24 participants was used in the analysis of the simulator portion of the study, all between the ages of 20 and 24 or 55 and 69. Participants were equally balanced across the age and gender groupings (younger M = 21.7, SD = 1.1; older M = 59.1, SD = 3.4). All reported having a valid driver's license for more than 3 years, driving on average one or more days per week, and were in self-reported reasonably good health for their age. The on-road analysis sample consisted of two groups of 24 participants each drawn from the earlier on-road study (*4, 5*) that were balanced across gender and representative of age characteristics similar to the simulator cohort (younger M = 22.9, SD = 2.3; older M = 58.28, SD = 5.4) although the age ranges were somewhat broader (20 to 28 and 46 to 69). All met the same qualifications as the simulator group, except all on-road participants were required to drive on average three or more days per week. One half of the on-road participants drove a 2013 Chevrolet Equinox, the other half drove a 2013 Volvo XC60.

Apparatus

Driving Simulator. This experiment used a high-fidelity driving simulator designed and fabricated by RTI (Realtime Technologies Inc., Ann Arbor, MI), which ran the SimCreator and SimVista software packages. The simulator consisted of the full cab of a 2001Volkswagen New Beetle. Three exterior screens provided a 180° simulated field of view. A rear projector screen hung behind the car displayed the rear roadway, which was viewable using the vehicle's standard rearview mirror. Two LCD displays were placed in the cab's side mirrors, which also presented the rear roadway. Images of the roadway were displayed at 60 frames per second. The steering wheel, gas, and brake pedals all provided force feedback to the participants. Internal and external speakers produced environmental noise during the experiment. The cab was instrumented with three cameras and one microphone to record the environment in the cab. One camera focused on the driver's face to capture head and eye movements, one camera recorded hand movements and interactions with a center-stack-mounted smartphone. The last camera recorded the forward screen to validate video and simulator data synchronization.

On-Road Vehicles. The two on-road vehicles (a 2013 Chevrolet Equinox and a 2013 Volvo XC60) were each instrumented with a data acquisition system designed for time-synchronized recordings. This system recorded information from the vehicle's controller area network bus, images from five video cameras monitoring the internal and external environments of the vehicles, and audio from a microphone inside the cab. Two cameras recorded the forward roadway (one narrow- and one wide-angle), one camera recorded the rear roadway, one focused on the center stack to monitor hand movements and interactions with a smartphone, and the last camera was focused on the driver's face to capture head and eye movements.

Smartphone. In each on-road vehicle and the simulator a Samsung Galaxy S4 (SCH-1545) running Android 4.3 (Jelly Bean) was mounted to the right side of the center stack. The Chevrolet Equinox had a mount attached to the dashboard above the center stack that allowed the phone to hang down to the right of the screen. The Volvo XC60 had a mount that clipped to the air vent below the center stack screen placing the phone to the right of the screen. The simulator had a mount that clipped to an air vent above the center stack screen placing the phone to the right of this screen. Voice input tasks were performed using Samsung's S-Voice interface with the hands-free mode turned on.

Tasks

Participants were instructed to place phone calls using a list of phone contacts stored in the smartphone, four using a visual-manual interface and four using a voice-based interface on the smartphone. Complete details are provided in Mehler et al. (*4*). The same four contact targets were used with each interface. In brief, to manually place a call, participants first had to wake up the phone by pressing the home button below the screen. Next they selected the "Contacts" icon along the bottom of the screen. This opened up a list of contacts that participants scrolled though until they came to the target name. For the first two calls, participants selected a contact's name, which initiated the call as each of these contacts had a single number listed. For the second two calls, selecting the contact name brought participants to another screen where they selected a specified phone (e.g., home, work) that was the target for that task. The voice-based calling tasks involved pressing the home button twice and waiting for the phone to give a greeting that indicated it was ready to accept a voice command. Participants then said "Call" followed by the name of the person they were instructed to call. For the second two calls, participants had to also say the name of the contact and a target phone (e.g., "Call Frank Scott at work"). After placing calls (both manual and voice), the phone connected to a voicemail message instructing the participant to hang up the call.

Procedure

Participants read and signed an informed consent form and completed demographic questionnaires. Next, participants were brought to the vehicle or simulator where they completed the remainder of the experiment. After being introduced to the operation of the vehicle or simulator, participants were trained on the tasks that they would perform during the evaluation portion of the experiment.

The driving simulation scenario consisted of a divided highway with two travel lanes in each direction, a posted speed limit of 65 mph, and low to moderate traffic; this scenario was modeled after I-495 north of Boston, MA used in the on-road study. Participants were instructed to follow the posted speed limit and drive in whichever lane they were comfortable.

On-road participants drove with a research assistant in the back seat of the vehicle. Participants drove approximately 20 min north of Boston on I-93 to I-495. On I-

495, participants were instructed to drive in whichever lane and at a speed that they were comfortable with and that was appropriate for the traffic conditions. The posted speed limit was 65 mph.

For analytic purposes, there was a 2-min baseline-driving period before initiation of the blocks of each task type (manual and voice-base phone contact calling). Within each block, individual calling tasks were separated by 30 s of just driving. The presentation order of the manual and voice blocks was counterbalanced across participants. All tasks began with a prerecorded audio prompt instructing the participant who to call and when to begin the calling task.

Data Reduction and Analysis

Given the importance of driver glance behaviors for safety, the majority of this analysis focuses on eye-glance measures. Following procedures detailed in Mehler et al. (*4*), two trained coders independently manually coded glances in each dataset and a third coder mediated any discrepancies. Open source software, which allowed for frame-by-frame review and annotation of driver eye movements, was used for this coding (*15*).

Values for the individual trials within each block (manual or voice task) were averaged across the block. The statistical tests used a between-subject design for the three environments (simulator, Chevrolet, Volvo) and a within-subject design for the two modalities (manual and voice) for a 3 × 2 design. ANOVAs were performed in R (*16*).

Results

Task Completion Time

Overall, there was a significant main effect of task modality on overall task completion times when the sample is considered as a whole, $F(1, 71) = 74.56, p < .001$. Task times were nominally lower during manual tasks compared with voice tasks in all three environments (Figure 1), generally supporting relative validity for this variable. However, a significant interaction between the three environments and task modality was also found, $F(2, 69) = 7.65, p < .001$. Although the difference by modality was statistically significant when testing individually for the two on-road vehicles (Volvo: $F[1, 23] = 69.44, p < .001$; Chevrolet: $F[1, 23] = 53.86, p < .001$), the difference in the simulator did not reach statistical significance, $F(1, 23) = 3.83, p = .062$; thus absolute validity is not supported. Looking at this another way, it can be observed that there was a significant effect of environment for manual tasks, $F(2, 69) = 9.6, p < .001$; participants took more time to complete the manual tasks in the simulator than participants took in the two on-road vehicles. However, there was no effect of environment on task completion time during voice tasks, $F(2, 69) = .91, p = .41$.

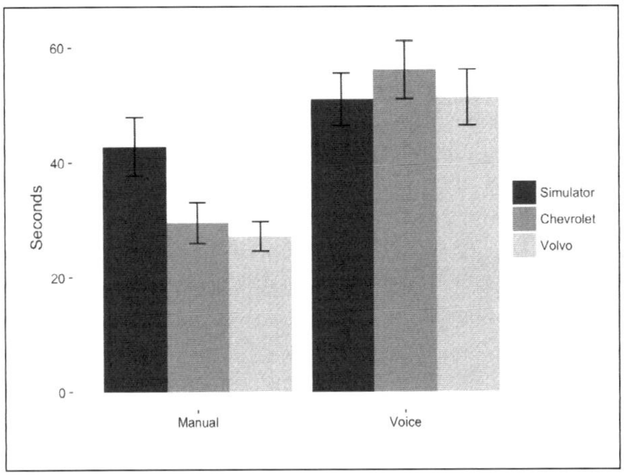

Figure 1. Mean total task completion time by environment and task modality. Error bars represent the standard error of the mean.

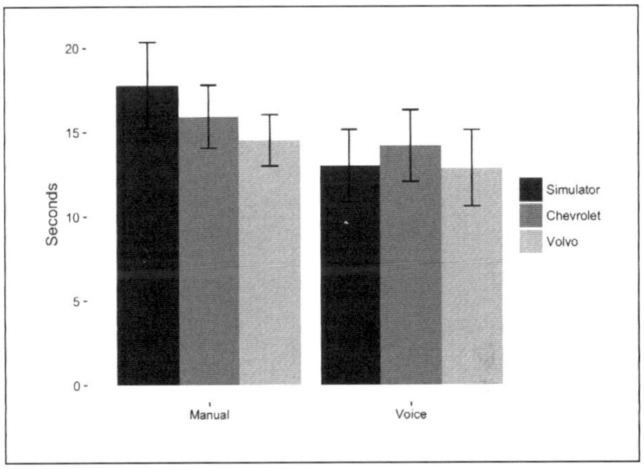

Figure 2. Mean total eyes-off-road time by environment and task modality. Error bars represent the standard error of the mean.

Total Eyes-Off-Road Time

The total amount of time that participants spent looking away from the forward road scene was coded and means across the trials making up each block were calculated for each participant. There was a significant main effect of task modality across the sample as a whole, $F(1, 71) = 8.23, p < .01$. On average, participants spent more time looking off-road during the manual tasks compared with voice tasks (Figure 2). There was not a significant interaction between environment and modality; however, it is evident that the modality effect was most prominent in the simulator, in which participants spent 4.8-s longer

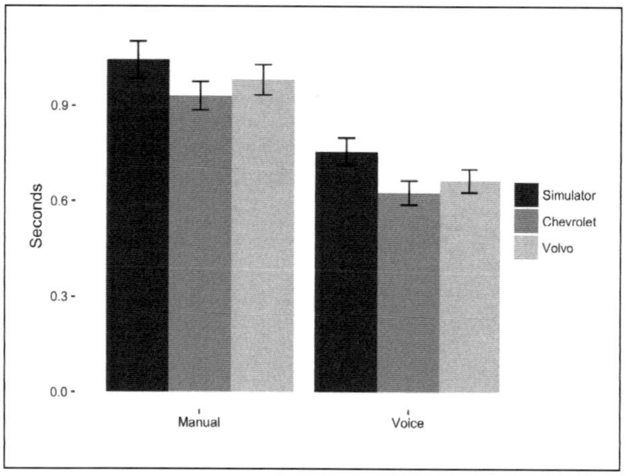

Figure 3. Mean single off-road glance duration by environment and task modality. Error bars represent the standard error of the mean.

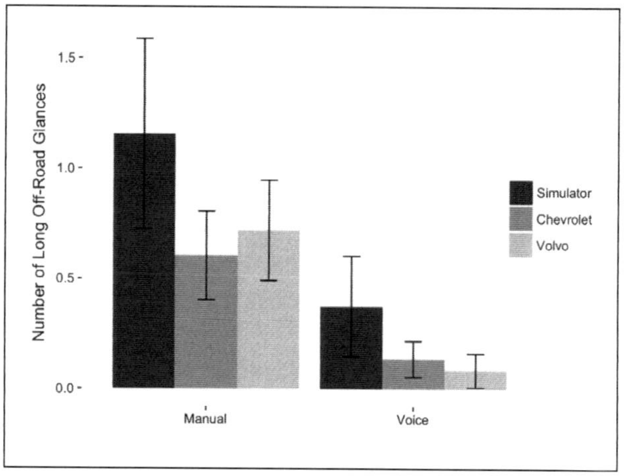

Figure 4. Mean number of long off-road glances by environment and task modality. Error bars represent the standard error of the mean.

looking off-road during manual versus voice-based calling. In the on-road vehicles total eyes-off-road time was 1.73-s longer in the Chevrolet and 1.65-s longer in the Volvo for the manual versus voice modality. When each of the samples was tested individually, only the difference in the simulator was statistically significant, $F(1, 23) = 7.27$, $p = .012$; the corresponding values for the Chevrolet are, $F(1, 23) = 0.94$, $p = .342$; and for the Volvo, $F(1,23) = 1.58$, $p = .222$). Thus, although the consistency in the patterning for nominally longer eyes off-road times for the manual interface across the environments is supportive of relative validity, absolute validity for the simulation was not demonstrated.

Mean Single-Glance Duration

Mean durations of single off-road glances showed a significant main effect of task modality, $F(1,71) = 180.15$, $p < .001$, in which individual glance durations were, on average, longer during the manual tasks than when using the voice-based interface (Figure 3). No interaction between environment and modality was present and this pattern of longer mean single-glance durations during manual tasks (M simulator = 1.04 s, M Chevrolet = .93 s, M Volvo = .98 s) than voice tasks (M simulator = .76 s, M Chevrolet = .63 s, M Volvo = .66 s) was consistent and statistically significant for all three samples: $F(1, 23) = 77.20$, $p < .001$; $F(1, 23) = 61.29$, $p < .001$; and $F(1,23) = 231.50$, $p < .001$, respectively. This patterning is consistent with relative validity.

Long Off-Road Glances

There was a significant main effect of task modality on the mean number of long-duration (longer than 2 s) off-road glances, $F(1, 71) = 32.24$, $p < .001$. On average, the number of long-duration glances was higher during manual tasks than during voice-based tasks (Figure 4). This finding was consistent and statistically significant across the three environments with participants making an average of 1.15 long off-road glances during manual tasks in the simulator, 0.6 in the Chevrolet, and 0.7 in the Volvo versus values for voice-based tasks of 0.37 in the simulator, 0.13 in the Chevrolet, and 0.08 in the Volvo. The corresponding post hoc tests within each environment were respectively: $F(1, 23) = 8.69$, $p = .007$; $F(1,23) = 10.86$, $p < .001$; and $F(1, 23) = 19.38$, $p < .001$. No interaction effects were present. These results can be considered consistent with both relative and absolute validity.

Standard Deviation of Velocity

The standard deviation of velocity during each task was recorded, and the percentage change between baseline periods and the task periods was calculated. A significant effect of modality was found across the sample, $F(1, 67) = 4.07$, $p = .047$. When considered by modality, there was a significant difference in the percentage change in standard deviation of velocity from the baseline-driving period for manual tasks, $F(2, 65) = 3.4$, $p = .039$), but not for voice tasks, $F(2, 65) = .58$, $p = .55$) (Figure 5). Both the vehicles and the simulator had smaller changes in the standard deviation of velocity from the baseline periods during voice tasks compared with manual tasks; however, the simulator showed overall increases in standard deviation of velocity from the baseline and the two vehicles showed decreases in the standard deviation of velocity from the baseline. Thus, neither absolute nor relative validity is apparent for this variable.

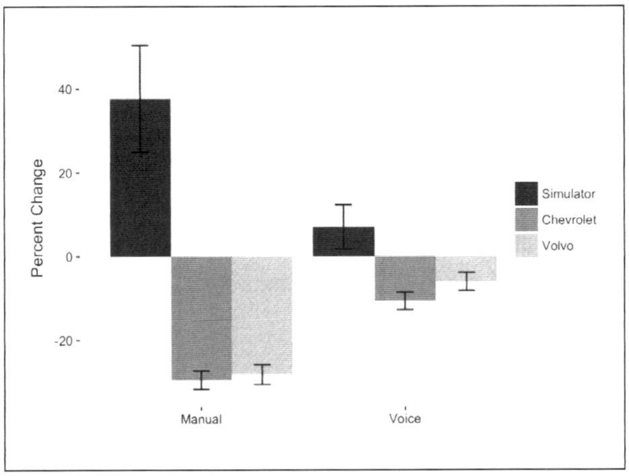

Figure 5. Mean percentage change in the standard deviation of velocity from baseline driving by environment and task modality. Error bars represent the standard error of the mean.

Discussion

The same basic behavior patterns appeared across the two on-road vehicles with regard to the relationship between task modality (manual versus voice-based interface) and each of the variables assessed. Except for the standard deviation of velocity variable, the simulator data largely conformed to these patterns in terms of relative validity, though some differences in absolute validity were present.

Task completion times in all environments were shorter for manual contact calling tasks compared with voice-based contact calling tasks with a mounted Samsung Galaxy S4. This is consistent with previous research (*4*, *5*) and is traceable to the duration of verbalizations by the driver, auditory prompts, and responses on the part of the voice system. The absolute difference in task completion time in the simulator was smaller between manual tasks and voice tasks than was observed in the two vehicles; this corresponded to simulator participants taking more time to complete manual calls relative to what was observed on-road, whereas task duration for voice-based calling was not significantly different across the three samples. Possible reasons for this finding are considered below.

A significant effect of task modality was found for total off-road glance time considering participants across the sample. The nominal mean values for each manual sample were greater than the mean value for each sample's corresponding voice-based calling task, thus in line with relative validity. The finding that a visual-manual task is more visually demanding than their corresponding voice-based alternative was expected and is consistent with numerous other studies (*17–19*). However, this difference was largely driven by the simulator, which was the only environment for which this difference was statistically significant. This will also be considered below.

Mean single-glance duration is often used to assess the demand required to interact with a system. All environments had a significant difference in mean single-glance time, with the primary visual-manual tasks eliciting longer glances. These results are consistent with previous literature (*4*, *5*). The consistency in patterning and lack of interaction between task modality and environment supports a finding of both relative and absolute validity in this simulator for this task, metric, and scenario.

Another glance metric that is often used to assess the demand brought on by using a device while driving is the number of off-road glances that last longer than 2 s. This measure is often used as a way to assess the risk involved with using a device as long glances are associated with increased accident risk (*12*, *13*). The significantly greater number of long off-road glances during manual tasks is consistent with previous findings (*20–22*), and is consistent with the idea that manual tasks place more visual demand on drivers. As the patterning of differences between task modalities is consistent across environments and there is no task modality by environment interaction, these findings are generally supportive of both relative and absolute validity.

Thus, the results for all three visual metrics are consistent with a finding of relative validity in the simulator for these variables. This is in line with findings for the medium-fidelity simulator that this high-fidelity simulator replaced, where the relative patterning of visual metrics characterizing how drivers looked off-road to interact with different visual-manual human machine interface (HMI) implementations aligned well with what was observed running the same tasks with the same interfaces under actual highway-driving conditions (*6*).

The fact that there were no significant differences by environment for these visual metrics when the modalities are considered independently could be taken as evidence for a degree of absolute validity as well. However, it is also worth noting that total off-road glance time during manual tasks, the mean single-glance duration for both task modalities, and the number of long-duration glances for both task modalities were all nominally greater in the simulator. Although not sufficient evidence to rule out absolute validity for these metrics, it is possible to hypothesize a trend toward participants being willing to visually engage somewhat more in off-road glance behavior during simulation than during the functionally more risky real-world driving environment. It can also be noted that participants engaged longer with the manual interface in the simulator than was observed on-road. This might also reflect on absolute validity in that participants may perhaps be comfortable taking their hands off the steering wheel somewhat longer in the simulator than they are on road. Nonetheless, these speculations do not affect the strong evidence for relative validity for the visual metrics.

Changes in a vehicle's velocity can suggest that the driver is experiencing added demand due the performance of a secondary task. Calculating the percentage change in the standard deviation of velocity from a designated baseline-driving period allows for a comparable measure that adjusts for driver-related individual differences and variation in environmental conditions. Again, there was an effect of task modality on the percentage change in standard deviation of velocity for each driving environment. Higher demand on drivers during manual tasks compared with voice tasks is the likely cause of this (*17*). The most striking aspect of these results is that both secondary tasks in the simulator resulted in an increase in the standard deviation of velocity, whereas engaging in the same tasks in the two on-road vehicles resulted in a decrease. It can be noted that there was only a statistical difference between the simulator and other vehicles during the manual tasks.

One major difference between simulator and highway driving is the consistency of the surrounding traffic. The simulator was programmed to have a consistent flow of traffic that would allow the driver to maintain a consistent speed. Not surprisingly, this is not the case on most highways. Though the highway-driving data was intentionally collected at times when there was a high likelihood of low to moderate traffic, this does not mean that traffic flow variation could be completely avoided, especially when things like accidents and construction can occur and affect traffic patterns. In addition to the variability of highway-driving conditions, participants in the simulator were instructed to drive at the posted speed limit of 65 mph. On the highway the posted speed limit was 65 mph, but participants were instructed to drive at a speed they were comfortable with that was also appropriate for the road and traffic conditions at the time. The divergence in the standard deviation of velocity between the simulator and highway-driving conditions suggests that the simulator did not achieve absolute validity in this case, and that it may continue to be a challenge to achieve in the future. However, the trend of standard deviations in velocity being closer to baseline driving during voice tasks and having an increased difference during manual tasks suggests that this measure may still demonstrate the relative validity of the simulator.

Summary

Previous work has argued for the importance of validating simulation systems for the use cases in which they are employed (*10*). The present study and analysis was run to assess the extent to which a simulator that was validated for studying driver glance behavior during interactions with HMIs (*6*) continued to be valid for this application following a major upgrade from a medium-fidelity to a high-fidelity configuration that also included a change to an entirely new software simulation package. Consistent with what was found in the previous validation effort, the new simulator demonstrated relative validity for studying HMI-related glance behavior. Specifically, the finding in the full on-road dataset (*5*) that the voice-based smartphone interface for phone contact calling was associated with shorter total eyes-off-road time, shorter mean single-glance durations, and fewer long-duration glances than the manual interface was also observed in the new simulator. Therefore, it seems appropriate to use this simulator as a tool to assess these aspects of secondary task demand on drivers in place of real-world driving. At the same time, other results reinforce the importance of recognizing that not all behaviors and metrics will show absolute validity, or even necessarily relative validity, between a given simulation and the real world. This is evident in participants' apparent willingness to spend more time engaged in the manual phone interface interaction in the simulator than was observed on-road, and the finding that a driving performance metric (standard deviation of velocity) changed in opposite directions under secondary task demand in the simulator versus actual on-road conditions.

Acknowledgments

This work was supported by the New England University Transportation Center and the Toyota Class Action Settlement Safety Research and Education Program.

Author Contributions

The authors confirm contribution to the paper as follows: study conception and design: TM, BM, BR; data collection: TM; analysis and interpretation of results: TM, NW, BM, BR; draft manuscript preparation: TM, NW, BM, BR. All authors reviewed the results and approved the final version of the manuscript.

References

1. De Silva, H. R. On an Investigation of Driving Skill. *Human Factors*, Vol. 10, 1936, pp. 1–13.
2. Fisher, D. L., M. Rizzo, J. Caird, and J. D. Lee. (eds.). *Handbook of Driving Simulation for Engineering, Medicine, and Psychology*. CRC Press, Boca Raton, FL, 2011.
3. Allen, R. W., G. D. Park, M. L. Cook, and D. Fiorentino. The Effect of Driving Simulator Fidelity on Training Effectiveness. *Proc., DSC 2007 North America, (DSC-NA)*, Dearborn, MI, 2007.
4. Mehler, B., D. Kidd, B. Reimer, I. Reagan, J. Dobres, and A. McCartt. Multi-Modal Assessment of On-Road Demand of Voice and Manual Phone Calling and Voice

Navigation Entry Across Two Embedded Vehicle Systems. *Ergonomics*, Vol. 59, No. 3, 2016, pp. 344–367.
5. Reimer, B., B. Mehler, I. Reagan, D. Kidd, and J. Dobres. Multi-Modal Demands of a Smartphone Used to Place Calls and Enter Addresses During Highway Driving Relative to Two Embedded Systems. *Ergonomics*, Vol. 59, No. 12, 2016, pp. 1565–1585.
6. Wang, Y., B. Mehler, B. Reimer, V. Lammers, L. D'Ambrosio, and J. Coughlin. The Validity of Driving Simulation for Assessing Differences between In-Vehicle Informational Interfaces: A Comparison with Field Testing. *Ergonomics*, Vol. 53, No. 3, 2010, pp. 404–420.
7. Blaauw, G. J. Driving Experience and Task Demands in Simulator and Instrumented Car: A Validation Study. *Human Factors*, Vol. 24, 1982, pp. 473–486.
8. Reimer, B., and B. Mehler. The Impact of Cognitive Workload on Physiological Arousal in Young Adult Drivers: A Field Study and Simulation Validation. *Ergonomics*, Vol. 54, No. 10, 2011, pp. 932–942.
9. Yan, X., M. Adbel-Aty, E. Radwan, X. Wang, and P. Chilakapati. Validating a Driving Simulator Using Surrogate Safety Measures. *Accident Analysis & Prevention*, Vol. 40, 2008, pp. 274–288.
10. Reimer, B., L. A. D'Ambrosio, J. F. Coughlin, M. E. Kafrissen, and J. Biederman. Using Self-Reported Data to Assess the Validity of Driving Simulation Data. *Behavior Research Methods*, Vol. 38, No. 2, 2006, pp. 314–324.
11. Perez, M., J. Owens, D. Viita, A. Angell, T. A. Ranney, G. H. S. Baldwin, E. Parmer, J. Martin, W. R. Garrott, and E. N. Mazzae. Radio Tuning Effects on Visual and Driving Performance – Simulator and Test Track Studies. Report No. DOT HS 811 781. National Highway Traffic Safety Administration, Washington, D.C., 2013.
12. Liang, Y., J. D. Lee, and L. Yekhshatyan. How Dangerous is Looking Away from the Road? Algorithms Predict Crash Risk from Glance Patterns in Naturalistic Driving. *Human Factors*, Vol. 54, No. 6, 2012, pp. 1104–1116.
13. Victor, T., J. Dozza, J. Bärgman, C. Boda, J. Engström, C. Flannagan, J. Lee, and G. Markkula. *SHRP 2 Report S2-S08A-RW-1: Analysis of Naturalistic Driving Study Data: Safer Glances, Driver Inattention, and Crash Risk*. Transportation Research Board of the National Academies, Washington, D.C., 2014.
14. National Highway Traffic Safety Administration. *Visual-Manual NHTSA Driver Distraction Guidelines for In-Vehicle Electronic Devices*. No. NHTSA-2010-0053. National Highway Traffic Safety Administration, Washington, D.C., 2013.
15. Reimer, B., P. Gruevski, and J. Coughlin. *MIT AgeLab Video Annotator*. Cambridge, MA, 2014. https://bitbucket.org/agelab/annotator.
16. R Core Team. *R: A Language and Environment for Statistical Computing. Vienna, Austria: R Foundation for Statistical Computing*. 2015. http://www.R-project.org/.
17. Lo, V. E-W., and P. A. Green. Development and Evaluation of Automotive Speech Interfaces: Useful Information from the Human Factors and the Related Literature. *International Journal of Vehicular Technology*, Vol 2013, Article ID 924170, pp. 1–13.
18. Maciej, J., and M. Vollrath. Comparison of Manual Versus Speech-Based Interaction with In-Vehicle Information Systems. *Accident Analysis & Prevention*, Vol. 41, No. 5, 2009, pp. 924–930.
19. Reimer, B., B. Mehler, J. Dobres, and J. F. Coughlin. *The Effects of a Production Level "Voice-Command" Interface on Driver Behavior: Reported Workload, Physiology, Visual Attention, and Driving Performance*. MIT AgeLab Technical Report No. 2013-17A. Massachusetts Institute of Technology, Cambridge, MA, 2013.
20. Mehler, B., B. Reimer, J. Dobres, and J. F. Coughlin. *Assessing the Demands of Voice Based In-Vehicle Based In-Vehicle Interfaces - Phase II Experiment 3 - 2015 Toyota Corolla*. MIT AgeLab Technical Report 2015-14. MIT, Cambridge, MA, 2015. http://doi.org/10.13140/RG.2.1.2244.3921.
21. Mehler, B., B. Reimer, J. Dobres, H. McAnulty, and J. F. Coughlin. *Assessing the Demands of Voice Based In-Vehicle Interfaces - Phase II Experiment 1 - 2014 Chevrolet Impala*. MIT AgeLab Technical Report 2015-6A. MIT, Cambridge, MA, 2015. http://doi.org/10.13140/RG.2.1.4996.9046.
22. Mehler, B., B. Reimer, H. McAnulty, J. Dobres, J. Lee, and J. F. Coughlin. *Assessing the Demands of Voice Based In-Vehicle Interfaces - Phase II Experiment 2 - 2014 Mercedes CLA*. MIT AgeLab Technical Report 2015-8. MIT, Cambridge, MA, 2015. http://doi.org/10.13140/RG.2.1.4996.9046.

The Standing Committee on Simulation and Measurement of Vehicle and Operator Performance (AND30) peer-reviewed this paper (18-06135).

The views and conclusions being expressed are those of the authors, and have not been sponsored, approved, or endorsed by Toyota or plaintiffs' class counsel.

Article

TRR
JOURNAL OF THE TRANSPORTATION RESEARCH BOARD

Transportation Research Record
2018, Vol. 2672(37) 172–180

DOI: 10.1177/0361198118783165
journals.sagepub.com/home/trr
SAGE

Exploring the Relationship between Electroencephalography (EEG) and Ordinary Driving Behavior: A Simulated Driving Study

Liu Yang[1], Zhengbing He[2], Wei Guan[1], and Shixiong Jiang[1]

Abstract

Driving behavior studies based on electroencephalography (EEG) have mostly investigated the relationship between various risky driving behaviors and brain activity, while only a few studies have discussed the relationship between ordinary driving behavior (drivers' behavior in normal situations) and brain activity. To bridge the gap, we conducted a driving simulator experiment to collect data on ordinary driving behavior, including acceleration, space headway, speed, time headway, lane deviation, and amplitude of steering wheel movements. At the same time, the amplitude, log-transformed power (LTP), and power spectral density of EEG were extracted as EEG features. The quantitative relationships between ordinary driving behavior features and EEG features were investigated, where power spectrum analysis was performed to process EEG signals and Pearson correlation analysis was utilized for statistical analysis. The results indicated that ordinary driving behavior relates to all four brain regions, especially the temporal, occipital, and frontal regions. β-LTP was found to be most relevant to ordinary driving behavior. Furthermore, acceleration, speed, and space headway may have potential correlation with EEG features (e.g., β-LTP). These findings improve our understanding of the correlation between brain activity and driving behavior, and show potential for application in transportation safety, such as advanced driver assistance systems design.

Driving behavior is widely studied in the field of transportation, especially driving behavior modeling, intelligent transportation systems, and transportation safety. As a noninvasive method, electroencephalography (EEG) is effective to detect drivers' cognitive states and to assist analysis of real-world driving. Hence, investigating the relationship between EEG and driving behavior may contribute to the development of advanced driver assistance systems. For example, in our recent research, EEG was utilized as the physiological index to recognize different driving behavior types (*1*). In the near future, once unsafe driving behavior has been identified, driver assistance systems will be able to send warnings to drivers and nearby vehicles.

In the past decades, with the development of EEG analysis technology, more and more researchers have applied EEG to vehicle driving behavior studies. As the most common driving behavior, ordinary driving behavior is defined as drivers' behavior in normal driving situations, where the driver is in good physical condition, and the vehicle, traffic (free flow), and weather (clear visibility) are all in good condition. The ordinary driving state has been involved in many EEG-based driving behavior studies. However, these studies have mainly focused on certain dangerous driving states, such as driving while fatigued or distracted, and the ordinary driving state is usually regarded as the baseline or the control group.

Early studies on fatigued driving aimed to explore the relationship between brain activity and fatigued driving behavior. The EEG recordings were used to measure the evolution of physiological vigilance level (*2*), and EEG indicators of driver fatigue were further analyzed (*3*, *4*). On the basis of fatigue indicator studies, EEG-based fatigue detection systems were constructed by using models such as artificial neural network, hierarchical gaussian mixture model, and parallel hidden Markov model (*5–7*). Then, five steps (data collection, data pre-processing,

[1]MOE Key Laboratory for Urban Transportation Complex Systems Theory and Technology, Beijing Jiaotong University, Beijing, P. R. China
[2]College of Metropolitan Transportation, Beijing University of Technology, Beijing, P. R. China

Corresponding Author:
Address correspondence to Wei Guan: weig@bjtu.edu.cn

fatigue indicator extraction, fatigue recognition and feedback) were proposed as the framework of such systems (*8–10*). As for distracted driving, primary driving tasks (e.g., lane keeping) and secondary driving tasks (e.g., auditory oddball task) were designed, and driving data (e.g., steering error) were collected to measure driving performance (*11*). By analyzing EEG signals, one can distinguish audio-distracted driving from ordinary driving (*12*). Other distracted driving events, such as unexpected car deviations and mathematical questions during normal driving process, were also studied, and the response time was analyzed to evaluate drivers' distraction (*13–15*).

As mentioned above, the ordinary driving state has always been treated as the baseline for the comparison of other risky driving conditions and seldom sufficiently discussed. In addition, few studies have investigated the correlation between different performance measures (e.g., space headway), associated EEG measures (e.g., power spectral density of α band), and brain regions (e.g., frontal region) with each one considered separately. To fill these gaps, the relationship between EEG (measured by EEG features in each brain lobe) and ordinary driving behavior (measured by multiple performance measures) was analyzed in the following aspects: (a) The correlation between different brain regions and driving behavior; (b) The correlation between various EEG bands, EEG features, and driving behavior; and (c) The correlation between different driving behavior features and EEG features.

The rest of this paper is organized as follows. The second section describes the methods of the experiment, including the experiment design, extraction of driving measures and EEG features, and data analysis process. The third section presents the results of the correlation analysis, and the results are discussed in the fourth section.

Methodology

Ethics Statement

The research involving human participants in this study has been approved by the Beijing Jiaotong University's research committee (per the Institutional Review Board [IRB]). The written informed consent form for the experiment was also signed by each participant in this study.

Subjects

A total of 29 subjects were recruited for this research. However, due to several of the subjects suffering motion sickness from using the driving simulator, 25 subjects (19 males, 6 females) completed the experiment successfully, a dropout rate of 13.79%. Gender differences are not discussed in this paper. All the subjects were asked to complete a form, which was used to collect their basic information, including name, age, gender, occupation, driving experience (years), number of days per week on which they drove a car, and average annual mileage driven. The age of the subjects ranged from 22 to 50 years (standard deviation: SD = 7.69). Each subject held a valid driver's license and their average driving experience was 6.28 years (SD = 3.74). The average number of driving days per week was 3.60 (SD = 2.85) and the annual driving mileage was between 5,000 and 40,000 kilometers.

Apparatus

The experiment was conducted using the BJTU (Beijing Jiaotong University) driving simulator (*16*). The BJTU simulator consists of a full-size vehicle cabin (Ford Focus) with a real operating interface, a digital video replay system, a vehicle dynamic simulation system, and an environmental noise and shaking simulation system. All these systems work together to ensure a high fidelity driving environment.

The EEG data collection equipment used in this study was Neuroscan system. The hardware of Neuroscan system is composed of a SynAmps^2 amplifier and an electrode cap. There were 64 channels in the electrode cap and the location followed the international 10–20 system. The BJTU driving simulator and the electrode cap are shown in Figure 1.

Scenario Design and Experimental Procedure

In this study, a 22-km two-way, four-lane ring road was designed. Each lane was 3.5 m wide. To make the driving scenario close to the real driving environment, the road was edged by an urban landscape and a free-flow traffic condition was established with an 80 km/h speed limit.

Figure 1. BJTU driving simulator and an electrode cap on the driver's head.

Before the experiment, all the subjects were required to undertake 10 minutes adaptive driving in the simulator. Then, each subject was asked to complete one lap of the ring road driving according to their daily driving habits. The average duration of each lap was about 20 minutes, with some variation. As it was easier to compare data records of the same length, the first 15 minutes of data for each subject were selected for further analysis. The EEG data and simulator data were collected simultaneously at sampling rates of 1000 Hz and 60 Hz, respectively.

Driving Measures

Driving behavior measures included acceleration (m/s^2), space headway (m), time headway (s), lane deviation (m), amplitude of steering wheel movements (o), and speed (m/s), all of which were extracted from the original simulator data. Additionally, only longitudinal acceleration was considered in this study. Since the sampling rate was 60 Hz, the driving performance data were averaged every one second.

Electroencephalographic Measures

Pre-processing. The raw EEG data usually contain noises and artifacts which come from the surroundings or subject's eye movements, etc. To this end, EEGlab toolbox (*17*) was utilized for pre-processing and the procedure followed the strategy suggested by Buiatti and Mognon (*18*). First, the original sampling rate was resampled down to 256 Hz to simplify the data processing. Next, the EEG recordings were band pass filtered between 1 and 30 Hz. Then, bad channels were removed and EEG data were epoched every one second, which resulted in 900 epochs. The automatic artifact identification was done by ADJUST (*18*). Finally, re-referencing of EEG data to the average reference and baseline correction was conducted.

EEG Measures. Power spectrum analysis is a well-established method utilized for the feature extraction of EEG signals. In this paper, the periodogram method (*19*) using fast Fourier transform (FFT) was used to obtain the power spectral density (PSD) estimation. The main procedure is presented as follows: first, the EEG data of each epoch were transformed into a frequency domain using FFT. Then, the amplitude, power, and PSD in each epoch of four waves (δ-band, θ-band, α-band, and β-band) were calculated. Each brain wave relates to different mental states: δ wave and θ wave usually appear when a person falls asleep; α wave comes out when a person is conscious and relaxed; β wave reflects the sober and nervous mental state.

The EEG power data were log-transformed for temporal correlation with the driving performance (*20*). Let P and P_{log} denote the power and log-transformed power (LTP), respectively. The formula is as follows:

$$P_{log} = 10 \times lg\ P \tag{1}$$

Then, the amplitude, LTP, and PSD of δ-band, θ-band, α-band, and β-band were extracted as the EEG features.

Division of EEG Channels. The human brain is usually divided into four regions,: the frontal, parietal, occipital, and temporal regions. Different brain regions have different functions: the frontal region is the motor center; the parietal region is the perception center; the occipital region is the vision center, and the temporal region is the acoustic center. According to the distribution of the channels on the electrode cap, the total of 64 channels were divided into four groups (see Table 1). After dividing the channels, EEG features were computed separately from the four EEG channel groups.

Data Analysis

Pearson correlation analysis was utilized for hypothesis tests with a significance level of 0.05. The correlation coefficient r was the basic evaluation index in this analysis; the formula is presented in Equation 2 (*21*):

$$r = \frac{\sum_{i=1}^{n}(x_i - \bar{x})(y_i - \bar{y})}{\sqrt{\sum_{i=1}^{n}(x_i - \bar{x})^2 \sum_{i=1}^{n}(y_i - \bar{y})^2}} \tag{2}$$

where $X = [x_1, x_2, ..., x_n]$ is one dataset, $Y = [y_1, y_2, ..., y_n]$ is another dataset; $\bar{x}$ and $\bar{y}$ is the mean of X and Y.

The correlation between the EEG and the driving behavior of each subject was analyzed from three separate aspects: brain regions, EEG features, and driving behavior features. Three indices were defined to evaluate the correlation degree in each aspect:

Table 1. Division of Channels among the Four Brain Regions

Brain region	Channels
Frontal	FP1, FP2, FPZ, FCZ, FZ, AF3, AF4, F1–F8, FC1–FC4
Parietal	C1–C4, CZ, CP1–CP4, CPZ, P1–P4, PZ
Occipital	PO3–PO8, POZ, O1, O2, OZ, CB1, CB2
Temporal	FT7, FT8, FC5, FC6, T7, T8, C5, C6, TP7, TP8, CP5, CP6, P5–P8

(a) The number of correlations (N): the number of correlations equals the number of p values that are smaller than 0.05.
(b) Average r value ($\overline{R}$): the average r value (only calculated when $p < .05$) of each subject in each aspect.
(c) Maximum r value (R_{max}): the maximum r value of each subject in each aspect.

The formulas of those indices are as follows:

$$N = \sum_{i=1}^{I} n_i; n_i = \begin{cases} 1 & p<0.05 \\ 0 & p \geq 0.05 \end{cases} \quad (3)$$

$$\overline{R} = \sum_{i=1}^{I} \frac{|r_i|}{N}; r_i = \begin{cases} r_i & p<0.05 \\ 0 & p \geq 0.05 \end{cases} \quad (4)$$

$$R_{max} = \max\{|r_i|\}; i \in [1, I] \quad (5)$$

where $i = 1, 2, ..., I$, and $I = 25$ is the total number of the subjects, n denotes the correlation of each subject in each aspect.

By calculating these indices, the comparison for the correlation of each EEG feature and driving behavior feature was performed. For example, let A be the α-PSD in the frontal region (one of the EEG features) of subject No. 1, let B be the space headway (one of the driving behavior features) of subject No. 1; since driving data and EEG data were collected synchronously, both A and B have the same length (900 epochs each). Then, the correlation coefficient r and the corresponding p value between A and B were calculated. Finally, the N, $\overline{R}$, and R_{max} were obtained. The process of the correlation analysis is presented as follows:

Step 1: Data classification. The data were classified by the different brain regions, the EEG features, and the driving behavior features.
Step 2: Correlation indices calculation. The correlation indices, i.e., N, $\bar{R}$ and R_{max} of each subject in each aspect, were calculated.
Step 3: Comparison of the indices. The indices of different aspects were compared, and the maximum value corresponds to the highest correlation.

Results

Brain Regions

Brain Regions and Driving Behavior. The Pearson correlation coefficients were computed for each subject in each brain region, and the correlation indices were extracted. The purpose was to find the brain regions significantly associated with particular features of ordinary driving behavior. To do this, N values of each brain region were compared first, and the maximum N value referred to the brain region most relevant to driving behavior. If the maximum N value was not unique, then compare $\overline{R}$ and R_{max}. Table 2 presents, for each subject, the brain region which is most active in their ordinary driving behavior. Through further analysis of the data presented in Table 2, the frequency of activity of the brain regions most relevant to driving behavior was obtained and the corresponding R_{max} extracted, as shown in Figure 2. The results shown in Figure 2 suggest that ordinary driving behavior may activate to all four brain regions, especially the temporal, frontal, and occipital regions. In addition, the differences in activity between these three brain regions are relatively small. Driving is a complex task, which requires perception, situation awareness, judgment, decision-making capacity, responsiveness, and reasonable physical capability (*22*). Hence, all the brain regions may cooperate with each other, and their contributions to the driving task may be similar.

Table 2. Brain Regions Most Active in Normal Driving Behavior

Subject no.	Brain region	N	$\bar{R}$	R_{max}*
1	Temporal	37	0.109	0.174
2	Temporal	15	0.087	0.113
3	Temporal	19	0.082	0.113
4	Frontal	46	0.122	0.222
5	Occipital	12	0.088	0.111
6	Occipital	35	0.146	0.371
7	Parietal	15	0.09	0.127
8	Frontal	26	0.082	0.11
9	Frontal	21	0.085	0.108
10	Occipital	35	0.127	0.279
11	Frontal	38	0.108	0.168
12	Temporal	42	0.114	0.184
13	Frontal	48	0.102	0.158
14	Temporal	31	0.204	0.523
15	Temporal	20	0.093	0.158
16	Temporal	29	0.094	0.161
17	Occipital	22	0.118	0.278
18	Temporal	3	0.082	0.087
19	Parietal	43	0.166	0.269
20	Frontal	38	0.131	0.248
21	Occipital	32	0.146	0.331
22	Occipital	25	0.114	0.188
23	Parietal	28	0.087	0.147
24	Occipital	25	0.113	0.22
25	Frontal	23	0.1	0.151

Further Analysis of the Driving Behavior Features. This subsection explores the most relevant brain regions with each driving behavior feature. Table 3 summarizes the results for each subject and Figure 3 shows the frequency of the most relevant brain regions in each driving behavior features. The results indicated that the frontal region and occipital region relate to all the driving behavior

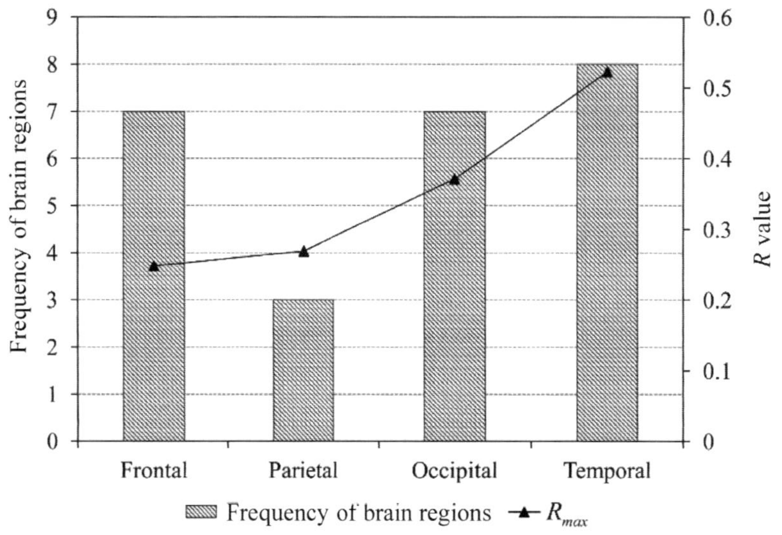

Figure 2. Frequency of the most relevant brain regions and corresponding R_{max}.

Table 3. Brain Regions Most Relevant to Each Driving Behavior Feature

No.	Acceleration	Space headway	Time headway	Lane deviation	Amplitude	Speed
1	Frontal	Occipital	Occipital	Frontal	–	Frontal
2	Occipital	–	–	Occipital	Temporal	Parietal
3	Temporal	Parietal	Temporal	Temporal	–	Temporal
4	Frontal	Frontal	Frontal	Frontal	–	Temporal
5	Occipital	Occipital	Occipital	–	Occipital	Parietal
6	Occipital	Occipital	Occipital	–	Occipital	Temporal
7	–	Parietal	Parietal	–	–	Temporal
8	Temporal	Temporal	Temporal	–	Frontal	Frontal
9	Frontal	Parietal	Frontal	–	Frontal	Occipital
10	Parietal	Frontal	Occipital	Occipital	Frontal	Occipital
11	Occipital	Parietal	Frontal	–	Frontal	Frontal
12	Frontal	Temporal	Frontal	Parietal	Temporal	Temporal
13	Frontal	Occipital	Frontal	Temporal	Frontal	Frontal
14	Temporal	Temporal	Temporal	Temporal	Frontal	Temporal
15	–	Temporal	Temporal	–	Temporal	Temporal
16	Parietal	Frontal	Occipital	Parietal	Occipital	Parietal
17	Frontal	Temporal	Temporal	Occipital	Occipital	Occipital
18	Temporal	–	–	–	–	Temporal
19	Parietal	Parietal	Frontal	Temporal	–	Frontal
20	Occipital	Frontal	Frontal	Temporal	–	Parietal
21	Occipital	Occipital	Occipital	Occipital	Temporal	Frontal
22	Occipital	Occipital	Occipital	Temporal	Frontal	Temporal
23	Frontal	Parietal	Parietal	Parietal	Temporal	Temporal
24	Occipital	Occipital	Occipital	Parietal	Occipital	Occipital
25	Parietal	Frontal	Frontal	Frontal	–	Occipital

features, especially to acceleration and time headway. Additionally, the temporal region is most relevant to speed compared with the other driving features.

EEG Bands and EEG Features

Different EEG frequency bands contain different information, which correspond to specific activities. The

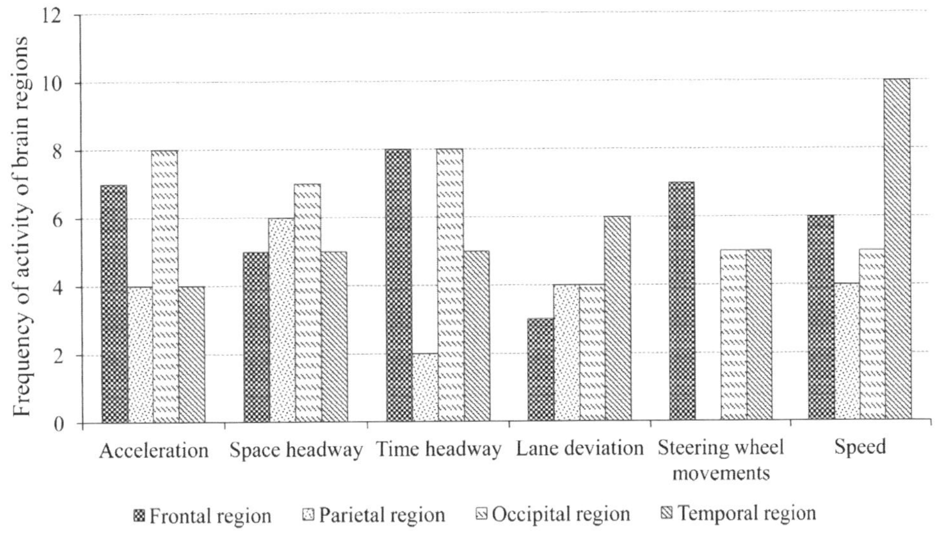

Figure 3. Brain regions most relevant in each driving behavior feature.

correlation coefficients between the EEG features and the driving behavior features were computed to find the EEG band and corresponding EEG feature that are most relevant to driving behavior. There were 48% of subjects whose β-band was significantly associated with driving behavior. Furthermore, θ-band and α-band were both the second most relevant bands with rates of 20% each, and δ-band had a weak relationship with driving behavior. Among the EEG features of β-band, β-LTP and β-PSD were found to be more relevant to driving behavior, with the rates of 50% and 30%, respectively. The analysis revealed that β-band relates to ordinary driving behavior, and β-LTP is the EEG feature most relevant to ordinary driving behavior. The results are shown in Table 4. Figure 4 presents the frequency of the most relevant EEG bands and the corresponding R_{max}.

Table 4. EEG Bands and EEG Features Most Relevant to Driving Behavior

No.	EEG bands	*N*	$\bar{R}$	R_{max}*
1	Beta	33	0.111	0.188
2	Alpha	17	0.104	0.2
3	Theta	15	0.085	0.124
4	Beta	42	0.141	0.256
5	Beta	10	0.089	0.111
6	Beta	41	0.152	0.371
7	Alpha	18	0.089	0.129
8	Theta	26	0.093	0.134
9	Theta	15	0.087	0.114
10	Beta	40	0.165	0.329
11	Alpha	37	0.105	0.168
12	Theta	45	0.106	0.17
13	Beta	49	0.111	0.184
14	Beta	51	0.195	0.523
15	Delta	27	0.106	0.158
16	Delta	36	0.106	0.165
17	Theta	24	0.088	0.141
18	Delta	4	0.078	0.087
19	Beta	45	0.189	0.378
20	Beta	39	0.152	0.313
21	Beta	43	0.138	0.331
22	Beta	30	0.113	0.188
23	Beta	37	0.113	0.197
24	Alpha	25	0.092	0.161
25	Alpha	31	0.107	0.153

Driving Behavior Features

Acceleration, space headway, time headway, lane deviation, amplitude of steering wheel movements, and speed were selected as driving behavior features. These features can describe drivers' performance and driving states in different ways. Thus, drivers' perceptions of each driving behavior feature may also differ. The results are shown in Table 5 and Figure 5. These results suggest that acceleration, space headway, and speed have closer relationships with brain activity than the other driving behavior features, with the rate of 72% in total (24% each), and only 4% of subjects whose most relevant driving behavior feature with brain activity is lane deviation. Thus, acceleration, space headway, and speed are significantly relevant to brain activity, while lane deviation has the weakest relationship with brain activity.

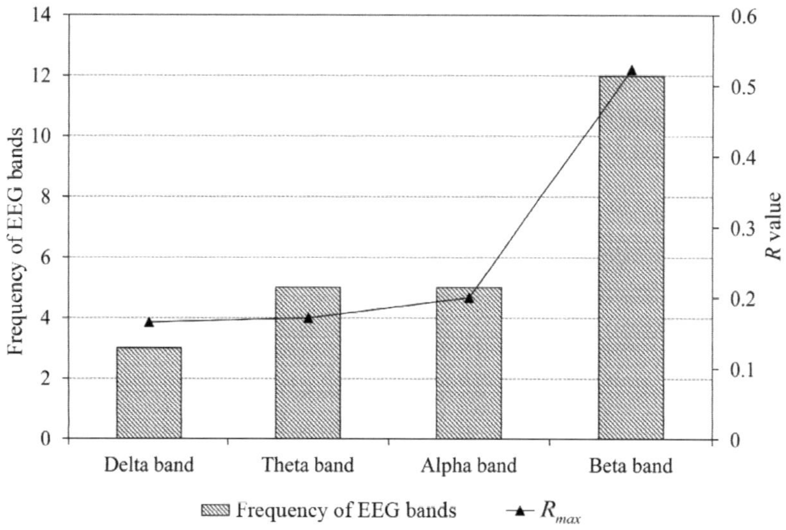

Figure 4. Frequency of most relevant EEG bands and corresponding R_{max}.

Table 5. Driving Behavior Features Most Relevant to Brain Activity

Subject no.	Driving behavior features	*N*	$\bar{R}$	$R_{max}{}^{*}$
1	Acceleration	38	0.126	0.176
2	Amplitude	32	0.085	0.113
3	Acceleration	28	0.086	0.124
4	Time headway	39	0.135	0.256
5	Acceleration	6	0.076	0.094
6	Speed	42	0.145	0.371
7	Speed	18	0.096	0.129
8	Space headway	20	0.086	0.132
9	Amplitude	15	0.089	0.114
10	Time headway	36	0.124	0.279
11	Space headway	45	0.103	0.183
12	Speed	48	0.127	0.184
13	Space headway	29	0.112	0.177
14	Time headway	18	0.254	0.523
15	Speed	20	0.09	0.141
16	Acceleration	25	0.116	0.165
17	Acceleration	19	0.14	0.307
18	Speed	3	0.072	0.082
19	Lane deviation	47	0.18	0.27
20	Acceleration	46	0.115	0.203
21	Speed	29	0.116	0.288
22	Space headway	34	0.115	0.17
23	Amplitude	22	0.118	0.267
24	Space headway	16	0.119	0.22
25	Space headway	34	0.093	0.147

Discussion

This paper investigates the relationship between EEG and ordinary driving behavior by treating acceleration, space headway, time headway, lane deviation, amplitude of steering wheel movements, and speed as the driving measures. As expected, EEG features correlate with driving behavior features during normal driving conditions. However, since features extracted from EEG data and driving behavior data are various, it is difficult to observe such correlations. For example, in the research of Otmani et al. (*20*) no significant correlation was found between any driving behavior indices and EEG data. The reason may be that both the EEG measures and driving measures were different from those in this study. The EEG feature analyzed in (*20*) was the power of ($\alpha + \beta$), while the EEG measures in this paper were the amplitude, LTP, and PSD of four EEG bands in four brain regions).

The regions of the brain have different functions. The current study indicates that all the regions of the brain may be involved in driving tasks, and that the temporal, frontal, and occipital regions are the most active in ordinary driving. As shown in Figure 3, the brain region most relevant to speed is the temporal region, and both the frontal region and occipital region have a closer relationship to acceleration, space headway, and time headway than the other regions. Since the temporal region deals with auditory tasks, drivers' perception of speed may relate to auditory perception. The frontal region is the motor center while the occipital region is the vision center, and the frontal region also influences drivers' cognitive distraction process (*23*). Hence, drivers' perceptions of acceleration, space headway, and time headway may be a kind of cognitive task which combines with their visual judgment.

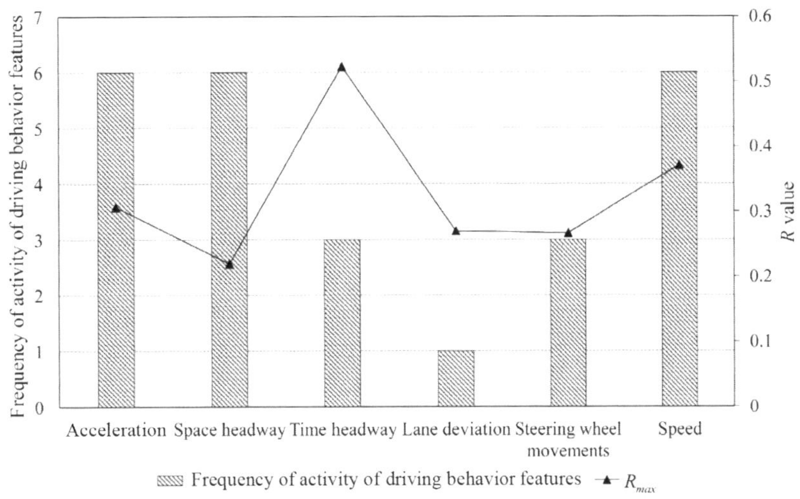

Figure 5. Frequency of most relevant driving behavior features and corresponding R_{max}.

As the results show in Figure 4, the EEG band most relevant to ordinary driving behavior was β-band. The β wave usually appears when people feel sober and nervous. It is sensitive to changes in alertness (*24*, *25*) and closely related to attention. Hence the results indicate that, during ordinary driving, drivers concentrate on their driving tasks, remain tense, and are vigilant to their surroundings. In Figure 5, the driving behavior features with most brain activity are acceleration, space headway, and speed, which are the minimal inputs required by car-following models to reproduce realistic traffic dynamics (*26*). The study confirms that these driving behavior features may be the link between brain activity and driving behavior at the level of human behavioral neuroscience.

Confirmation of the correlation between driving behavior and brain activity is a fundamental research need for transportation studies. Most studies, however, focus on risky driving conditions, such as fatigue and distraction situations, while normal driving conditions are seldom fully discussed and always treated as the baseline. Furthermore, to the best of our knowledge, the proposed method is the first of its kind to isolate different driving tasks (e.g., following traffic) and performance measures (e.g., space headway), and to associate EEG patterns and brain lobes with each one. Thus, the major contribution of this study is to fill these gaps by investigating the relationship between EEG and ordinary driving behavior in a comprehensive perspective. The major finding of this study is to demonstrate a clear relationship between driving behavior and the brain activity from three aspects: brain regions, EEG measures, and driving features. The results indicate that driving behavior is related to all the brain regions, and β-band along with β-LTP is closely related to ordinary driving behavior. In addition, acceleration, space headway, and speed were found to be related to brain activity.

Some limitations faced by the current study are also acknowledged. This study is the first attempt of its kind, to our knowledge, and a sample size of 25 may not provide adequate data for strong conclusions. Additionally, as a general issue for all driving simulator studies, there still exists some differences between simulated driving and actual driving, and this may influence both drivers' behavior and their brain activity.

Future work will be required to expand the current studies by recruiting more subjects and considering demographic variables (e.g., age, gender, and occupational differences). Real-world driving experiments should be conducted to eliminate the potential differences between simulated scenarios and actual driving scenarios.

Acknowledgments

This work was supported by the National Natural Science Foundation of China (NFSC, No. 71471014 and No. 71621001) and the Fundamental Research Funds for the Central Universities (No. 2017YJS093).

Author Contributions

The authors confirm contribution to the paper as follows: study conception and design: Liu Yang, Wei Guan; data collection: Liu Yang, Shixiong Jiang; analysis and interpretation of results: Liu Yang, Wei Guan, Zhengbing He; draft manuscript preparation: Liu Yang, Zhengbing He, Shixiong Jiang. All authors reviewed the results and approved the final version of the manuscript.

References

1. Yang, L., R. Ma, M. Zhang, W. Guan, and S. Jiang. Driving Behavior Recognition Using EEG Data from a Simulated Car-Following Experiment. *Accident Analysis and Prevention*, Vol. 116, 2018, pp. 30–40.
2. Campagne, A., T. Pebayle, and A. Muzet. Correlation Between Driving Errors and Vigilance Level: Influence of the Driver's Age. *Physiology & Behavior*, Vol. 80, No. 4, 2004, pp. 515–524.
3. Kar, S., M. Bhagat, and A. Routray. EEG Signal Analysis for the Assessment and Quantification of Driver's Fatigue. *Transportation Research Part F: Traffic Psychology and Behaviour*, Vol. 13, No. 5, 2010, pp. 297–306.
4. Jap, B. T., S. Lal, and P. Fischer. Comparing Combinations of EEG Activity in Train Drivers During Monotonous Driving. *Expert Systems with Applications*, Vol. 38, No. 1, 2011, pp. 996–1003.
5. Vuckovic, A., V. Radivojevic, A. C. N. Chen, and D. Popovic. Automatic Recognition of Alertness and Drowsiness from EEG by an Artificial Neural Network. *Medical Engineering and Physics*, Vol. 24, 2002, pp. 349–360.
6. Rosipal, R., B. Peters, G. Kecklund, and T. Åkerstedt. EEG-Based Drivers' Drowsiness Monitoring Using a Hierarchical Gaussion Mixture Model. In *Lecture Notes in Computer Science, Foundations of Augmented Cognition*, Springer, Berlin, 2007, pp. 294–303.
7. Sun, H., and B.-L. Lu. EEG-Based Fatigue Classification by Using Parallel Hidden Markov Model and Pattern Classifier Combination. *Proc., 19th International Conference on Neural Information Processing—Volume Part IV*, Springer, Berlin, 2012, pp. 484–491.
8. Lin, C. T., Y. C. Chen, R. C. Wu, S. F. Liang, and T. Y. Huang. Assessment of Driver's Driving Performance and Alertness Using EEG-based Fuzzy Neural Networks. *Proc., IEEE International Symposium on Circuits & Systems*, IEEE, 2005, pp. 152–155.
9. Raut, S. R., and S. M. Kulkarni. A Real Time Drowsiness Detection System for Safe Driving. *International Journal of Electronics, Electrical and Computational Systems*, Vol. 3, No. 5, 2014, pp. 15–19.
10. Chuang, C. H., C. S. Huang, L. W. Ko, and C. T. Lin. An EEG-Based Perceptual Function Integration Network for Application to Drowsy Driving. *Knowledge-Based Systems*, Vol. 80, 2015, pp. 143–152.
11. Wester, A. E., K. B. E. Böcker, E. R. Volkerts, J. C. Verster, and J. L. Kenemans. Event-Related Potentials and Secondary Task Performance During Simulated Driving. *Accident Analysis and Prevention*, Vol. 40, No. 1, 2008, pp. 1–7.
12. Dahal, N., N. Nandagopal, B. Cocks, N. Dasari, R. Vijayalakshmi, and P. Gaertner. Event Related Synchronization and Desynchronization of EEG Data During Distracted Virtual Driving. *Proc., 7th Asia Modelling Symposium*, IEEE, 2013, pp. 23–28.
13. Lin, C. T., T. W. Lin, H. Z. Chiu, C. F. Chao, Y. C. Chen, S. F. Liang, and L. W. Ko. Distraction-Related EEG Dynamics in Virtual Reality Driving Simulation. *Proc., 2008 International Symposium on Circuits and Systems*, IEEE, 2008, pp. 1088–1091.
14. Lin, C. T., S. A. Chen, T. T. Chiu, H. Z. Lin, and L. W. Ko. Spatial and Temporal EEG Dynamics of Dual-Task Driving Performance. *Journal of Neuroengineering and Rehabilitation*, Vol. 8, 2011, pp. 1–13.
15. Lin, C. T., S. A. Chen, L. W. Ko, and Y. K. Wang. EEG-Based Brain Dynamics of Driving Distraction. *The 2011 International Joint Conference on Neural Networks*, IEEE, 2011, pp. 1497–1500. https://doi.org/10.1109/IJCNN.2011.6033401
16. Li, X., X. Yan, and S. C. Wong. Effects of Fog, Driver Experience and Gender on Driving Behavior on S-Curved Road Segments. *Accident Analysis and Prevention*, Vol. 77, 2015, pp. 91–104.
17. Delorme, A., and S. Makeig. EEGLAB: An Open Source Toolbox for Analysis of Single-Trial EEG Dynamics Including Independent Component Analysis. *Journal of Neuroscience Methods*, Vol. 134, No. 1, 2004, pp. 9–21.
18. Mognon, A., J. Jovicich, L. Bruzzone, and M. Buiatti. *ADJUST: An Automatic EEG Artifact Detector based on the Joint Use of Spatial and Temporal Features. Psychophysiology*, Vol. 28, No. 2, 2011, pp. 229–240.
19. Zhao, L., and Y. He. Power Spectrum Estimation of the Welch Method Based on Imagery EEG. *Applied Mechanics and Materials*, Vol. 278–280, 2013, pp. 1260–1264.
20. Otmani, S., T. Pebayle, J. Roge, and A. Muzet. Effect of Driving Duration and Partial Sleep Deprivation on Subsequent Alertness and Performance of Car Drivers. *Physiology & Behavior*, Vol. 84, No. 5, 2005, pp. 715–724.
21. Ahlgren, P., B. Jarneving, and R. Rousseau. Requirements for a Cocitation Similarity Measure, with Special Reference to Pearson's Correlation Coefficient. *Journal of the American Society for Information Science and Technology*, Vol. 54, No. 6, 2003, pp. 550–560.
22. Chapman, J. Health Initiatives to Target Obesity in Surface Transport Industries: Review and Implications for Action. *Evidence Base*, Vol. 2, 2015, pp. 1–32.
23. Almahasneh, H., W. T. Chooi, N. Kamel, and A. S. Malik. Deep in Thought While Driving: An EEG Study on Drivers' Cognitive Distraction. *Transportation Research Part F: Traffic Psychology and Behaviour*, Vol. 26, 2014, pp. 218–226.
24. Eoh, H. J., M. K. Chung, and S.-H. Kim. Electroencephalographic Study of Drowsiness in Simulated Driving with Sleep Deprivation. *International Journal of Industrial Ergonomics*, Vol. 35, No. 4, 2005, pp. 307–320.
25. Subasi, A. Automatic Recognition of Alertness Level from EEG by Using Neural Network and Wavelet Coefficients. *Expert Systems with Applications*, Vol. 28, No. 4, 2005, pp. 701–711.
26. He, Z., L. Zheng, and W. Guan. A Simple Nonparametric Car-Following Model Driven by Field Data. *Transportation Research Part B: Methodological*, Vol. 80, 2015, pp. 185–201.

The Standing Committee on Simulation and Measurement of Vehicle and Operator Performance (AND30) peer-reviewed this paper (18-05856).

Sage Publications USA
sagepub.com

CURRAN ASSOCIATES INC.
proceedings.com

ISBN 978-1-5108-8245-4
9 781510 882454